EMS Textbooks in Mathematics

EMS Textbooks in Mathematics is a series of books aimed at students or professional mathematicians seeking an introduction into a particular field. The individual volumes are intended not only to provide relevant techniques, results, and applications, but also to afford insight into the motivations and ideas behind the theory. Suitably designed exercises help to master the subject and prepare the reader for the study of more advanced and specialized literature.

Jørn Justesen and Tom Høholdt, *A Course In Error-Correcting Codes*
Markus Stroppel, *Locally Compact Groups*
Peter Kunkel and Volker Mehrmann, *Differential-Algebraic Equations*
Dorothee D. Haroske and Hans Triebel, *Distributions, Sobolev Spaces, Elliptic Equations*
Thomas Timmermann, *An Invitation to Quantum Groups and Duality*
Oleg Bogopolski, *Introduction to Group Theory*
Marek Jarnicki and Peter Pflug, *First Steps in Several Complex Variables: Reinhardt Domains*
Tammo tom Dieck, *Algebraic Topology*
Mauro C. Beltrametti et al., *Lectures on Curves, Surfaces and Projective Varieties*
Wolfgang Woess, *Denumerable Markov Chains*
Eduard Zehnder, *Lectures on Dynamical Systems. Hamiltonian Vector Fields and Symplectic Capacities*
Andrzej Skowroński and Kunio Yamagata, *Frobenius Algebras I. Basic Representation Theory*
Piotr W. Nowak and Guoliang Yu, *Large Scale Geometry*
Joaquim Bruna and Juliá Cufí, *Complex Analysis*
Eduardo Casas-Alvero, *Analytic Projective Geometry*
Fabrice Baudoin, *Diffusion Processes and Stochastic Calculus*
Olivier Lablée, *Spectral Theory in Riemannian Geometry*

Dietmar A. Salamon

Measure and Integration

European Mathematical Society

Author:

Dietmar A. Salamon
Departement Mathematik
ETH Zürich
Rämistrasse 101
8092 Zürich
Switzerland

E-mail: dietmar.salamon@math.ethz.ch

2010 Mathematics Subject Classification: Primary: 28-01; Secondary: 28A05, 28A10, 28A12, 28A15, 28A20, 28A25, 28A30, 28A33, 28A35, 28C05, 28C10, 28C15, 35J05, 43A05, 44A35, 46B22, 46C05, 46E27, 46E30

Key words: sigma-Algebra, Lebesgue monotone convergence, Caratheodory criterion, Lebesgue measure, Borel measure, Dieudonné's measure, Riesz Representation Theorem, Alexandrov Double Arrow Space, Sorgenfrey Line, separability, Cauchy–Schwarz inequality, Jensen's inequality, Egoroff's theorem, Hardy's inequality, absolutely continuous measure, truly continuous measure, singular measure, signed measure, Radon–Nikodym Theorem, Lebesgue Decomposition Theorem, Hahn Decomposition Theorem, Jordan Decomposition Theorem, Hardy–Littlewood maximal inequality, Vitali's Covering Lemma, Lebesgue point, Lebesgue Differentiation Theorem, Banach-Zarecki Theorem, Vitali–Caratheodory Theorem, Cantor function, product sigma-algebra, Fubini's Theorem, convolution, Young's inequality, mollifier, Marcinciewicz interpolation, Poisson identity, Green's formula, Calderon–Zygmund inequality, Haar measure, modular character.

ISBN 978-3-03719-159-0

The Swiss National Library lists this publication in The Swiss Book, the Swiss national bibliography, and the detailed bibliographic data are available on the Internet at http://www.helveticat.ch.

Contact address:

European Mathematical Society Publishing House
Seminar for Applied Mathematics
ETH-Zentrum SEW A27
CH-8092 Zürich
Switzerland

Phone: +41 (0)44 632 34 36
Email: info@ems-ph.org
Homepage: www.ems-ph.org

Typeset using the author's TEX files: M. Zunino, Stuttgart, Germany
Printing and binding: Beltz Bad Langensalza GmbH, Bad Langensalza, Germany
∞ Printed on acid free paper
9 8 7 6 5 4 3 2 1

Preface

This book is based on notes for the lecture course *Measure and Integration* held at ETH Zürich in the spring semester 2014. Prerequisites are the first-year courses on analysis and linear algebra, including the Riemann integral [9, 18, 19, 21], as well as some basic knowledge of metric and topological spaces. The course material is based in large parts on Chapters 1–8 of the textbook *real and complex analysis* by Walter Rudin [17]. In addition to Rudin's book, the lecture notes by Urs Lang [10, 11], the five volumes on measure theory by David H. Fremlin [4], the paper by Heinz König [8] on the generalized Radon–Nikodým theorem, the lecture notes by C. E. Heil [7] on absolutely continuous functions, Dan Ma's topology blog [12] on exotic examples of topological spaces, and the paper by Gert K. Pedersen [16] on the Haar measure were very helpful in preparing this manuscript.

The text also contains some material that was not covered in the lecture course, namely some of the results in Sections 4.5 and 5.2 (concerning the dual space of $L^p(\mu)$ in the non-σ-finite case), Section 5.4 on the generalized Radon–Nikodým theorem, Sections 7.6 and 7.7 on Marcinkiewicz interpolation and the Calderón–Zygmund inequality, and Chapter 8 on the Haar measure.

I am grateful to many people who helped to improve this manuscript. Thanks to the students at ETH who pointed out typos or errors in earlier drafts. Thanks to Andreas Leiser for his careful proofreading. Thanks to Theo Buehler for many enlightening discussions and for pointing out the book by Fremlin, Dan Ma's topology blog, and the paper by Pedersen. Thanks to Urs Lang for his insightful comments on the construction of the Haar measure.

ETH, Zürich
1 August 2015

Dietmar A. Salamon

Contents

Introduction

We learn already in high school that integration plays a central role in mathematics and physics. One encounters integrals in the notions of area or volume, when solving a differential equation, in the fundamental theorem of calculus, in Stokes' theorem, or in classical and quantum mechanics. The first year analysis course at ETH includes an introduction to the Riemann integral, which is satisfactory for many applications. However, it has certain drawbacks, in that some very basic functions are not Riemann integrable, that the pointwise limit of a sequence of Riemann integrable functions need not be Riemann integrable, and that the space of Riemann integrable functions is not complete with respect to the L^1-norm. One purpose of this book is to introduce the *Lebesgue integral,* which does not suffer from these drawbacks and agrees with the Riemann integral whenever the latter is defined. Chapter 1 introduces abstract integration theory for functions on measure spaces. It includes proofs of the Lebesgue monotone convergence theorem, the Fatou lemma, and the Lebesgue dominated convergence theorem. In Chapter 2 we move on to outer measures and introduce the Lebesgue measure on Euclidean space. Borel measures on locally compact Hausdorff spaces are the subject of Chapter 3. Here the central result is the Riesz representation theorem. In Chapter 4 we encounter L^p spaces and show that the compactly supported continuous functions form a dense subspace of L^p for a regular Borel measure on a locally compact Hausdorff space when $p < \infty$. Chapter 5 is devoted to the proof of the Radon–Nikodým theorem about absolutely continuous measures and to the proof that L^q is naturally isomorphic to the dual space of L^p when $1/p + 1/q = 1$ and $1 < p < \infty$. Chapter 6 deals with differentiation. Chapter 7 introduces product measures and contains a proof of Fubini's Theorem, an introduction to the convolution product on $L^1(\mathbb{R}^n)$, and a proof of the Calderón–Zygmund inequality. Chapter 8 constructs Haar measures on locally compact Hausdorff groups.

Despite the overlap with the book of Rudin [17], there are some differences in exposition and content. A small expository difference is that in Chapter 1 measurable functions are defined in terms of pre-images of (Borel) measurable sets rather than pre-images of open sets. The Lebesgue measure in Chapter 2 is introduced in terms of the Lebesgue outer measure instead of as a corollary of the Riesz representation theorem. The notion of a Radon measure on a locally compact Hausdorff space in Chapter 3 is defined in terms of inner regularity, rather than outer regularity together with inner regularity on open sets. This leads to a somewhat different formulation of the Riesz representation theorem (which includes the result as formulated by Rudin). In Chapters 4 and 5 it is shown that $L^q(\mu)$ is isomorphic to the dual space of $L^p(\mu)$ for all measure spaces (not just the σ-finite ones) whenever $1 < p < \infty$ and $1/p + 1/q = 1$. It is also shown that $L^\infty(\mu)$ is isomorphic to the dual space of $L^1(\mu)$ if and only if the measure space is localizable. Chapter 5 includes a generalized version of the Radon–Nikodým theorem for signed measures, due to Fremlin [4], which does not require that the underlying measure μ is σ-finite. In the formulation of König [8] it asserts that a signed measure admits a μ-density if and only if it is both absolutely continuous and inner regular with respect to μ. In addition, the present book includes a self-contained proof of the Calderón–Zygmund inequality in Chapter 7 and an existence and uniqueness proof for (left and right) Haar measures on locally compact Hausdorff groups in Chapter 8.

The book is intended as a companion for a foundational one-semester lecture course on measure and integration theory and there are many topics that it does not cover. For example, the subject of probability theory is only touched upon briefly at the end of Chapter 1 and the interested reader is referred to the book of Malliavin [13] which covers many additional topics including Fourier analysis, limit theorems in probability theory, Sobolev spaces, and the stochastic calculus of variations. Many other fields of mathematics require the basic notions of measure and integration. They include functional analysis and partial differential equations (see, e.g., Gilbarg–Trudinger [5]), geometric measure theory, geometric group theory, ergodic theory and dynamical systems, and differential topology and geometry.

There are many other textbooks on measure theory that cover most or all of the material in the present book, as well as much more, perhaps from somewhat different view points. They include the book of Bogachev [2] which also contains many historical references, the book of Halmos [6], and the aforementioned books of Fremlin [4], Malliavin [13], and Rudin [17].

Chapter 1

Abstract measure theory

The purpose of this first chapter is to introduce integration on abstract measure spaces. The basic idea is to assign to a real valued function on a given domain a number that gives a reasonable meaning to the notion of *area under the graph*. For example, to the characteristic function of a subset of the domain one would want to assign the *length* or *area* or *volume* of that subset. To carry this out one needs a sensible notion of *measuring the size* of the subsets of a given domain. Formally this can take the form of a function which assigns a nonnegative real number, possibly also infinity, to each subset of our domain. This function should have the property that the measure of a disjoint union of subsets is the sum of the measures of the individual subsets. However, as is the case with many beautiful ideas, this naive approach does not work. Consider for example the notion of the length of an interval of real numbers. In this situation each single point has measure zero. With the additivity requirement it would then follow that every subset of the reals, when expressed as the disjoint union of all its elements, must also have measure zero, thus defeating the original purpose of defining the *length* of an arbitrary subset of the reals. This reasoning carries over to any dimension and makes it impossible to define the familiar notions of *area* or *volume* in the manner outlined above. To find a way around this, it helps to recall the basic observation that any uncountable sum of positive real numbers must be infinity. Namely, if we are given a collection of positive real numbers whose sum is finite, then only finitely many of these numbers can be bigger than $1/n$ for each natural number n, and so it can only be a countable collection. Thus it makes sense to demand additivity only for countable collections of disjoint sets.

Even with the restricted concept of *countable additivity* it will not be possible to assign a measure to every subset of the reals and recover the notion of the length of an interval. For example, call two real numbers equivalent if their difference is rational, and let E be a subset of the half unit interval that contains precisely one element of each equivalence class. Since each equivalence class has a nonempty intersection with the half unit interval, such a set exists by the Axiom of Choice. Assume that all translates of E have the same measure. Then countable additivity would imply that the unit interval has measure zero or infinity.

One way out of this dilemma is to give up on the idea of countable additivity and replace it by the weaker requirement of *countable subadditivity*. This leads to the notion of an *outer measure* which will be discussed in Chapter 2. Another way out is to retain the requirement of countable additivity, but give up on the idea of assigning a measure to *every* subset of a given domain. Instead, one assigns a measure only to *some* subsets which are then called *measurable*. This idea will be pursued in the present chapter. A subtlety of this approach is that in some important cases it is not possible to give an explicit description of those subsets of a given domain that one wants to measure, and instead one can only impose certain axioms that the collection of *all measurable sets* must satisfy. By contrast, in topology the open sets can often be described explicitly. For example, the open subsets of the real line are countable unions of open intervals, while there is no such explicit description for the *Borel measurable* subsets of the real line.

The precise formulation of this approach leads to the notion of a *σ-algebra* which is discussed in Section 1.1. Section 1.2 introduces *measurable functions* and examines their basic properties. *Measures* and the *integrals* of positive measurable functions are the subject of Section 1.3. Here the nontrivial part is to establish additivity of the integral and the proof is based on the *Lebesgue monotone convergence theorem.* An important inequality is the *Fatou lemma.* It is needed to prove the *Lebesgue dominated convergence theorem* in Section 1.4 for real valued *integrable functions.* Section 1.5 deals with *sets of measure zero,* which are negligible for many purposes. For example, it is often convenient to identify two measurable functions if they agree *almost everywhere*, i.e., on the complement of a set of measure zero. This defines an equivalence relation. The quotient of the space of integrable functions by this equivalence relation is a Banach space and is denoted by L^1. Section 1.6 discusses the *completion* of a measure space. Here the idea is to declare every subset of a set of measure zero to be measurable as well.

1.1 σ-algebras

For any fixed set X denote by 2^X the set of all subsets of X and, for any subset $A \subset X$, denote by

$$A^c := X \setminus A$$

its complement.

Definition 1.1 (measurable space). Let X be a set. A collection $\mathcal{A} \subset 2^X$ of subsets of X is called a *σ-algebra* if it satisfies the following axioms:

(a) $X \in \mathcal{A}$;

(b) if $A \in \mathcal{A}$, then $A^c \in \mathcal{A}$;

(c) every countable union of elements of $\mathcal{A}$ is again an element of $\mathcal{A}$, i.e., if $A_i \in \mathcal{A}$ for $i = 1, 2, 3, \ldots$, then $\bigcup_{i=1}^{\infty} A_i \in \mathcal{A}$.

A *measurable space* is a pair $(X, \mathcal{A})$ consisting of a set X and a σ-algebra $\mathcal{A} \subset 2^X$. The elements of a σ-algebra $\mathcal{A}$ are called *measurable sets*.

Lemma 1.2. *Every σ-algebra $\mathcal{A} \subset 2^X$ satisfies the following:*

(d) $\emptyset \in \mathcal{A}$;

(e) *if $n \in \mathbb{N}$ and $A_1, \ldots, A_n \in \mathcal{A}$, then $\bigcup_{i=1}^{n} A_i \in \mathcal{A}$;*

(f) *every finite or countable intersection of elements of $\mathcal{A}$ is an element of $\mathcal{A}$;*

(g) *if $A, B \in \mathcal{A}$, then $A \setminus B \in \mathcal{A}$.*

Proof. Condition (d) follows from (a), (b) because $X^c = \emptyset$, and (e) follows from (c), (d) by taking $A_i := \emptyset$ for $i > n$. Condition (f) follows from (b), (c), (e) because $\left(\bigcap_i A_i\right)^c = \bigcup_i A_i^c$, and (g) follows from (b), (f) because $A \setminus B = A \cap B^c$. This proves Lemma 1.2. □

Example 1.3. The sets $\mathcal{A} := \{\emptyset, X\}$ and $\mathcal{A} := 2^X$ are σ-algebras.

Example 1.4. Let X be an uncountable set. Then the collection $\mathcal{A} \subset 2^X$ of all subsets $A \subset X$ such that either A or A^c is countable is a σ-algebra. (Here *countable* means finite or countably infinite.)

Example 1.5. Let X be a set and let $\{A_i\}_{i \in I}$ be a *partition of* X, i.e., A_i is a nonempty subset of X for each $i \in I$, $A_i \cap A_j = \emptyset$ for $i \neq j$, and $X = \bigcup_{i \in I} A_i$. Then

$$\mathcal{A} := \{A_J := \bigcup_{j \in J} A_j \mid J \subset I\}$$

is a σ-algebra.

Exercise 1.6. (i) Let X be a set and let $A, B \subset X$ be subsets such that the four sets $A \setminus B, B \setminus A, A \cap B, X \setminus (A \cup B)$ are nonempty. What is the cardinality of the smallest σ-algebra $\mathcal{A} \subset X$ containing A and B?

(ii) How many σ-algebras on X are there when $\#X = k$ for $k = 0, 1, 2, 3, 4$?

(iii) Is there an infinite σ-algebra with countable cardinality?

Exercise 1.7. Let X be any set and let I be any nonempty index set. Suppose that for every $i \in I$ a σ-algebra $\mathcal{A}_i \subset 2^X$ is given. Prove that the intersection $\mathcal{A} := \bigcap_{i \in I} \mathcal{A}_i = \{A \subset X \mid A \in \mathcal{A}_i \text{ for all } i \in I\}$ is a σ-algebra.

Lemma 1.8. *Let X be a set and $\mathcal{E} \subset 2^X$ be any set of subsets of X. Then there is a unique smallest σ-algebra $\mathcal{A} \subset 2^X$ containing $\mathcal{E}$ (i.e., $\mathcal{A}$ is a σ-algebra, $\mathcal{E} \subset \mathcal{A}$, and if $\mathcal{B}$ is any other σ-algebra with $\mathcal{E} \subset \mathcal{B}$, then $\mathcal{A} \subset \mathcal{B}$).*

Proof. Uniqueness follows directly from the definition. Namely, if $\mathcal{A}$ and $\mathcal{B}$ are two smallest σ-algebras containing $\mathcal{E}$, we have both $\mathcal{B} \subset \mathcal{A}$ and $\mathcal{A} \subset \mathcal{B}$ and hence $\mathcal{A} = \mathcal{B}$. To prove existence, denote by $\mathcal{S} \subset 2^{2^X}$ the collection of all σ-algebras $\mathcal{B} \subset 2^X$ that contain $\mathcal{E}$ and define

$$\mathcal{A} := \bigcap_{\mathcal{B} \in \mathcal{S}} \mathcal{B} = \left\{ A \subset X \;\middle|\; \begin{array}{l} \text{if } \mathcal{B} \subset 2^X \text{ is a } \sigma\text{-algebra} \\ \text{such that } \mathcal{E} \subset \mathcal{B}, \text{ then } A \in \mathcal{B} \end{array} \right\}.$$

Thus $\mathcal{A}$ is a σ-algebra by Exercise 1.7. Moreover, it follows directly from the definition of $\mathcal{A}$ that $\mathcal{E} \subset \mathcal{A}$ and that every σ-algebra $\mathcal{B}$ that contains $\mathcal{E}$ also contains $\mathcal{A}$. This proves Lemma 1.8. □

Lemma 1.8 is a useful tool to construct nontrivial σ-algebras. Before doing that, let us first take a closer look at Definition 1.1. The letter "σ" stands for *countable* and the crucial observation is that axiom (c) allows for countable unions. On the one hand, this is a lot more general than only allowing for finite unions, which would be the subject of *Boolean algebra.* On the other hand, it is a lot more restrictive than allowing for arbitrary unions, which one encounters in the subject of *topology.* Topological spaces will play a central role in this book and we recall here the formal definition.

Definition 1.9 (topological space). Let X be a set. A collection $\mathcal{U} \subset 2^X$ of subsets of X is called a *topology on* X if it satisfies the following axioms:

(a) $\emptyset, X \in \mathcal{U}$;

(b) if $n \in \mathbb{N}$ and $U_1, \dots, U_n \in \mathcal{U}$, then $\bigcap_{i=1}^n U_i \in \mathcal{U}$;

(c) if I is any index set and $U_i \in \mathcal{U}$ for $i \in I$, then $\bigcup_{i \in I} U_i \in \mathcal{U}$.

A *topological space* is a pair $(X, \mathcal{U})$ consisting of a set X and a topology $\mathcal{U} \subset 2^X$. If $(X, \mathcal{U})$ is a topological space, the elements of $\mathcal{U}$ are called *open sets,* and a subset $F \subset X$ is called *closed* if its complement is open, i.e., $F^c \in \mathcal{U}$. Thus finite intersections of open sets are open and arbitrary unions of open sets are open. Likewise, finite unions of closed sets are closed and arbitrary intersections of closed sets are closed.

Conditions (a) and (b) in Definition 1.9 are also properties of every σ-algebra. However, condition (c) in Definition 1.9 is not shared by σ-algebras because it permits arbitrary unions. On the other hand, complements of open sets are typically not open. Many of the topologies used in this book arise from metric spaces and are familiar from first-year analysis courses. Here is a recollection of the definition.

Definition 1.10 (metric space). A *metric space* is a pair (X, d) consisting of a set X and a function

$$d: X \times X \longrightarrow \mathbb{R}$$

satisfying the following axioms:

(a) $d(x, y) \geq 0$ for all $x, y \in X$, with equality if and only if $x = y$;

(b) $d(x, y) = d(y, x)$ for all $x, y \in X$;

(c) $d(x, z) \leq d(x, y) + d(y, z)$ for all $x, y, z \in X$.

A function $d: X \times X \to \mathbb{R}$ that satisfies these axioms is called a *distance function* and the inequality in (c) is called the *triangle inequality*. A subset $U \subset X$ of a metric space (X, d) is called *open* (or *d-open*) if for every $x \in U$ there exists a constant $\varepsilon > 0$ such that the *open ball*

$$B_\varepsilon(x) := B_\varepsilon(x, d) := \{y \in X \mid d(x, y) < \varepsilon\}$$

(centered at x and of radius ε) is contained in U. The collection of d-open subsets of X will be denoted by

$$\mathcal{U}(X, d) := \{U \subset X \mid U \text{ is } d\text{-open}\}.$$

It follows directly from the definitions that the collection $\mathcal{U}(X, d) \subset 2^X$ of d-open sets in a metric space (X, d) satisfies the axioms of a topology in Definition 1.9. A subset F of a metric space (X, d) is closed if and only if the limit point of every convergent sequence in F is itself contained in F.

Example 1.11. A *normed vector space* is a pair $(X, \|\cdot\|)$ consisting of a real vector space X and a function

$$X \longrightarrow \mathbb{R}: x \longmapsto \|x\|$$

satisfying the following:

(a) $\|x\| \geq 0$ for all $x \in X$, with equality if and only if $x = 0$;

(b) $\|\lambda x\| = |\lambda| \, \|x\|$ for all $x \in X$ and $\lambda \in \mathbb{R}$;

(c) $\|x + y\| \leq \|x\| + \|y\|$ for all $x, y \in X$.

Let $(X, \|\cdot\|)$ be a normed vector space. Then the formula

$$d(x, y) := \|x - y\|$$

defines a distance function on X. X is called a *Banach space* if the metric space (X, d) is *complete,* i.e., if every Cauchy sequence in X converges.

Example 1.12. The set $X = \mathbb{R}$ of real numbers is a metric space with the standard distance function

$$d(x, y) := |x - y|.$$

The topology on $\mathbb{R}$ induced by this distance function is called the *standard topology on* $\mathbb{R}$. The open sets in the standard topology are unions of open intervals.

Exercise. Every union of open intervals is a *countable* union of open intervals.

Exercise 1.13. Consider the set

$$\overline{\mathbb{R}} := [-\infty, \infty] := \mathbb{R} \cup \{-\infty, \infty\}.$$

For $a, b \in \mathbb{R}$ define

$$(a, \infty] := (a, \infty) \cup \{\infty\}, \quad [-\infty, b) := (-\infty, b) \cup \{-\infty\}.$$

Call a subset $U \subset \overline{\mathbb{R}}$ *open* if it is a countable union of open intervals in $\mathbb{R}$ and sets of the form $(a, \infty]$ or $[-\infty, b)$ for $a, b \in \mathbb{R}$.

(i) Show that the set of open subsets of $\overline{\mathbb{R}}$ satisfies the axioms of a topology. This is called the *standard topology on* $\overline{\mathbb{R}}$.

(ii) Prove that the standard topology on $\overline{\mathbb{R}}$ is induced by the distance function

$$d : \overline{\mathbb{R}} \times \overline{\mathbb{R}} \longrightarrow \mathbb{R}$$

defined by the following formulas for $x, y \in \mathbb{R}$:

$$d(x, y) := \frac{2|e^{x-y} - e^{y-x}|}{e^{x+y} + e^{x-y} + e^{y-x} + e^{-x-y}},$$

$$d(x, \infty) := d(\infty, x) := \frac{2e^{-x}}{e^x + e^{-x}},$$

$$d(x, -\infty) := d(-\infty, x) := \frac{2e^x}{e^x + e^{-x}},$$

$$d(-\infty, \infty) := d(\infty, -\infty) := 2.$$

(iii) Prove that the map

$$f: \overline{\mathbb{R}} \longrightarrow [-1, 1]$$

defined by

$$f(x) := \tanh(x) := \frac{e^x - e^{-x}}{e^x + e^{-x}}, \qquad x \in \mathbb{R},$$

$$f(\pm\infty) := \pm 1,$$

is a homeomorphism. Prove that it is an isometry with respect to the metric in (ii) on $\overline{\mathbb{R}}$ and the standard metric on the interval $[-1, 1]$. Deduce that $(\overline{\mathbb{R}}, d)$ is a compact metric space.

Exercise 1.14. Extend the total ordering of $\mathbb{R}$ to $\overline{\mathbb{R}}$ by $-\infty \leq a \leq \infty$ for $a \in \overline{\mathbb{R}}$. Extend addition by

$$\infty + a := \infty \qquad \text{for } -\infty < a \leq \infty,$$

and by

$$-\infty + a := -\infty \qquad \text{for } -\infty \leq a < \infty.$$

(The sum $a + b$ is undefined when $\{a, b\} = \{-\infty, \infty\}$.) Let $a_1, a_2, a_3, \ldots$ and $b_1, b_2, b_3, \ldots$ be sequences in $\overline{\mathbb{R}}$.

(i) Define $\limsup_{n\to\infty} a_n$ and $\liminf_{n\to\infty} a_n$ and show that they always exist.

(ii) Show that

$$\limsup_{n\to\infty}(-a_n) = -\liminf_{n\to\infty} a_n.$$

(iii) Assume $\{a_n, b_n\} \neq \{-\infty, \infty\}$ so the sum $a_n + b_n$ is defined for $n \in \mathbb{N}$. Prove the inequality

$$\limsup_{n\to\infty}(a_n + b_n) \leq \limsup_{n\to\infty} a_n + \limsup_{n\to\infty} b_n,$$

whenever the right-hand side exists. Find an example where the inequality is strict.

(iv) If $a_n \leq b_n$ for all $n \in \mathbb{N}$, show that

$$\liminf_{n\to\infty} a_n \leq \liminf_{n\to\infty} b_n.$$

Definition 1.15. Let $(X, \mathcal{U})$ be a topological space and let $\mathcal{B} \subset 2^X$ be the smallest σ-algebra containing $\mathcal{U}$. Then $\mathcal{B}$ is called the *Borel σ-algebra of* $(X, \mathcal{U})$ and the elements of $\mathcal{B}$ are called *Borel (measurable) sets*.

Lemma 1.16. *Let $(X, \mathfrak{U})$ be a topological space. Then the following holds.*

(i) *Every closed subset $F \subset X$ is a Borel set.*

(ii) *Every countable union $\bigcup_{i=1}^{\infty} F_i$ of closed subsets $F_i \subset X$ is a Borel set. (These are sometimes called **F_σ-sets**.)*

(iii) *Every countable intersection $\bigcap_{i=1}^{\infty} U_i$ of open subsets $U_i \subset X$ is a Borel set. (These are sometimes called **G_δ-sets**.)*

Proof. Part (i) follows from the definition of Borel sets and condition (b) in Definition 1.1, part (ii) follows from (i) and (c), and part (iii) follows from (ii) and (b), because the complement of an F_σ-set is a G_δ-set. □

Consider for example the Borel σ-algebra on the real axis $\mathbb{R}$ with its standard topology. In view of Lemma 1.16 it is a legitimate question whether there is any subset of $\mathbb{R}$ at all that is not a Borel set. The answer to this question is positive, which may not be surprising; however, the proof of the existence of subsets that are not Borel sets is surprisingly nontrivial. It will only appear much later in this book, after we have introduced the Lebesgue measure (see Lemma 2.15). For now it is useful to note that, roughly speaking, every set that one can construct in terms of some explicit formula, will be a Borel set, and one can only prove with the Axiom of Choice that subsets of $\mathbb{R}$ must exist that are not Borel sets.

Recollections about point set topology

We close this section with a digression into some basic notions in topology that, at least for metric spaces, are familiar from first-year analysis and will be used throughout this book. The two concepts we recall here are *compactness* and *continuity*. A subset $K \subset X$ of a metric space (X, d) is called *compact* if every sequence in K has a subsequence that converges to some element of K. Thus, in particular, every compact subset is closed. The notion of compactness carries over to general topological spaces as follows.

Let $(X, \mathfrak{U})$ be a topological space and let $K \subset X$. An *open cover* of K is a collection of open sets $\{U_i\}_{i \in I}$, indexed by a set I, such that

$$K \subset \bigcup_{i \in I} U_i.$$

The set K is called *compact* if every open cover of K has a finite subcover, i.e., if for every open cover $\{U_i\}_{i \in I}$ of K there exist finitely many indices $i_1, \dots, i_n \in I$ such that

$$K \subset U_{i_1} \cup \dots \cup U_{i_n}.$$

When (X, d) is a metric space and $\mathcal{U} = \mathcal{U}(X, d)$ is the topology induced by the distance function (Definition 1.10), the two notions of compactness agree. Thus, for every subset $K \subset X$, every sequence in K has a subsequence converging to an element of K if and only if every open cover of K has a finite subcover. For a proof see, for example, Munkres [14] or [20, Appendix C.1]. We emphasize that when K is a compact subset of a general topological space $(X, \mathcal{U})$ it does not follow that K is closed. For example, a finite subset of X is always compact, but need not be closed or, if $\mathcal{U} = \{\emptyset, X\}$, then every subset of X is compact, but only the empty set and X itself are closed subsets of X. If, however, $(X, \mathcal{U})$ is a *Hausdorff space* (i.e., for any two distinct points $x, y \in X$ there exist open sets $U, V \in \mathcal{U}$ such that $x \in U$, $y \in V$, and $U \cap V = \emptyset$), then every compact subset of X is closed (Lemma A.2).

Next, recall that a map $f: X \to Y$ between two metric spaces (X, d_X) and (Y, d_Y) is continuous (i.e., for every $x \in X$ and every $\varepsilon > 0$ there is a $\delta > 0$ such that $f(B_\delta(x, d_X)) \subset B_\varepsilon(f(x), d_Y))$ if and only if the pre-image

$$f^{-1}(V) := \{x \in X \mid f(x) \in V\}$$

of every open subset of Y is an open subset of X. This second notion carries over to general topological spaces, i.e., a map $f: X \to Y$ between topological spaces $(X, \mathcal{U}_X)$ and $(Y, \mathcal{U}_Y)$ is called *continuous* if

$$V \in \mathcal{U}_Y \quad \Longrightarrow \quad f^{-1}(V) \in \mathcal{U}_X.$$

It follows directly from the definition that topological spaces form a *category,* in that the composition $g \circ f: X \to Z$ of two continuous maps $f: X \to Y$ and $g: Y \to Z$ between topological spaces is again continuous. Another basic observation is that if $f: X \to Y$ is a continuous map between topological spaces and K is a compact subset of X, then its image $f(K)$ is a compact subset of Y.

1.2 Measurable functions

In analogy to continuous maps between topological spaces, one can define measurable maps between measurable spaces as those maps under which pre-images of measurable sets are again measurable. A slightly different approach is taken by Rudin [17], who defines a measurable map from a measurable space to a topological space as one under which pre-images of open sets are measurable. Both definitions agree whenever the target space is equipped with its Borel σ-algebra.

As a warmup we begin with some recollections about pre-images of sets that are also relevant for the above discussion of continuity. For any map $f: X \to Y$ between two sets X and Y and any subset $B \subset Y$, the *pre-image*

$$f^{-1}(B) := \{x \in X \mid f(x) \in B\}$$

of B under f is a well-defined subset of X, whether or not the map f is bijective, i.e., even if there does not exist any map $f^{-1}: Y \to X$. The pre-image defines a map from 2^Y to 2^X. It satisfies

$$f^{-1}(Y) = X, \quad f^{-1}(\emptyset) = \emptyset, \tag{1.1}$$

and preserves union, intersection, and complement. Thus

$$f^{-1}(Y \setminus B) = X \setminus f^{-1}(B) \tag{1.2}$$

for every subset $B \subset Y$, and

$$f^{-1}\Big(\bigcup_{i \in I} B_i\Big) = \bigcup_{i \in I} f^{-1}(B_i), \quad f^{-1}\Big(\bigcap_{i \in I} B_i\Big) = \bigcap_{i \in I} f^{-1}(B_i) \tag{1.3}$$

for every collection of subsets $B_i \subset Y$, indexed by a set I.

Definition 1.17 (measurable function). (i) Let $(X, \mathcal{A}_X)$ and $(Y, \mathcal{A}_Y)$ be measurable spaces. A map $f: X \to Y$ is called *measurable* if the pre-image of every measurable subset of Y under f is a measurable subset of X, i.e.,

$$B \in \mathcal{A}_Y \implies f^{-1}(B) \in \mathcal{A}_X.$$

(ii) Let $(X, \mathcal{A}_X)$ be a measurable space. A function $f: X \to \overline{\mathbb{R}}$ is called *measurable* if it is measurable with respect to the Borel σ-algebra on $\overline{\mathbb{R}}$ associated to the standard topology in Exercise 1.13 (see Definition 1.15).

(iii) Let $(X, \mathcal{U}_X)$ and $(Y, \mathcal{U}_Y)$ be topological spaces. A map $f: X \to Y$ is called *Borel measurable* if the pre-image of every Borel measurable subset of Y under f is a Borel measurable subset of X.

Example 1.18. Let X be a set. The *characteristic function* of a subset $A \subset X$ is the function

$$\chi_A: X \longrightarrow \mathbb{R}$$

defined by

$$\chi_A(x) := \begin{cases} 1 & \text{if } x \in A, \\ 0 & \text{if } x \notin A. \end{cases} \tag{1.4}$$

Now assume $(X, \mathcal{A})$ is a measurable space, consider the Borel σ-algebra on $\mathbb{R}$, and let $A \subset X$ be any subset. Then χ_A is a measurable function if and only if A is a measurable set.

Part (iii) in Definition 1.17 is the special case of part (i), where $\mathcal{A}_X \subset 2^X$ and $\mathcal{A}_Y \subset 2^Y$ are the σ-algebras of Borel sets (see Definition 1.15). Theorem 1.20 below shows that every continuous function between topological spaces is Borel measurable. It also shows that a function from a measurable space to a topological space is measurable with respect to the Borel σ-algebra on the target space if and only if the pre-image of every open set is measurable. Since the collection of Borel sets is in general much larger than the collection of open sets, the collection of measurable functions is then also much larger than the collection of continuous functions.

Theorem 1.19 (measurable maps). *Let $(X, \mathcal{A}_X)$, $(Y, \mathcal{A}_Y)$, and $(Z, \mathcal{A}_Z)$ be measurable spaces.*

(i) *The identity map* $\mathrm{id}_X \colon X \to X$ *is measurable.*

(ii) *If $f \colon X \to Y$ and $g \colon Y \to Z$ are measurable maps, then so is the composition $g \circ f \colon X \to Z$.*

(iii) *Let $f \colon X \to Y$ be any map. Then the set*

$$f_*\mathcal{A}_X := \{B \subset Y \mid f^{-1}(B) \in \mathcal{A}_X\} \tag{1.5}$$

is a σ-algebra on Y, called the ***pushforward of $\mathcal{A}_X$ under f***.

(iv) *A map $f \colon X \to Y$ is measurable if and only if $\mathcal{A}_Y \subset f_*\mathcal{A}_X$.*

Proof. Parts (i) and (ii) follow directly from the definitions. That $f_*\mathcal{A}_X \subset 2^Y$ defined by (1.5) is a σ-algebra follows from (1.1) (for axiom (a)), (1.2) (for axiom (b)), and (1.3) (for axiom (c)). This proves part (iii). Moreover by Definition 1.17, f is measurable if and only if $f^{-1}(B) \in \mathcal{A}_X$ for every $B \in \mathcal{A}_Y$, and this means that $\mathcal{A}_Y \subset f_*\mathcal{A}_X$. This establishes (iv) and completes the proof of Theorem 1.19. □

Theorem 1.20 (measurable and continuous maps). *Let $(X, \mathcal{A}_X)$ and $(Y, \mathcal{A}_Y)$ be measurable spaces. Assume $\mathcal{U}_Y \subset 2^Y$ is a topology on Y such that $\mathcal{A}_Y$ is the Borel σ-algebra of $(Y, \mathcal{U}_Y)$.*

(i) *A map $f \colon X \to Y$ is measurable if an only if the pre-image of every open subset $V \subset Y$ under f is measurable, i.e.,*

$$V \in \mathcal{U}_Y \implies f^{-1}(V) \in \mathcal{A}_X.$$

(ii) *Assume $\mathcal{U}_X \subset 2^X$ is a topology on X such that $\mathcal{A}_X$ is the Borel σ-algebra of $(X, \mathcal{U}_X)$. Then every continuous map $f \colon X \to Y$ is (Borel) measurable.*

Proof. By part (iv) of Theorem 1.19, a map $f: X \to Y$ is measurable if and only if $\mathcal{A}_Y \subset f_*\mathcal{A}_X$. Since $f_*\mathcal{A}_X$ is a σ-algebra on Y by part (iii) of Theorem 1.19, and the Borel σ-algebra $\mathcal{A}_Y$ is the smallest σ-algebra on Y containing the collection of open sets $\mathcal{U}_Y$ by Definition 1.15, it follows that $\mathcal{A}_Y \subset f_*\mathcal{A}_X$ if and only if $\mathcal{U}_Y \subset f_*\mathcal{A}_X$. By the definition of $f_*\mathcal{A}_X$ in (1.5), this translates into the condition

$$V \in \mathcal{U}_Y \implies f^{-1}(V) \in \mathcal{A}_X.$$

This proves part (i). If, in addition, $\mathcal{A}_X$ is the Borel σ-algebra of a topology $\mathcal{U}_X$ on X and $f: (X, \mathcal{U}_X) \to (Y, \mathcal{U}_Y)$ is a continuous map, then the pre-image of every open subset $V \subset Y$ under f is an open subset of X and hence is a Borel subset of X; thus it follows from part (i) that f is Borel measurable. This proves part (ii) and Theorem 1.20. □

Theorem 1.21 (characterization of measurable functions). *Let $(X, \mathcal{A})$ be a measurable space and let $f: X \to \overline{\mathbb{R}}$ be any function. Then the following are equivalent*:

(i) *f is measurable*;

(ii) *$f^{-1}((a, \infty])$ is a measurable subset of X for every $a \in \mathbb{R}$*;

(iii) *$f^{-1}([a, \infty])$ is a measurable subset of X for every $a \in \mathbb{R}$*;

(iv) *$f^{-1}([-\infty, b))$ is a measurable subset of X for every $b \in \mathbb{R}$*;

(v) *$f^{-1}([-\infty, b])$ is a measurable subset of X for every $b \in \mathbb{R}$*.

Proof. That (i) implies (ii), (iii), (iv), and (v) follows directly from the definitions. We prove that (ii) implies (i). Thus let $f: X \to \overline{\mathbb{R}}$ be such that $f^{-1}((a, \infty]) \in \mathcal{A}_X$ for every $a \in \mathbb{R}$ and define

$$\mathcal{B} := f_*\mathcal{A}_X = \{B \subset \overline{\mathbb{R}} \mid f^{-1}(B) \in \mathcal{A}_X\} \subset 2^{\overline{\mathbb{R}}}.$$

Then $\mathcal{B}$ is a σ-algebra on $\overline{\mathbb{R}}$ by part (iii) of Theorem 1.19 and $(a, \infty] \in \mathcal{B}$ for every $a \in \mathbb{R}$ by assumption. Therefore $[-\infty, b] = \overline{\mathbb{R}} \setminus (b, \infty] \in \mathcal{B}$ for every $b \in \mathbb{R}$ by axiom (b) and hence

$$[-\infty, b) = \bigcup_{n \in \mathbb{N}} [-\infty, b - \tfrac{1}{n}] \in \mathcal{B}$$

by axiom (c) in Definition 1.1. Hence it follows from (f) in Lemma 1.2 that

$$(a, b) = [-\infty, b) \cap (a, \infty] \in \mathcal{B}$$

for every pair of real numbers $a < b$. Since every open subset of $\overline{\mathbb{R}}$ is a countable union of sets of the form (a, b), $(a, \infty]$, $[-\infty, b)$, it follows from axiom (c) in Definition 1.1 that every open subset of $\overline{\mathbb{R}}$ is an element of $\mathcal{B}$. Hence it follows from Theorem 1.20 that f is measurable. This shows that (ii) implies (i). That either of the conditions (iii), (iv), and (v) also implies (i) is shown by a similar argument which is left as an exercise for the reader. This proves Theorem 1.21. □

Our next goal is to show that sums, products, and limits of measurable functions are again measurable. The next two results are useful for the proofs of these fundamental facts.

Theorem 1.22 (vector valued measurable functions). *Let $(X, \mathcal{A})$ be a measurable space and let $f = (f_1, \dots, f_n): X \to \mathbb{R}^n$ be a function. Then f is measurable if and only if $f_i: X \to \mathbb{R}$ is measurable for each i.*

Proof. For $i = 1, \dots, n$ define the projection $\pi_i: \mathbb{R}^n \to \mathbb{R}$ by $\pi_i(x) := x_i$ for $x = (x_1, \dots, x_n) \in \mathbb{R}$. Since π_i is continuous, it follows from Theorems 1.19 and 1.20 that if f is measurable, so is $f_i = \pi_i \circ f$ for all i. Conversely, suppose that f_i is measurable for $i = 1, \dots, n$. Let $a_i < b_i$ for $i = 1, \dots, n$ and define

$$Q(a, b) := \{x \in \mathbb{R}^n \mid a_i < x_i < b_i, \text{for all } i\} = (a_1, b_1) \times \cdots \times (a_n, b_n).$$

Then

$$f^{-1}(Q(a, b)) = \bigcap_{i=1}^{n} f_i^{-1}((a_i, b_i)) \in \mathcal{A}$$

by property (f) in Lemma 1.2. Now every open subset of $\mathbb{R}^n$ can be expressed as a countable union of sets of the form $Q(a, b)$. (Prove this!) Hence it follows from axiom (c) in Definition 1.1 that $f^{-1}(U) \in \mathcal{A}$ for every open set $U \subset \mathbb{R}^n$, and so f is measurable. This proves Theorem 1.22. □

Lemma 1.23. *Let $(X, \mathcal{A})$ be a measurable space and let $u, v: X \to \mathbb{R}$ be measurable functions. If $\phi: \mathbb{R}^2 \to \mathbb{R}$ is continuous, then the function $h: X \to \mathbb{R}$ defined by*

$$h(x) := \phi(u(x), v(x)) \quad \textit{for } x \in X$$

is measurable.

Proof. The map $f := (u, v): X \to \mathbb{R}^2$ is measurable (with respect to the Borel σ-algebra on $\mathbb{R}^2$) by Theorem 1.22 and the map $\phi: \mathbb{R}^2 \to \mathbb{R}$ is Borel measurable by Theorem 1.20. Hence the composition $h = \phi \circ f: X \to \mathbb{R}$ is measurable by Theorem 1.19. This proves Lemma 1.23. □

Theorem 1.24 (properties of measurable functions). *Let $(X, \mathcal{A})$ be a measurable space.*

(i) *If $f, g: X \to \mathbb{R}$ are measurable functions, then so are the functions*

$$f + g, \quad fg, \quad \max\{f, g\}, \quad \min\{f, g\}, \quad |f|.$$

(ii) *Let $f_k: X \to \overline{\mathbb{R}}$, $k = 1, 2, 3, \dots$, be a sequence of measurable functions. Then the following functions from X to $\overline{\mathbb{R}}$ are measurable:*

$$\inf_k f_k, \quad \sup_k f_k, \quad \limsup_{k\to\infty} f_k, \quad \liminf_{k\to\infty} f_k.$$

Proof. (i) The functions $\phi: \mathbb{R}^2 \to \mathbb{R}$ defined by $\phi(s,t) := s + t$, $\phi(s,t) := st$, $\phi(s,t) := \max\{s,t\}$, $\phi(s,t) := \min\{s,t\}$, or $\phi(s,t) := |s|$ are all continuous. Hence assertion (i) follows from Lemma 1.23.

(ii) Define

$$g := \sup_k f_k: X \longrightarrow \overline{\mathbb{R}}$$

and let $a \in \mathbb{R}$. Then the set

$$\begin{aligned} g^{-1}((a,\infty]) &= \{x \in X \mid \sup_k f_k(x) > a\} \\ &= \{x \in X \mid \text{there exists } k \in \mathbb{N} \text{ such that } f_k(x) > a\} \\ &= \bigcup_{k\in\mathbb{N}} \{x \in X \mid f_k(x) > a\} \\ &= \bigcup_{k\in\mathbb{N}} f_k^{-1}((a,\infty]) \end{aligned}$$

is measurable. Hence it follows from Theorem 1.21 that g is measurable. It also follows from part (i) (already proved) that $-f_k$ is measurable, hence so is $\sup_k(-f_k)$ by what we have just proved, and then so is the function $\inf_k f_k = -\sup_k(-f_k)$. With this understood, we conclude that the functions

$$\limsup_{k\to\infty} f_k = \inf_{\ell\in\mathbb{N}} \sup_{k\geq\ell} f_k, \quad \liminf_{k\to\infty} f_k = \sup_{\ell\in\mathbb{N}} \inf_{k\geq\ell} f_k$$

are also measurable. This proves Theorem 1.24. □

It follows from Theorem 1.24 that the pointwise limit of a sequence of measurable functions, if it exists, is again measurable. This is in sharp contrast to Riemann integrable functions.

Step functions

We close this section with a brief discussion of measurable step functions. Such functions will play a central role throughout this book. In particular, they are used in the definition of the Lebesgue integral.

Definition 1.25 (step function). Let X be a set. A function $s\colon X \to \mathbb{R}$ is called a *step function* (or *simple function*) if it takes on only finitely many values, i.e., the image $s(X)$ is a finite subset of $\mathbb{R}$.

Let $s\colon X \to \mathbb{R}$ be a step function, write

$$s(X) = \{\alpha_1, \dots, \alpha_\ell\} \quad \text{with } \alpha_i \neq \alpha_j \text{ for } i \neq j,$$

and define

$$A_i := s^{-1}(\alpha_i) = \{x \in X \mid s(x) = \alpha_i\}$$

for $i = 1, \dots, \ell$. Then the sets $A_1, \dots, A_\ell$ form a partition of X, i.e.,

$$X = \bigcup_{i=1}^{\ell} A_i, \qquad A_i \cap A_j = \emptyset \text{ for } i \neq j. \tag{1.6}$$

(See Example 1.5.) Moreover,

$$s = \sum_{i=1}^{\ell} \alpha_i \chi_{A_i}, \tag{1.7}$$

where $\chi_{A_i}\colon X \to \mathbb{R}$ is the characteristic function of the set A_i for $i = 1, \dots, \ell$ (see (1.4)). In this situation s is measurable if and only if the set $A_i \subset X$ is measurable for each i. For later reference we prove the following.

Theorem 1.26 (approximation). *Let $(X, \mathcal{A})$ be a measurable space and let*

$$f\colon X \longrightarrow [0, \infty]$$

be a function. Then f is measurable if and only if there exists a sequence of measurable step functions $s_n\colon X \to [0, \infty)$ such that

$$0 \le s_1(x) \le s_2(x) \le \cdots \le f(x), \qquad f(x) = \lim_{n\to\infty} s_n(x) \quad \textit{for all } x \in X.$$

Proof. If f can be approximated by a sequence of measurable step functions, then f is measurable by Theorem 1.24. Conversely, suppose that f is measurable. For $n \in \mathbb{N}$ define

$$\phi_n\colon [0,\infty] \longrightarrow \mathbb{R}$$

by

$$\phi_n(t) := \begin{cases} k2^{-n} & \text{if } k2^{-n} \le t < (k+1)2^{-n},\ k = 0,1,\dots,n2^n - 1, \\ n & \text{if } t \ge n. \end{cases} \tag{1.8}$$

These functions are Borel measurable and satisfy $\phi_n(0) = 0$ and $\phi_n(\infty) = n$ for all n, as well as $t - 2^{-n} \le \phi_n(t) \le \phi_{n+1}(t) \le t$ whenever $n \ge t > 0$. Thus

$$\lim_{n\to\infty} \phi_n(t) = t \quad \text{for all } t \in [0,\infty].$$

Hence the functions $s_n := \phi_n \circ f$ satisfy the requirements of the theorem. □

1.3 Integration of nonnegative functions

Our next goal is to define the integral of a measurable step function and then the integral of a general nonnegative measurable function via approximation. This requires the notion of *volume* or *measure* of a measurable set. The definitions of measure and integral will require some arithmetic on the space $[0,\infty]$. Addition to ∞ and multiplication by ∞ are defined by

$$a + \infty := \infty + a := \infty, \qquad a \cdot \infty := \infty \cdot a := \begin{cases} \infty & \text{if } a \ne 0, \\ 0 & \text{if } a = 0. \end{cases}$$

With this convention, addition and multiplication are commutative, associative, and distributive. Moreover, if a_i and b_i are nondecreasing sequences in $[0,\infty]$, then the limits

$$a := \lim_{i\to\infty} a_i \quad \text{and} \quad b := \lim_{i\to\infty} b_i$$

exists in $[0,\infty]$ and satisfy the familiar rules

$$a + b = \lim_{i\to\infty}(a_i + b_i) \quad \text{and} \quad ab = \lim_{i\to\infty}(a_i b_i).$$

These rules must be treated with caution. The product rule does not hold when the sequences are not nondecreasing. For example $a_i := i$ converges to $a = \infty$, $b_i := 1/i$ converges to $b = 0$, but $a_i b_i = 1$ does not converge to $ab = 0$. (*Exercise.* Show that the sum of two convergent sequences in $[0, \infty]$ always converges to the sum of the limits.) Also, for all $a, b, c \in [0, \infty]$,

$$a + b = a + c, \qquad a < \infty \implies b = c,$$

$$ab = ac, \qquad 0 < a < \infty \implies b = c.$$

Neither of these assertions extend to the case $a = \infty$.

Definition 1.27 (measure). Let $(X, \mathcal{A})$ be a measurable space. A *measure on* $(X, \mathcal{A})$ is a function

$$\mu \colon \mathcal{A} \longrightarrow [0, \infty]$$

satisfying the following axioms.

(a) μ is *σ-additive*, i.e., if $A_i \in \mathcal{A}$, $i = 1, 2, 3, \dots$, is a sequence of pairwise disjoint measurable sets, then

$$\mu\Big(\bigcup_{i=1}^{\infty} A_i\Big) = \sum_{i=1}^{\infty} \mu(A_i).$$

(b) There exists a measurable set $A \in \mathcal{A}$ such that $\mu(A) < \infty$.

A *measure space* is a triple $(X, \mathcal{A}, \mu)$ consisting of a set X, a σ-algebra $\mathcal{A} \subset 2^X$, and a measure $\mu \colon \mathcal{A} \to [0, \infty]$.

The basic properties of measures are summarized in the next theorem.

Theorem 1.28 (properties of measures). *Let $(X, \mathcal{A}, \mu)$ be a measure space. Then the following holds.*

(i) $\mu(\emptyset) = 0$.

(ii) *If $n \in \mathbb{N}$ and $A_1, \dots, A_n \in \mathcal{A}$ are such that $A_i \cap A_j = \emptyset$ for $i \neq j$, then*

$$\mu(A_1 \cup \dots \cup A_n) = \mu(A_1) + \dots + \mu(A_n).$$

(iii) *If $A, B \in \mathcal{A}$ are such that $A \subset B$, then*

$$\mu(A) \leq \mu(B).$$

(iv) *Let $A_i \in \mathcal{A}$ be a sequence such that $A_i \subset A_{i+1}$ for all i. Then*

$$\mu\Big(\bigcup_{i=1}^{\infty} A_i\Big) = \lim_{i\to\infty} \mu(A_i).$$

(v) *Let $A_i \in \mathcal{A}$ be a sequence such that $A_i \supset A_{i+1}$ for all i. Then*

$$\mu(A_1) < \infty \implies \mu\Big(\bigcap_{i=1}^{\infty} A_i\Big) = \lim_{i\to\infty} \mu(A_i).$$

Proof. We prove (i). Choose $A_1 \in \mathcal{A}$ such that $\mu(A_1) < \infty$ and define $A_i := \emptyset$ for $i > 1$. Then it follows from σ-additivity that

$$\mu(A_1) = \mu(A_1) + \sum_{i>1} \mu(\emptyset),$$

and hence $\mu(\emptyset) = 0$. This proves part (i).

Part (ii) follows from (i) and σ-additivity by choosing $A_i := \emptyset$ for $i > n$.

We prove (iii). If $A, B \in \mathcal{A}$ are such that $A \subset B$, then $B \setminus A \in \mathcal{A}$ by property (g) in Lemma 1.2 and hence $\mu(B) = \mu(A) + \mu(B \setminus A) \geq \mu(A)$ by part (ii). This proves part (iii).

We prove (iv). Assume $A_i \subset A_{i+1}$ for all i and set

$$B_i := \begin{cases} A_1 & \text{for } i = 1, \\ A_i \setminus A_{i-1} & \text{for } i > 1. \end{cases}$$

Then B_i is measurable for all i and, for $n \in \mathbb{N}$,

$$A_n = \bigcup_{i=1}^{n} B_i, \quad A := \bigcup_{i=1}^{\infty} A_i = \bigcup_{i=1}^{\infty} B_i.$$

Since $B_i \cap B_j = \emptyset$ for $i \neq j$, it follows from σ-additivity that

$$\mu(A) = \sum_{i=1}^{\infty} \mu(B_i) = \lim_{n\to\infty} \sum_{i=1}^{n} \mu(B_i) = \lim_{n\to\infty} \mu(A_n).$$

Here the last equality follows from part (ii). This proves part (iv).

We prove (v). Assume $A_i \supset A_{i+1}$ for all i and define $C_i := A_i \setminus A_{i+1}$. Then C_i is measurable for all i and, for $n \in \mathbb{N}$,

$$A_n = A \cup \bigcup_{i=n}^{\infty} C_i, \quad A := \bigcap_{i=1}^{\infty} A_i.$$

Since $C_i \cap C_j = \emptyset$ for $i \neq j$, σ-additivity yields

$$\mu(A_n) = \mu(A) + \sum_{i=n}^{\infty} \mu(C_i)$$

for all $n \in \mathbb{N}$. Since $\mu(A_1) < \infty$. It follows that $\sum_{i=1}^{\infty} \mu(C_i) < \infty$ and hence

$$\lim_{n\to\infty} \mu(A_n) = \mu(A) + \lim_{n\to\infty} \sum_{i=n}^{\infty} \mu(C_i) = \mu(A).$$

This proves part (v) and Theorem 1.28. □

Exercise 1.29. Let $(X, \mathcal{A}, \mu)$ be a measure space and let $A_i \in \mathcal{A}$ be a sequence of measurable sets. Prove that $\mu\big(\bigcup_i A_i\big) \leq \sum_i \mu(A_i)$.

Example 1.30. Let $(X, \mathcal{A})$ be a measurable space. The *counting measure*

$$\mu \colon \mathcal{A} \longrightarrow [0, \infty]$$

is defined by

$$\mu(A) := \#A \quad \text{for } A \in \mathcal{A}.$$

As an example, consider the counting measure $\mu \colon 2^{\mathbb{N}} \to [0, \infty]$ on the natural numbers. Then the sets $A_n := \{n, n+1, \cdots\}$ all have infinite measure and their intersection is the empty set and hence has measure zero. Thus the hypothesis $\mu(A_1) < \infty$ cannot be removed in part (v) of Theorem 1.28.

Example 1.31. Let $(X, \mathcal{A})$ be a measurable space and fix an element $x_0 \in X$. The *Dirac measure at* x_0 is the measure $\delta_{x_0} \colon \mathcal{A} \to [0, \infty]$ defined by

$$\delta_{x_0}(A) := \begin{cases} 1 & \text{if } x_0 \in A, \\ 0 & \text{if } x_0 \notin A, \end{cases} \qquad \text{for } A \in \mathcal{A}.$$

Example 1.32. Let X be an uncountable set and let $\mathcal{A}$ be the σ-algebra of all subsets of X that are either countable or have countable complements (Example 1.4). Then the function $\mu \colon \mathcal{A} \to [0, 1]$ defined by $\mu(A) := 0$ when A is countable and by $\mu(A) := 1$ when A^{c} is countable is a measure.

Example 1.33. Let $X = \bigcup_{i\in I} A_i$ be a partition and let $\mathcal{A} \subset 2^X$ be the σ-algebra in Example 1.5. Then any function $I \to [0, \infty] \colon i \mapsto \alpha_i$ determines a measure $\mu \colon \mathcal{A} \to [0, \infty]$ via $\mu(A_J) := \sum_{j\in J} \alpha_j$ for $J \subset I$ and $A_J = \bigcup_{j\in J} A_j \in \mathcal{A}$.

With these preparations in place we are now ready to introduce the Lebesgue integral of a nonnegative measurable function

Definition 1.34 (Lebesgue integral). Let $(X, \mathcal{A}, \mu)$ be a measure space and let $E \in \mathcal{A}$ be a measurable set.

(i) Let $s: X \to [0, \infty)$ be a measurable step function of the form

$$s = \sum_{i=1}^{n} \alpha_i \chi_{A_i} \tag{1.9}$$

with $\alpha_i \in [0, \infty)$ and $A_i \in \mathcal{A}$ for $i = 1, \dots, n$. The *(Lebesgue) integral of* s *over* E is the number $\int_E s \, d\mu \in [0, \infty]$ defined by

$$\int_E s \, d\mu := \sum_{i=1}^{n} \alpha_i \mu(E \cap A_i). \tag{1.10}$$

(ii) Let $f: X \to [0, \infty]$ be a measurable function. The *(Lebesgue) integral of* f *over* E is the number $\int_E f \, d\mu \in [0, \infty]$ defined by

$$\int_E f \, d\mu := \sup_{s \leq f} \int_E s \, d\mu,$$

where the supremum is taken over all measurable step function $s: X \to [0, \infty)$ that satisfy $s(x) \leq f(x)$ for all $x \in X$.

The reader may verify that the right-hand side of (1.10) depends only on s and not on the choice of α_i and A_i. The same definition can be used if f is only defined on the measurable set $E \subset X$. Then $\mathcal{A}_E := \{A \in \mathcal{A} \mid A \subset E\}$ is a σ-algebra on E and $\mu_E := \mu|_{\mathcal{A}_E}$ is a measure. So $(E, \mathcal{A}_E, \mu_E)$ is a measure space and the integral $\int_E f \, d\mu_E$ is well defined. It agrees with the integral of the extended function on X defined by $f(x) := 0$ for $x \in X \setminus E$.

Theorem 1.35 (basic properties of the Lebesgue integral). *Let $(X, \mathcal{A}, \mu)$ be a measure space, $f, g: X \to [0, \infty]$ be measurable functions, and $E \in \mathcal{A}$. Then the following holds:*

(i) *if $f \leq g$ on E, then $\int_E f \, d\mu \leq \int_E g \, d\mu$;*

(ii) $\int_E f \, d\mu = \int_X f \chi_E \, d\mu$;

(iii) *if $f(x) = 0$ for all $x \in E$, then $\int_E f \, d\mu = 0$;*

(iv) *if $\mu(E) = 0$, then $\int_E f \, d\mu = 0$;*

(v) *if $A \in \mathcal{A}$ and $E \subset A$, then $\int_E f \, d\mu \leq \int_A f \, d\mu$;*

(vi) *if $c \in [0, \infty)$, then $\int_E c f \, d\mu = c \int_E f \, d\mu$.*

Proof. To prove (i), assume $f \le g$ on E. If $s\colon X \to [0,\infty)$ is a measurable step function such that $s \le f$, then $s\chi_E \le g$, so

$$\int_E s\,d\mu = \int_E s\chi_E\,d\mu \le \int_E g\,d\mu$$

by the definition of the integral of g. Now take the supremum over all measurable step functions $s \le f$ to obtain

$$\int_E f\,d\mu \le \int_E g\,d\mu.$$

This proves (i).

We prove (ii). It follows from the definitions that

$$\begin{aligned}\int_E f\,d\mu &= \sup_{s\le f}\int_E s\,d\mu \\ &= \sup_{s\le f}\int_X s\chi_E\,d\mu \\ &= \sup_{t\le f\chi_E}\int_X t\,d\mu \\ &= \int_X f\chi_E\,d\mu.\end{aligned}$$

Here the supremum is over all measurable step functions $s\colon X \to [0,\infty)$, respectively $t\colon X \to [0,\infty)$, that satisfy $s \le f$, respectively $t \le f\chi_E$. The second equality follows from the fact that every measurable step function $s\colon X \to [0,\infty)$ satisfies $\int_E s\,d\mu = \int_X s\chi_E\,d\mu$ by the definition of the integral. The third equation follows from the fact that a measurable step function $t\colon X \to [0,\infty)$ satisfies $t \le f\chi_E$ if and only if it has the form $t = s\chi_E$ for some measurable step function $s\colon X \to [0,\infty)$ such that $s \le f$.

Part (iii) is a consequence of (i) with $g = 0$ and the fact that $\int_E f\,d\mu \ge 0$ by definition. Part (iv) follows from the fact that $\int_E s\,d\mu = 0$ for every measurable step function s when $\mu(E) = 0$. Part (v) follows from parts (i) and (ii) and the fact that $f\chi_E \le f\chi_A$ whenever $E \subset A$. Part (vi) follows from the fact that $\int_E cs\,d\mu = c\int_E s\,d\mu$ for every $c \in [0,\infty)$ and every measurable step function s, by the commutativity, associativity, and distributivity rules for calculations with numbers in $[0,\infty]$. This proves Theorem 1.35. □

Notably absent from the statements of Theorem 1.35 is the assertion that the integral of a sum is the sum of the integrals. This is a fundamental property that any integral should have. The proof that the integral in Definition 1.34 indeed satisfies this crucial condition requires some preparation. First one verifies this property for integrals of step functions and then one uses the Lebesgue monotone convergence Theorem 1.37.

Lemma 1.36 (additivity for step functions). *Let $(X, \mathcal{A}, \mu)$ be a measure space and let $s, t\colon X \to [0, \infty)$ be measurable step functions.*

(i) *For every measurable set $E \in \mathcal{A}$*

$$\int_E (s+t)\,d\mu = \int_E s\,d\mu + \int_E t\,d\mu.$$

(ii) *If $E_1, E_2, E_3, \dots$ is a sequence of pairwise disjoint measurable sets, then*

$$\int_E s\,d\mu = \sum_{k=1}^{\infty} \int_{E_k} s\,d\mu, \qquad E := \bigcup_{k\in\mathbb{N}} E_k.$$

Proof. Write the functions s and t in the form

$$s = \sum_{i=1}^{m} \alpha_i \chi_{A_i}, \qquad t = \sum_{j=1}^{n} \beta_j \chi_{B_j}$$

where $\alpha_i, \beta_j \in [0, \infty)$ and $A_i, B_j \in \mathcal{A}$ are such that $A_i \cap A_{i'} = \emptyset$ for $i \neq i'$, $B_j \cap B_{j'} = \emptyset$ for $j \neq j'$, and $X = \bigcup_{i=1}^m A_i = \bigcup_{j=1}^n B_j$. Then

$$s + t = \sum_{i=1}^{m}\sum_{j=1}^{n} (\alpha_i + \beta_j)\chi_{A_i \cap B_j},$$

and hence

$$\begin{aligned}
\int_E (s+t)\,d\mu &= \sum_{i=1}^{m}\sum_{j=1}^{n} (\alpha_i + \beta_j)\mu(A_i \cap B_j \cap E) \\
&= \sum_{i=1}^{m} \alpha_i \sum_{j=1}^{n} \mu(A_i \cap B_j \cap E) + \sum_{j=1}^{n} \beta_j \sum_{i=1}^{m} \mu(A_i \cap B_j \cap E) \\
&= \sum_{i=1}^{m} \alpha_i \mu(A_i \cap E) + \sum_{j=1}^{n} \beta_j \mu(B_j \cap E) \\
&= \int_E s\,d\mu + \int_E t\,d\mu.
\end{aligned}$$

To prove (ii), let $E_1, E_2, E_3, \dots$ be a sequence of pairwise disjoint measurable sets and define $E := \bigcup_{k=1}^{\infty} E_k$. Then

$$\begin{aligned}
\int_E s\,d\mu &= \sum_{i=1}^{m} \alpha_i \mu(E \cap A_i) \\
&= \sum_{i=1}^{m} \alpha_i \sum_{k=1}^{\infty} \mu(E_k \cap A_i) \\
&= \sum_{i=1}^{m} \alpha_i \lim_{n\to\infty} \sum_{k=1}^{n} \mu(E_k \cap A_i) \\
&= \lim_{n\to\infty} \sum_{i=1}^{m} \alpha_i \sum_{k=1}^{n} \mu(E_k \cap A_i) \\
&= \lim_{n\to\infty} \sum_{k=1}^{n} \sum_{i=1}^{m} \alpha_i \mu(E_k \cap A_i) \\
&= \lim_{n\to\infty} \sum_{k=1}^{n} \int_{E_k} s\,d\mu \\
&= \sum_{k=1}^{\infty} \int_{E_k} s\,d\mu.
\end{aligned}$$

This proves Lemma 1.36. □

Theorem 1.37 (Lebesgue monotone convergence theorem). *Let $(X, \mathcal{A}, \mu)$ be a measure space and let*

$$f_n \colon X \longrightarrow [0, \infty]$$

be a sequence of measurable functions such that

$$f_n(x) \le f_{n+1}(x) \quad \text{for all } x \in X \text{ and all } n \in \mathbb{N}.$$

Define $f \colon X \to [0, \infty]$ by

$$f(x) := \lim_{n\to\infty} f_n(x) \quad \text{for } x \in X.$$

Then f is measurable and

$$\lim_{n\to\infty} \int_X f_n\,d\mu = \int_X f\,d\mu.$$

Proof. By part (i) of Theorem 1.35, we have

$$\int_X f_n \, d\mu \le \int_X f_{n+1} \, d\mu$$

for all $n \in \mathbb{N}$ and hence the limit

$$\alpha := \lim_{n \to \infty} \int_X f_n \, d\mu \tag{1.11}$$

exists in $[0, \infty]$. Moreover, $f = \sup_n f_n$ is a measurable function on X, by part (ii) of Theorem 1.24, and satisfies $f_n \le f$ for all $n \in \mathbb{N}$. Thus it follows from part (i) of Theorem 1.35 that

$$\int_X f_n \, d\mu \le \int_X f \, d\mu \quad \text{for all } n \in \mathbb{N},$$

and hence

$$\alpha \le \int_X f \, d\mu. \tag{1.12}$$

Now fix a measurable step function $s\colon X \to [0, \infty)$ such that $s \le f$. Define

$$\mu_s\colon \mathcal{A} \longrightarrow [0, \infty]$$

by

$$\mu_s(E) := \int_E s \, d\mu \quad \text{for } E \in \mathcal{A}. \tag{1.13}$$

This function is a measure by part (ii) of Lemma 1.36 (which asserts that μ_s is σ-additive) and by part (iv) of Theorem 1.35 (which asserts that $\mu_s(\emptyset) = 0$). Now fix a constant $0 < c < 1$ and define

$$E_n := \{x \in X \mid cs(x) \le f_n(x)\} \quad \text{for } n \in \mathbb{N}.$$

Then $E_n \in \mathcal{A}$ is a measurable set and $E_n \subset E_{n+1}$ for all $n \in \mathbb{N}$. Moreover,

$$\bigcup_{n=1}^{\infty} E_n = X. \tag{1.14}$$

(To spell it out, choose an element $x \in X$. If $f(x) = \infty$, then $f_n(x) \to \infty$ and hence $cs(x) \le s(x) \le f_n(x)$ for some $n \in \mathbb{N}$, which means that x belongs to one of the sets E_n. If $f(x) < \infty$, then $f_n(x)$ converges to $f(x) > cf(x)$, hence $f_n(x) > cf(x) \ge cs(x)$ for some $n \in \mathbb{N}$, and for this n we have $x \in E_n$.) Since $cs \le f_n$ on E_n, it follows from parts (i) and (vi) of Theorem 1.35 that

$$c\mu_s(E_n) = c \int_{E_n} s \, d\mu = \int_{E_n} cs \, d\mu \le \int_{E_n} f_n \, d\mu \le \int_X f_n \, d\mu \le \alpha.$$

Here the last inequality follows from the definition of α in (1.11). Hence

$$\mu_s(E_n) \le \frac{\alpha}{c} \quad \text{for all } n \in \mathbb{N}. \tag{1.15}$$

Since $\mu_s: \mathcal{A} \to [0, \infty]$ is a measure, by part (i) of Theorem 1.35, it follows from (1.14) and part (iv) of Theorem 1.28 that

$$\int_X s\, d\mu = \mu_s(X) = \lim_{n\to\infty} \mu_s(E_n) \le \frac{\alpha}{c}. \tag{1.16}$$

Here the last inequality follows from (1.15). Since (1.16) holds for every constant $0 < c < 1$, we have $\int_X s\, d\mu \le \alpha$ for every measurable step function $s: X \to [0, \infty)$ such that $s \le f$. Take the supremum over all such s to obtain

$$\int_X f\, d\mu = \sup_{s \le f} \int_X s\, d\mu \le \alpha.$$

Combining this with (1.12) we obtain $\int_X f\, d\mu = \alpha$ and hence the assertion of Theorem 1.37 follows from the definition of α in (1.11). □

Theorem 1.38 (σ-additivity of the Lebesgue integral). *Let $(X, \mathcal{A}, \mu)$ be a measure space.*

(i) *If $f, g: X \to [0, \infty]$ are measurable and $E \in \mathcal{A}$, then*

$$\int_E (f + g)\, d\mu = \int_E f\, d\mu + \int_E g\, d\mu. \tag{1.17}$$

(ii) *Let $f_n: X \to [0, \infty]$ be a sequence of measurable functions and define*

$$f(x) := \sum_{n=1}^{\infty} f_n(x) \quad \textit{for } x \in X.$$

Then $f: X \to [0, \infty]$ is measurable and, for every $E \in \mathcal{A}$,

$$\int_E f\, d\mu = \sum_{n=1}^{\infty} \int_E f_n\, d\mu. \tag{1.18}$$

(iii) *If $f: X \to [0, \infty]$ is measurable and $E_1, E_2, E_3, \ldots$ is a sequence of pairwise disjoint measurable sets, then*

$$\int_E f\, d\mu = \sum_{k=1}^{\infty} \int_{E_k} f\, d\mu, \quad E := \bigcup_{k\in\mathbb{N}} E_k. \tag{1.19}$$

Proof. (i) By Theorem 1.26, there exist sequences of measurable step functions $s_n, t_n\colon X \to [0,\infty)$ such that

$$s_n \le s_{n+1} \quad \text{and} \quad t_n \le t_{n+1}$$

for all $n \in \mathbb{N}$ and

$$f(x) = \lim_{n\to\infty} s_n(x) \quad \text{and} \quad g(x) = \lim_{n\to\infty} t_n(x)$$

for all $x \in X$. Then $s_n + t_n$ is a monotonically nondecreasing sequence of measurable step functions converging pointwise to $f + g$. Hence,

$$\begin{aligned}
\int_X (f+g)\,d\mu &= \lim_{n\to\infty} \int_X (s_n + t_n)\,d\mu \\
&= \lim_{n\to\infty} \left(\int_X s_n\,d\mu + \int_X t_n\,d\mu \right) \\
&= \lim_{n\to\infty} \int_X s_n\,d\mu + \lim_{n\to\infty} \int_X t_n\,d\mu \\
&= \int_X f\,d\mu + \int_X g\,d\mu.
\end{aligned}$$

Here the first and last equalities follow from Theorem 1.37, while the second equality follows from part (i) of Lemma 1.36. This proves (i) for $E = X$. To prove it in general, replace f, g by $f\chi_E, g\chi_E$ and use part (ii) of Theorem 1.35.

(ii) Define $g_n\colon X \to [0,\infty]$ by

$$g_n := \sum_{k=1}^{n} f_k.$$

This is a nondecreasing sequence of measurable functions, by part (i) of Theorem 1.24, and it converges pointwise to f by definition. Hence it follows from part (ii) of Theorem 1.24 that f is measurable and how the Lebesgue monotone convergence theorem (Theorem 1.37) shows that

$$\begin{aligned}
\int_X f\,d\mu &= \lim_{n\to\infty} \int_X g_n\,d\mu \\
&= \lim_{n\to\infty} \int_X \sum_{k=1}^{n} f_k\,d\mu \\
&= \lim_{n\to\infty} \sum_{k=1}^{n} \int_X f_k\,d\mu \\
&= \sum_{n=1}^{\infty} \int_X f_n\,d\mu.
\end{aligned}$$

Here the second equality follows from the definition of g_n while the third one follows from part (i) of the present theorem (already proved). This proves (ii) for $E = X$. To prove it in general, replace f, f_n by $f\chi_E, f_n\chi_E$ and use part (ii) of Theorem 1.35.

(iii) Let $f\colon X \to [0,\infty]$ be a measurable function and let $E_k \in \mathcal{A}$ be a sequence of pairwise disjoint measurable sets. Define

$$E := \bigcup_{k=1}^{\infty} E_k, \qquad f_n := \sum_{k=1}^{n} f\chi_{E_k}.$$

Then it follows from part (i) of the present theorem (already proved) and part (ii) of Theorem 1.35 that

$$\begin{aligned}\int_X f_n\,d\mu &= \int_X \sum_{k=1}^{n} f\chi_{E_k}\,d\mu \\ &= \sum_{k=1}^{n}\int_X f\chi_{E_k}\,d\mu \\ &= \sum_{k=1}^{n}\int_{E_k} f\,d\mu.\end{aligned}$$

Now $f_n\colon X \to [0,\infty]$ is a nondecreasing sequence of measurable functions converging pointwise to $f\chi_E$. Hence it follows from the Lebesgue monotone convergence theorem (Theorem 1.37) that

$$\begin{aligned}\int_E f\,d\mu &= \int_X f\chi_E\,d\mu \\ &= \lim_{n\to\infty}\int_X f_n\,d\mu \\ &= \lim_{n\to\infty}\sum_{k=1}^{n}\int_{E_k} f\,d\mu \\ &= \sum_{k=1}^{\infty}\int_{E_k} f\,d\mu.\end{aligned}$$

This proves Theorem 1.38. □

Exercise 1.39. Let $\mu: 2^{\mathbb{N}} \to [0, \infty]$ be the counting measure on the natural numbers. Show that in this case equation (1.18) in part (ii) of Theorem 1.38 is equivalent to the formula

$$\sum_{i=1}^{\infty} \Big(\sum_{j=1}^{\infty} a_{ij} \Big) = \sum_{j=1}^{\infty} \Big(\sum_{i=1}^{\infty} a_{ij} \Big) \tag{1.20}$$

for every map $\mathbb{N} \times \mathbb{N} \to [0, \infty]: (i, j) \mapsto a_{ij}$.

The next theorem shows that every measurable function $f: X \to [0, \infty]$ induces another measure μ_f on $(X, \mathcal{A})$.

Theorem 1.40. *Let $(X, \mathcal{A}, \mu)$ be a measure space and let $f: X \to [0, \infty]$ be a measurable function. Then the function*

$$\mu_f: \mathcal{A} \longrightarrow [0, \infty],$$

defined by

$$\mu_f(E) := \int_E f \, d\mu \quad \textit{for } E \in \mathcal{A} \tag{1.21}$$

is a measure, and

$$\int_E g \, d\mu_f = \int_E fg \, d\mu \tag{1.22}$$

for every measurable function $g: X \to [0, \infty]$ and every $E \in \mathcal{A}$.

Proof. μ_f is σ-additive by part (iii) of Theorem 1.38 and $\mu_f(\emptyset) = 0$ by part (iv) of Theorem 1.35. Hence μ_f is a measure (see Definition 1.27). Now let $g := \chi_A$ be the characteristic function of a measurable set $A \in \mathcal{A}$. Then

$$\int_X \chi_A \, d\mu_f = \mu_f(A) = \int_A f \, d\mu = \int_X f \chi_A \, d\mu.$$

Here the first equality follows from the definition of the integral for measurable step functions in Definition 1.34, the second equality from the definition of μ_f, and the last equality from part (ii) of Theorem 1.35. Thus equation (1.22) (with $E = X$) holds for characteristic functions of measurable sets. Taking finite sums and using part (vi) of Theorem 1.35 and part (i) of Theorem 1.38 we find that (1.22) (with $E = X$) continues to hold for all measurable step functions $g = s: X \to [0, \infty)$. Now approximate an arbitrary measurable function $g: X \to [0, \infty]$ by a sequence of measurable step functions via Theorem 1.26 and use the Lebesgue monotone convergence theorem (Theorem 1.37) to deduce that (1.22) holds with $E = X$ for all measurable functions $g: X \to [0, \infty]$. Now replace g by $g\chi_E$ and use part (ii) of Theorem 1.35 to obtain equation (1.22) in general. This proves Theorem 1.40. □

It is one of the central questions in measure theory under which conditions a measure $\lambda: \mathcal{A} \to [0, \infty]$ can be expressed in the form μ_f for some measurable function $f: X \to [0, \infty]$. We return to this question in Chapter 5. The final result in this section is an inequality which will be used in the proof of the Lebesgue dominated convergence theorem (Theorem 1.45).

Theorem 1.41 (Fatou's lemma). *Let $(X, \mathcal{A}, \mu)$ be a measure space. Consider a sequence of measurable functions $f_n: X \to [0, \infty]$. Then*

$$\int_X \liminf_{n\to\infty} f_n \, d\mu \le \liminf_{n\to\infty} \int_X f_n \, d\mu.$$

Proof. For $n \in \mathbb{N}$ define $g_n: X \to [0, \infty]$ by

$$g_n(x) := \inf_{i \ge n} f_i(x)$$

for $x \in X$. Then g_n is measurable, by Theorem 1.24, and

$$g_1(x) \le g_2(x) \le g_3(x) \le \cdots, \qquad \lim_{n\to\infty} g_n(x) = \liminf_{n\to\infty} f_n(x) =: f(x)$$

for all $x \in X$. Moreover, $g_n \le f_i$ for all $i \ge n$. By part (i) of Theorem 1.35 this implies

$$\int_X g_n \, d\mu \le \int_X f_i \, d\mu$$

for all $i \ge n$, and hence

$$\int_X g_n \, d\mu \le \inf_{i \ge n} \int_X f_i \, d\mu$$

for all $n \in \mathbb{N}$. Thus, by the Lebesgue monotone convergence theorem (Theorem 1.37),

$$\int_X f \, d\mu = \lim_{n\to\infty} \int_X g_n \, d\mu \le \lim_{n\to\infty} \inf_{i \ge n} \int_X f_i \, d\mu = \liminf_{n\to\infty} \int_X f_n \, d\mu.$$

This proves Theorem 1.41. □

Example 1.42. Let $(X, \mathcal{A}, \mu)$ be a measure space and $E \in \mathcal{A}$ be a measurable set such that $0 < \mu(E) < \mu(X)$. Define $f_n := \chi_E$ when n is even and $f_n := 1 - \chi_E$ when n is odd. Then $\liminf_{n\to\infty} f_n = 0$ and so

$$\int_X \liminf_{n\to\infty} f_n d\mu = 0 < \min\{\mu(E), \mu(X \setminus E)\} = \liminf_{n\to\infty} \int_X f_n \, d\mu.$$

Thus the inequality in Theorem 1.41 can be strict.

1.4 Integration of real valued functions

The integral of a real valued measurable function is defined as the difference of the integrals of its positive and negative parts. This definition makes sense whenever at least one of these numbers is not equal to infinity. It leads naturally to the concepts of *integrability* and the *Lebesgue integral.* The basic properties of the Lebesgue integral are summarized in Theorem 1.44 below. The main result of this section is the Lebesgue dominated convergence theorem (Theorem 1.45).

Definition 1.43 (Lebesgue integrable functions). Let $(X, \mathcal{A}, \mu)$ be a measure space. A function $f: X \to \mathbb{R}$ is called *(Lebesgue) integrable* or *μ-integrable* if f is measurable and $\int_X |f| \, d\mu < \infty$. Denote the set of μ-integrable functions by

$$\mathcal{L}^1(\mu) := \mathcal{L}^1(X, \mathcal{A}, \mu) := \{f: X \to \mathbb{R} \mid f \text{ is } \mu\text{-integrable}\}.$$

The *Lebesgue integral* of $f \in \mathcal{L}^1(\mu)$ over a set $E \in \mathcal{A}$ is the real number

$$\int_E f \, d\mu := \int_E f^+ \, d\mu - \int_E f^- \, d\mu, \tag{1.23}$$

where the functions $f^\pm: X \to [0, \infty)$ are defined by

$$f^+(x) := \max\{f(x), 0\}, \quad f^-(x) := \max\{-f(x), 0\} \tag{1.24}$$

The functions $f^\pm$ are measurable by Theorem 1.24 and $0 \le f^\pm \le |f|$. Hence their integrals over E are finite by part (i) of Theorem 1.35.

Theorem 1.44 (properties of the Lebesgue integral). *Let $(X, \mathcal{A}, \mu)$ be a measure space. Then the following holds.*

(i) *The set $\mathcal{L}^1(\mu)$ is a real vector space and, for every $E \in \mathcal{A}$, the function $\mathcal{L}^1(\mu) \to \mathbb{R}: f \mapsto \int_E f \, d\mu$ is linear, i.e., if $f, g \in \mathcal{L}^1(\mu)$ and $c \in \mathbb{R}$, then $f + g, cf \in \mathcal{L}^1(\mu)$, and*

$$\int_E (f+g) \, d\mu = \int_E f \, d\mu + \int_E g \, d\mu, \quad \int_E cf \, d\mu = c \int_E f \, d\mu. \tag{1.25}$$

(ii) *For all $f, g \in \mathcal{L}^1(\mu)$ and all $E \in \mathcal{A}$,*

$$f \le g \text{ on } E \implies \int_E f \, d\mu \le \int_E g \, d\mu. \tag{1.26}$$

(iii) *If $f \in \mathcal{L}^1(\mu)$, then $|f| \in \mathcal{L}^1(\mu)$ and, for all $E \in \mathcal{A}$,*

$$\left| \int_E f \, d\mu \right| \le \int_E |f| \, d\mu. \tag{1.27}$$

(iv) *If $f \in \mathcal{L}^1(\mu)$ and $E_1, E_2, E_3, \dots$ is a sequence of pairwise disjoint measurable sets, then*

$$\int_E f\,d\mu = \sum_{k=1}^{\infty} \int_{E_k} f\,d\mu, \quad E := \bigcup_{k\in\mathbb{N}} E_k. \tag{1.28}$$

(v) *For all $E \in \mathcal{A}$ and all $f \in \mathcal{L}^1(\mu)$,*

$$\int_E f\,d\mu = \int_X f\chi_E\,d\mu. \tag{1.29}$$

(vi) *Let $E \in \mathcal{A}$ and $f \in \mathcal{L}^1(\mu)$. If $\mu(E) = 0$ or $f|_E = 0$, then*

$$\int_E f\,d\mu = 0.$$

Proof. (i) Let $f, g \in \mathcal{L}^1(\mu)$ and let $c \in \mathbb{R}$. Then $f + g \in \mathcal{L}^1(\mu)$ because $|f + g| \le |f| + |g|$ and hence $\int_X |f + g|\,d\mu < \infty$ by part (i) of Theorem 1.38. Likewise, $cf \in \mathcal{L}^1(\mu)$ because $|cf| = |c||f|$ and hence $\int_X |cf|\,d\mu < \infty$ by part (vi) of Theorem 1.35. To prove the second equation in (1.25) assume first that $c \ge 0$. Then $(cf)^{\pm} = cf^{\pm}$ and hence

$$\begin{aligned}\int_E cf\,d\mu &= \int_E cf^+\,d\mu - \int_E cf^-\,d\mu \\ &= c\int_E f^+\,d\mu - c\int_E f^-\,d\mu \\ &= c\int_E f\,d\mu.\end{aligned}$$

Here the second equation follows from part (vi) of Theorem 1.35. If $c < 0$, then $(cf)^+ = (-c)f^-$ and $(cf)^- = (-c)f^+$ and hence, again using part (iv) of Theorem 1.35, we obtain

$$\begin{aligned}\int_E cf\,d\mu &= \int_E (-c)f^-\,d\mu - \int_E (-c)f^+\,d\mu \\ &= (-c)\int_E f^-\,d\mu - (-c)\int_E f^+\,d\mu \\ &= c\int_E f\,d\mu.\end{aligned}$$

Now let $h := f + g$. Then $h^+ - h^- = f^+ - f^- + g^+ - g^-$ and hence

$$h^+ + f^- + g^- = h^- + f^+ + g^+.$$

Hence it follows from part (i) of Theorem 1.38 that

$$\int_E h^+\,d\mu + \int_E f^-\,d\mu + \int_E g^-\,d\mu = \int_E h^-\,d\mu + \int_E f^+\,d\mu + \int_E g^+\,d\mu.$$

Therefore,

$$\begin{aligned}
\int_E h\,d\mu &= \int_E h^+\,d\mu - \int_E h^-\,d\mu \\
&= \int_E f^+\,d\mu + \int_E g^+\,d\mu - \int_E f^-\,d\mu - \int_E g^-\,d\mu \\
&= \int_E f\,d\mu + \int_E g\,d\mu
\end{aligned}$$

and this proves (i).

(ii) Assume $f = f^+ - f^- \le g = g^+ - g^-$ on E. Then $f^+ + g^- \le g^+ + f^-$ on E and hence $\int_E (f^+ + g^-)\,d\mu \le \int_E (g^+ + f^-)\,d\mu$ by part (i) of Theorem 1.35. Now use the additivity of the integral in part (i) of Theorem 1.38 to obtain

$$\int_E f^+\,d\mu + \int_E g^-\,d\mu \le \int_E g^+\,d\mu + \int_E f^-\,d\mu.$$

This implies (1.26).

(iii) Since $-|f| \le f \le |f|$ it follows from (1.25) and (1.26) that

$$-\int_E |f|\,d\mu = \int_E (-|f|)\,d\mu \le \int_E f\,d\mu \le \int_E |f|\,d\mu$$

and this implies (1.27).

(iv) Equation (1.28) holds for $f^\pm$ by part (iii) of Theorem 1.38 and hence holds for f by Definition 1.43.

(v) The formula $\int_E f\,d\mu = \int_X f\chi_E\,d\mu$ in (1.29) follows from part (ii) of Theorem 1.35 since $f^\pm \chi_E = (f\chi_E)^\pm$.

(vi) If f vanishes on E, then $f^\pm$ also vanish on E and hence $\int_E f^\pm\,d\mu = 0$ by part (iii) of Theorem 1.35. If $\mu(E) = 0$, then $\int_E f^\pm\,d\mu = 0$ by part (iv) of Theorem 1.35. In either case it follows from the definition of the integral in Definition 1.43 that $\int_E f\,d\mu = 0$. This proves Theorem 1.44. □

Theorem 1.45 (Lebesgue dominated convergence theorem). *Let $(X, \mathcal{A}, \mu)$ be a measure space, $g: X \to [0, \infty)$ be an integrable function, let $f_n: X \to \mathbb{R}$ be a sequence of integrable functions satisfying*

$$|f_n(x)| \leq g(x) \quad \textit{for all } x \in X \textit{ and } n \in \mathbb{N}, \tag{1.30}$$

and converging pointwise to $f: X \to \mathbb{R}$, i.e.,

$$f(x) = \lim_{n\to\infty} f_n(x) \quad \textit{for all } x \in X. \tag{1.31}$$

Then f is integrable and, for every $E \in \mathcal{A}$,

$$\int_E f \, d\mu = \lim_{n\to\infty} \int_E f_n \, d\mu. \tag{1.32}$$

Proof. f is measurable by part (ii) of Theorem 1.24 and $|f(x)| \leq g(x)$ for all $x \in X$ by (1.30) and (1.31). Hence it follows from part (i) of Theorem 1.35 that

$$\int_X |f| \, d\mu \leq \int_X g \, d\mu < \infty,$$

and so f is integrable. Moreover

$$|f_n - f| \leq |f_n| + |f| \leq 2g.$$

Therefore, by Fatou's lemma (Theorem 1.41), that

$$\begin{aligned}
\int_X 2g \, d\mu &= \int_X \liminf_{n\to\infty} (2g - |f_n - f|) \, d\mu \\
&\leq \liminf_{n\to\infty} \int_X (2g - |f_n - f|) \, d\mu \\
&= \liminf_{n\to\infty} \left(\int_X 2g \, d\mu - \int_X |f_n - f| \, d\mu \right) \\
&= \int_X 2g \, d\mu - \limsup_{n\to\infty} \int_X |f_n - f| \, d\mu.
\end{aligned}$$

Here the penultimate step follows from part (i) of Theorem 1.44. This implies

$$\limsup_{n\to\infty} \int_X |f_n - f| \, d\mu \leq 0.$$

Hence

$$\lim_{n\to\infty} \int_X |f_n - f| \, d\mu = 0.$$

Since

$$\left| \int_E f_n \, d\mu - \int_E f \, d\mu \right| \le \int_E |f_n - f| \, d\mu \le \int_X |f_n - f| \, d\mu,$$

part (iii) of Theorem 1.44 yields

$$\lim_{n\to\infty} \left| \int_E f_n \, d\mu - \int_E f \, d\mu \right| = 0,$$

which is equivalent to (1.32). This completes the proof of Theorem 1.45. □

1.5 Sets of measure zero

Assume throughout this section that $(X, \mathcal{A}, \mu)$ is a measure space. A *set of measure zero* (or *null set*) is a measurable set $N \in \mathcal{A}$ such that $\mu(N) = 0$. Let $\mathcal{P}$ stand for some property that a point $x \in X$ may have, or not have, depending on x. For example, if $f: X \to [0, \infty]$ is a measurable function on X, then $\mathcal{P}$ could stand for the condition $f(x) > 0$, or for the condition $f(x) = 0$, or for the condition $f(x) = \infty$. Or if $f_n: X \to \mathbb{R}$ is a sequence of measurable functions, $\mathcal{P}$ could stand for the statement "the sequence $f_n(x)$ converges." In such a situation we say that $\mathcal{P}$ holds *almost everywhere* if there exists a set $N \subset X$ of measure zero such that every element $x \in X \setminus N$ has the property $\mathcal{P}$. It is not required that the set of all elements $x \in X$ that have the property $\mathcal{P}$ is measurable, although that may often be the case.

Example 1.46. Let $f_n: X \to \mathbb{R}$ be any sequence of measurable functions. Then the set

$$\begin{aligned} E &:= \{x \in X \mid (f_n(x))_{n=1}^{\infty} \text{ is a Cauchy sequence}\} \\ &= \bigcap_{k\in\mathbb{N}} \bigcup_{n_0\in\mathbb{N}} \bigcap_{n,m\ge n_0} \{x \in X \mid |f_n(x) - f_m(x)| < 2^{-k}\} \end{aligned}$$

is measurable. If $N := X \setminus E$ is a set of measure zero, then f_n converges almost everywhere to a function $f: X \to \mathbb{R}$. This function can be chosen measurable by defining

$$f(x) := \lim_{n\to\infty} f_n(x) \quad \text{for } x \in E$$

and

$$f(x) := 0 \qquad\qquad \text{for } x \in N.$$

This is the pointwise limit of the sequence of measurable functions $g_n := f_n \chi_E$ and hence is measurable by part (ii) of Theorem 1.24.

The first observation is that every nonnegative function with finite integral is almost everywhere finite.

Lemma 1.47. *Let $f: X \to [0, \infty]$ be a measurable function. If $\int_X f \, d\mu < \infty$, then $f < \infty$ almost everywhere.*

Proof. Define $N := \{x \in X \mid f(x) = \infty\}$ and $h := \infty \chi_N$. Then $h \leq f$ and so $\infty \mu(N) = \int_X h \, d\mu \leq \int_X f \, d\mu < \infty$ by Theorem 1.35. Hence $\mu(N) = 0$. □

The second observation is that if two integrable, or nonnegative measurable, functions agree almost everywhere, then their integrals agree over every measurable set.

Lemma 1.48. *Assume either that $f, g: X \to [0, \infty]$ are measurable functions that agree almost everywhere, or that $f, g: X \to \mathbb{R}$ are μ-integrable functions that agree almost everywhere. Then*

$$\int_A f \, d\mu = \int_A g \, d\mu \quad \textit{for all } A \in \mathcal{A}. \tag{1.33}$$

Proof. Fix a measurable set $A \in \mathcal{A}$ and define $N := \{x \in X \mid f(x) \neq g(x)\}$. Then N is measurable and $\mu(N) = 0$ by assumption. Hence $\mu(A \cap N) = 0$ by part (iii) of Theorem 1.28. This implies

$$\begin{aligned} \int_A f \, d\mu &= \int_{A \setminus N} f \, d\mu + \int_{A \cap N} f \, d\mu \\ &= \int_{A \setminus N} f \, d\mu \\ &= \int_X f \chi_{A \setminus N} \, d\mu. \end{aligned}$$

Here the first equality follows from part (iii) of Theorem 1.38 in the nonnegative case and from part (iv) of Theorem 1.44 in the integrable case. The second equality follows from part (iv) of Theorem 1.35 in the nonnegative case and from part (vi) of Theorem 1.44 in the integrable case. The third equality follows from part (ii) of Theorem 1.35 in the nonnegative case and from part (v) of Theorem 1.44 in the integrable case. Since $f \chi_{A \setminus N} = g \chi_{A \setminus N}$, it follows that the integrals of f and g over A agree. This proves Lemma 1.48. □

The converse of Lemma 1.48 fails for nonnegative measurable functions. For example, if X is a singleton and $\mu(X) = \infty$, then the integrals of any two positive functions agree over every measurable set. However, the converse of Lemma 1.48 does hold for integrable functions. Since the difference of two integrable functions is again integrable, it suffices to assume $g = 0$, and in this case the converse also holds for nonnegative measurable functions. This is the content of the next lemma.

Lemma 1.49. *Assume either that $f: X \to [0, \infty]$ is measurable, or that $f: X \to \mathbb{R}$ is μ-integrable. Then the following are equivalent:*

(i) $f = 0$ *almost everywhere;*

(ii) $\int_A f \, d\mu = 0$ *for all* $A \in \mathcal{A}$;

(iii) $\int_X |f| \, d\mu = 0$.

Proof. That (i) implies (ii) is the content of Lemma 1.48. That (ii) implies (iii) is obvious in the nonnegative case. In the integrable case, define

$$A^+ := \{x \in X \mid f(x) \geq 0\}, \quad A^- := \{x \in X \mid f(x) < 0\}.$$

Then $f^+ = f\chi_{A^+}$ and $f^- = -f\chi_{A^-}$ by (1.24). Hence

$$\int_X |f| \, d\mu = \int_X f^+ \, d\mu + \int_X f^- \, d\mu = \int_{A^+} f \, d\mu - \int_{A^-} f \, d\mu = 0$$

by Theorem 1.44 and (ii).

It remains to prove that (iii) implies (i). Let $f: X \to [0, \infty]$ be a measurable function such that $\int_X f = 0$ and define the measurable sets

$$A_n := \{x \in X \mid f(x) > 2^{-n}\} \quad \text{for } n \in \mathbb{N}.$$

Then

$$2^{-n}\mu(A_n) = \int_X 2^{-n}\chi_{A_n} \, d\mu \leq \int_X f \, d\mu = 0$$

for all $n \in \mathbb{N}$ by Theorem 1.35. Hence $\mu(A_n) = 0$ for all $n \in \mathbb{N}$, and so

$$N := \{x \in X \mid f(x) \neq 0\} = \bigcup_{n=1}^{\infty} A_n$$

is a set of measure zero. In the integrable case apply this argument to the function $|f|: X \to [0, \infty)$. This proves Lemma 1.49. □

Lemma 1.50. *Let $f \in \mathcal{L}^1(\mu)$. Then*

$$\left| \int_X f \, d\mu \right| = \int_X |f| \, d\mu \tag{1.34}$$

if and only if $f = |f|$ almost everywhere or $f = -|f|$ almost everywhere.

Proof. Assume (1.34). Then $\int_X f \, d\mu = \int_X |f| \, d\mu$ or $\int_X f \, d\mu = -\int_X |f| \, d\mu$. In the first case $\int_X (|f| - f) \, d\mu = 0$, and so $|f| - f = 0$ almost everywhere by Lemma 1.49. In the second case $\int_X (|f| + f) \, d\mu = 0$, and so $|f| + f = 0$ almost everywhere. This proves Lemma 1.50. □

Definition 1.51 (the Banach space $L^1(\mu)$). Define an equivalence relation on the real vector space of all measurable function from X to $\mathbb{R}$ by

$$f \overset{\mu}{\sim} g \quad \overset{\text{def}}{\Longleftrightarrow} \quad \text{the set } \{x \in X \mid f(x) \neq g(x)\} \text{ has measure zero.} \tag{1.35}$$

Thus two functions are equivalent if and only if they agree almost everywhere. (Verify that this is an equivalence relation!) By Lemma 1.48 the subspace $\mathcal{L}^1(\mu)$ is invariant under this equivalence relation, i.e., if $f, g \colon X \to \mathbb{R}$ are measurable, $f \in \mathcal{L}^1(\mu)$, and $f \overset{\mu}{\sim} g$, then $g \in \mathcal{L}^1(\mu)$. Moreover, the set $\{f \in \mathcal{L}^1(\mu) \mid f \overset{\mu}{\sim} 0\}$ is a linear subspace of $\mathcal{L}^1(\mu)$ and hence the quotient space

$$L^1(\mu) := \mathcal{L}^1(\mu)/\overset{\mu}{\sim}$$

is again a real vector space. It is the space of all equivalence classes in $\mathcal{L}^1(\mu)$ under the equivalence relation (1.35). Thus an element of $L^1(\mu)$ is not a function on X, but a set of functions on X. By Lemma 1.48, the map

$$\mathcal{L}^1(\mu) \longrightarrow \mathbb{R} \colon f \longmapsto \int_X |f|\, d\mu =: \|f\|_{L^1}$$

takes on the same value on all the elements in a given equivalence class and so descends to the quotient space $L^1(\mu)$. By Lemma 1.49 it defines a norm on $L^1(\mu)$ and Theorem 1.53 below shows that $L^1(\mu)$ is a Banach space with this norm (i.e., a complete normed vector space).

Theorem 1.52 (convergent series of integrable functions). *Let $(X, \mathcal{A}, \mu)$ be a measure space and let $f_n \colon X \to \mathbb{R}$ be a sequence of μ-integrable functions such that*

$$\sum_{n=1}^{\infty} \int_X |f_n|\, d\mu < \infty. \tag{1.36}$$

Then there are a set N of measure zero and a function $f \in \mathcal{L}^1(\mu)$ such that

$$\sum_{n=1}^{\infty} |f_n(x)| < \infty \quad \textit{and} \quad f(x) = \sum_{n=1}^{\infty} f_n(x) \quad \textit{for all } x \in X \setminus N, \tag{1.37}$$

$$\int_A f\, d\mu = \sum_{n=1}^{\infty} \int_A f_n\, d\mu \quad \textit{for all } A \in \mathcal{A}, \tag{1.38}$$

and

$$\lim_{n\to\infty} \int_X \left| f - \sum_{k=1}^{n} f_k \right| d\mu = 0. \tag{1.39}$$

Proof. Define

$$\phi(x) := \sum_{k=1}^{\infty} |f_k(x)|$$

for $x \in X$. This function is measurable by part (ii) of Theorem 1.24. Moreover, it follows from the Lebesgue monotone convergence theorem (Theorem 1.37) and from part (i) of Theorem 1.38 that

$$\begin{aligned}
\int_X \phi \, d\mu &= \lim_{n\to\infty} \int_X \sum_{k=1}^{n} |f_k| \, d\mu \\
&= \lim_{n\to\infty} \sum_{k=1}^{n} \int_X |f_k| \, d\mu \\
&= \sum_{k=1}^{\infty} \int_X |f_k| \, d\mu \\
&< \infty.
\end{aligned}$$

Hence the set

$$N := \{x \in X \mid \phi(x) = \infty\}$$

has measure zero by Lemma 1.47 and

$$\sum_{k=1}^{\infty} |f_k(x)| < \infty \quad \text{for all } x \in X \setminus N.$$

Define the function $f \colon X \to \mathbb{R}$ by

$$f(x) := \begin{cases} 0 & \text{for } x \in N, \\ \displaystyle\sum_{k=1}^{\infty} f_k(x) & \text{for } x \in X \setminus N. \end{cases}$$

Then f satisfies (1.37). Define the functions $g \colon X \to \mathbb{R}$ and $g_n \colon X \to \mathbb{R}$ by

$$g := \phi \chi_{X \setminus N}, \quad g_n := \sum_{k=1}^{n} f_k \chi_{X \setminus N} \quad \text{for } n \in \mathbb{N}.$$

These functions are measurable by part (i) of Theorem 1.24. Moreover,

$$\int_X g\,d\mu = \int_X \phi\,d\mu < \infty$$

by Lemma 1.48. Since $|g_n(x)| \le g(x)$ for all $n \in \mathbb{N}$ and g_n converges pointwise to f, it follows from the Lebesgue dominated convergence theorem (Theorem 1.45) that $f \in \mathcal{L}^1(\mu)$ and, for all $A \in \mathcal{A}$,

$$\int_A f\,d\mu = \lim_{n\to\infty}\int_A g_n\,d\mu = \lim_{n\to\infty}\int_A \sum_{k=1}^n f_k\,d\mu = \sum_{n=1}^\infty \int_A f_n\,d\mu.$$

Here the second step follows from Lemma 1.48 because $g_n = \sum_{k=1}^n f_k$ almost everywhere. The last step follows by interchanging sum and integral, using part (i) of Theorem 1.44. This proves (1.38). To prove (1.39) note that

$$f - \sum_{k=1}^n f_k = f - g_n$$

almost everywhere, that $f(x) - g_n(x)$ converges to zero for all $x \in X$, and that $|f - g_n| \le |f| + g$, where $|f| + g$ is integrable. Hence, by Lemma 1.48 and the Lebesgue dominated convergence theorem (Theorem 1.45),

$$\lim_{n\to\infty}\int_X \left| f - \sum_{k=1}^n f_k \right| d\mu = \lim_{n\to\infty}\int_X |f - g_n|\,d\mu = 0,$$

This proves (1.39) and Theorem 1.52. □

Theorem 1.53 (completeness of L^1). *Let $(X, \mathcal{A}, \mu)$ be a measure space and let $f_n \in \mathcal{L}^1(\mu)$ be a sequence of integrable functions. Assume f_n is a Cauchy sequence with respect to the L^1-norm, i.e., for every $\varepsilon > 0$ there is an $n_0 \in \mathbb{N}$ such that, for all $m, n \in \mathbb{N}$,*

$$n, m \ge n_0 \implies \int_X |f_n - f_m|\,d\mu < \varepsilon. \tag{1.40}$$

Then there exists a function $f \in \mathcal{L}^1(\mu)$ such that

$$\lim_{n\to\infty}\int_X |f_n - f|\,d\mu = 0. \tag{1.41}$$

Moreover, there is a subsequence f_{n_i} that converges almost everywhere to f.

Proof. By assumption, there is a sequence $n_i \in \mathbb{N}$ such that

$$\int_X |f_{n_{i+1}} - f_{n_i}|\, d\mu < 2^{-i}, \quad n_i < n_{i+1}, \quad \text{for all } i \in \mathbb{N}.$$

Then the sequence

$$g_i := f_{n_{i+1}} - f_{n_i} \in \mathcal{L}^1(\mu)$$

satisfies (1.36). Hence, by Theorem 1.52, there exists a function $g \in \mathcal{L}^1(\mu)$ such that

$$g = \sum_{i=1}^{\infty} g_i = \sum_{i=1}^{\infty} (f_{n_{i+1}} - f_{n_i})$$

almost everywhere and

$$0 = \lim_{k\to\infty} \int_X \left| \sum_{i=1}^{k-1} g_i - g \right| d\mu = \lim_{k\to\infty} \int_X |f_{n_k} - f_{n_1} - g|\, d\mu. \tag{1.42}$$

Define

$$f := f_{n_1} + g.$$

Then

$$f_{n_i} = f_{n_1} + \sum_{j=1}^{i-1} g_j$$

converges almost everywhere to f. We prove (1.41). Let $\varepsilon > 0$. By (1.42), there is an $\ell \in \mathbb{N}$ such that

$$\int_X |f_{n_k} - f|\, d\mu < \varepsilon/2 \quad \text{for all } k \geq \ell.$$

By (1.40), the integer ℓ can be chosen such that

$$\int_X |f_n - f_m|\, d\mu < \varepsilon/2 \quad \text{for all } n, m \geq n_\ell.$$

Then

$$\int_X |f_n - f|\, d\mu \leq \int_X |f_n - f_{n_\ell}|\, d\mu + \int_X |f_{n_\ell} - f|\, d\mu < \varepsilon$$

for all $n \geq n_\ell$. This proves (1.41) and Theorem 1.53. □

1.6 Completion of a measure space

The discussion in Section 1.5 shows that sets of measure zero are *negligible* in the sense that the integral of a measurable function remains the same if the function is modified on a set of measure zero. Thus also subsets of sets of measure zero can be considered *negligible.* However, such subsets need not be elements of our σ-algebra $\mathcal{A}$. It is sometimes convenient to form a new σ-algebra by including all subsets of sets of measure zero. This leads to the notion of a *completion* of a measure space $(X, \mathcal{A}, \mu)$.

Definition 1.54. A measure space $(X, \mathcal{A}, \mu)$ is called *complete* if

$$N \in \mathcal{A}, \quad \mu(N) = 0, \quad E \subset N \implies E \in \mathcal{A}.$$

Theorem 1.55. *Let $(X, \mathcal{A}, \mu)$ be a measure space and define*

$$\mathcal{A}^* := \left\{ E \subset X \;\middle|\; \begin{array}{l} \text{there exist measurable sets } A, B \in \mathcal{A} \text{ such that} \\ A \subset E \subset B \text{ and } \mu(B \setminus A) = 0 \end{array} \right\}.$$

Then the following holds.

(i) *$\mathcal{A}^*$ is a σ-algebra and $\mathcal{A} \subset \mathcal{A}^*$.*

(ii) *There exists a unique measure $\mu^* \colon \mathcal{A}^* \to [0, \infty]$ such that*

$$\mu^*|_{\mathcal{A}} = \mu.$$

(iii) *The triple $(X, \mathcal{A}^*, \mu^*)$ is a complete measure space. It is called the* **completion** *of $(X, \mathcal{A}, \mu)$.*

(iv) *If $f \colon X \to \mathbb{R}$ is μ-integrable, then f is μ^*-integrable and, for $E \in \mathcal{A}$,*

$$\int_E f \, d\mu^* = \int_E f \, d\mu \tag{1.43}$$

This continues to hold for all $\mathcal{A}$-measurable functions $f \colon X \to [0, \infty]$.

(v) *If $f^* \colon X \to \overline{\mathbb{R}}$ is $\mathcal{A}^*$-measurable, then there exists an $\mathcal{A}$-measurable function $f \colon X \to \overline{\mathbb{R}}$ such that the set*

$$N^* := \{x \in X \mid f(x) \neq f^*(x)\} \in \mathcal{A}^*$$

has measure zero, i.e., $\mu^*(N^*) = 0$.

Proof. (i) First, $X \in \mathcal{A}^*$ because $\mathcal{A} \subset \mathcal{A}^*$. Second, let $E \in \mathcal{A}^*$ and choose $A, B \in \mathcal{A}$ such that $A \subset E \subset B$ and $\mu(B \setminus A) = 0$. Then $B^c \subset E^c \subset A^c$ and $A^c \setminus B^c = A^c \cap B = B \setminus A$. Hence $\mu(A^c \setminus B^c) = 0$ and so $E^c \in \mathcal{A}^*$. Third, let $E_i \in \mathcal{A}^*$ for $i \in \mathbb{N}$ and choose $A_i, B_i \in \mathcal{A}$ such that $A_i \subset E_i \subset B_i$ and $\mu(B_i \setminus A_i) = 0$. Define

$$A := \bigcup_i A_i, \quad E := \bigcup_i E_i, \quad B := \bigcup_i B_i.$$

Then $A \subset E \subset B$ and $B \setminus A = \bigcup_i (B_i \setminus A) \subset \bigcup_i (B_i \setminus A_i)$. Hence

$$\mu(B \setminus A) \leq \sum_i \mu(B_i \setminus A_i) = 0$$

and this implies $E \in \mathcal{A}^*$. Thus we have proved (i).

(ii) For $E \in \mathcal{A}^*$ define

$$\mu^*(E) := \mu(A) \quad \text{where } \begin{aligned} & A, B \in \mathcal{A}, \\ & A \subset E \subset B, \\ & \mu(B \setminus A) = 0. \end{aligned} \tag{1.44}$$

This is the only possibility for defining a measure $\mu^* : \mathcal{A}^* \to [0, \infty]$ that agrees with μ on $\mathcal{A}$ because $\mu(A) = \mu(B)$ whenever $A, B \in \mathcal{A}$ such that $A \subset B$ and $\mu(B \setminus A) = 0$. To prove that μ^* is well defined, let $E \in \mathcal{A}^*$ and $A, B \in \mathcal{A}$ as in (1.44). If $A', B' \in \mathcal{A}$ is another pair such that $A' \subset E \subset B'$ and $\mu(B' \setminus A') = 0$, then $A \setminus A' \subset E \setminus A' \subset B' \setminus A'$ and hence $\mu(A \setminus A') = 0$. This implies $\mu(A) = \mu(A \cap A') = \mu(A')$, where the last equality follows by interchanging the roles of the pairs (A, B) and (A', B'). Thus the map $\mu^* : \mathcal{A}^* \to [0, \infty]$ in (1.44) is well defined.

We prove that μ^* is a measure. Let $E_i \in \mathcal{A}^*$ be a sequence of pairwise disjoint sets and choose sequences $A_i, B_i \in \mathcal{A}$ such that $A_i \subset E_i \subset B_i$ for all i. Then the A_i are pairwise disjoint and $\mu^*(E_i) = \mu(A_i)$ for all i. Moreover, $A := \bigcup_i A_i \in \mathcal{A}$, $B := \bigcup_i B_i \in \mathcal{A}$, $A \subset E \subset B$, and $\mu(B \setminus A) = 0$ as we have seen in the proof of part (i). Hence $\mu^*(E) = \mu(A) = \sum_i \mu(A_i) = \sum_i \mu^*(E_i)$. This proves (ii).

(iii) Let $E \in \mathcal{A}^*$ such that $\mu^*(E) = 0$ and let $E' \subset E$. Choose $A, B \in \mathcal{A}$ such that $A \subset E \subset B$ and $\mu(B \setminus A) = 0$. Then $\mu(A) = \mu^*(E) = 0$ and hence $\mu(B) = \mu(A) + \mu(B \setminus A) = 0$. Since $E' \subset E \subset B$, this implies that $E' \in \mathcal{A}^*$ (by choosing $B' := B$ and $A' := \emptyset$). This shows that $(X, \mathcal{A}^*, \mu^*)$ is a complete measure space.

(iv) Assume $f: X \to [0, \infty]$ is $\mathcal{A}$-measurable. By Theorem 1.26, there exists a sequence of $\mathcal{A}$-measurable step functions $s_n: X \to \mathbb{R}$ such that

$$0 \le s_1 \le s_2 \le \cdots \le f$$

and

$$f(x) = \lim_{n\to\infty} s_n(x) \quad \text{for all } x \in X.$$

Since $\mu^*|_{\mathcal{A}} = \mu$ we have

$$\int_X s_n \, d\mu = \int_X s_n \, d\mu^*$$

for all n and hence it follows from the Lebesgue monotone convergence theorem (Theorem 1.37) for both μ and μ^* that

$$\int_X f \, d\mu = \lim_{n\to\infty} \int_X s_n \, d\mu = \lim_{n\to\infty} \int_X s_n \, d\mu^* = \int_X f \, d\mu^*.$$

This proves (1.43) for $E = X$ and all $\mathcal{A}$-measurable functions $f: X \to [0, \infty]$. To prove it for all E replace f by $f\chi_E$ and use part (ii) of Theorem 1.35. This proves equation (1.43) for all $\mathcal{A}$-measurable functions $f: X \to [0, \infty]$. That it continues to hold for all $f \in \mathcal{L}^1(\mu)$ follows directly from Definition 1.43. This proves (iv).

(v) If $f^* = \chi_E$ for $E \in \mathcal{A}^*$, choose $A, B \in \mathcal{A}$ such that

$$A \subset E \subset B, \quad \mu(B \setminus A) = 0,$$

and define $f := \chi_A$. Then

$$N^* = \{x \in X \mid f^*(x) \ne f(x)\} = E \setminus A \subset B \setminus A.$$

Hence $\mu^*(N^*) \le \mu^*(B \setminus A) = \mu(B \setminus A) = 0$. This proves (v) for characteristic functions of $\mathcal{A}^*$-measurable sets. For $\mathcal{A}^*$-measurable step functions the assertion follows by multiplication with real numbers and taking finite sums. Now let $f^*: X \to [0, \infty]$ be an arbitrary $\mathcal{A}^*$-measurable function. By Theorem 1.26, there exists a sequence of $\mathcal{A}^*$-measurable step functions $s_i^*: X \to [0, \infty)$ such that s_i^* converges pointwise to f^*. For each $i \in \mathbb{N}$ choose an $\mathcal{A}$-measurable step function $s_i: X \to [0, \infty)$ and a set $N_i^* \in \mathcal{A}^*$ such that $s_i = s_i^*$ on $X \setminus N_i^*$ and $\mu^*(N_i^*) = 0$. Then there is a sequence of sets $N_i \in \mathcal{A}$ such that $N_i^* \subset N_i$ and $\mu(N_i) = 0$ for all i. Define $f: X \to [0, \infty]$ by

$$f(x) := \begin{cases} f^*(x) & \text{if } x \notin N, \\ 0 & \text{if } x \in N, \end{cases} \qquad N := \bigcup_i N_i.$$

Then $N \in \mathcal{A}$, $\mu(N) = 0$, and the sequence of $\mathcal{A}$-measurable functions $s_i \chi_{X \setminus N}$ converges pointwise to f as i tends to infinity. Hence f is $\mathcal{A}$-measurable by part (ii) of Theorem 1.24 and, agrees with f^* on $X \setminus N$ by definition. Now let $f^*: X \to \overline{\mathbb{R}}$ be $\mathcal{A}^*$-measurable. Then so are $(f^*)^\pm := \max\{\pm f^*, 0\}$. Construct $f^\pm: X \to [0, \infty]$ as above. Then $f^-(x) = 0$ whenever $f^+(x) > 0$ and vice versa. Thus $f := f^+ - f^-$ is well defined, $\mathcal{A}$-measurable, and agrees with f^* on the complement of a μ-null set. This proves Theorem 1.55. □

Corollary 1.56. *Let $(X, \mathcal{A}, \mu)$ be a measure space and let $(X, \mathcal{A}^*, \mu^*)$ be its completion. Denote the equivalence class of a μ-integrable function $f \in \mathcal{L}^1(\mu)$ under the equivalence relation* (1.35) *in Definition* 1.51 *by*

$$[f]_\mu := \{g \in \mathcal{L}^1(\mu) \mid \mu(\{x \in X \mid f(x) \neq g(x)\}) = 0\}.$$

Then the map

$$L^1(\mu) \longrightarrow L^1(\mu^*): [f]_\mu \longmapsto [f]_{\mu^*} \tag{1.45}$$

is a Banach space isometry.

Proof. The map (1.45) is linear and injective by definition. It preserves the L^1-norm by part (iv) of Theorem 1.55 and is surjective by part (v) of Theorem 1.55. □

As noted in Section 1.5, sets of measure zero can be neglected when integrating functions. Hence it is sometimes convenient to enlarge the notion of integrability. It is not even necessary that the function be defined on all of X, as long as it is defined on the complement of a set of measure zero.

Thus, let $(X, \mathcal{A}, \mu)$ be a measure space and call a function $f: E \to \mathbb{R}$, defined on a measurable subset $E \subset X$, *measurable* if $\mu(X \setminus E) = 0$ and the set $f^{-1}(B) \subset E$ is measurable for every Borel set $B \subset \mathbb{R}$. Call f *integrable* if the function on all of X, obtained by setting $f|_{X \setminus E} = 0$, is integrable.

If $(X, \mathcal{A}, \mu)$ is complete our integrable function $f: E \to \mathbb{R}$ can be extended in any manner whatsoever to all of X, and the extended function on X is then integrable in the original sense, regardless of the choice of the extension. Moreover, its integral over any measurable set $A \in \mathcal{A}$ is unaffected by the choice of the extension (see Lemma 1.48).

With this extended notion of integrability we see that the Lebesgue dominated convergence theorem (Theorem 1.45) continues to hold if (1.31) is replaced by the weaker assumption that f_n only converges to f almost everywhere.

That such an extended terminology might be useful can also be seen in Theorem 1.52, where the series $\sum_{n=1}^\infty f_n$ only converges on the complement of a set N of measure zero, and the function f can only be naturally defined on $E := X \setminus N$. Our choice in the proof of Theorem 1.52 was to define $f|_N := 0$, but this choice does not affect any of the statements of the theorem. Moreover, when working with the quotient space $L^1(\mu) = \mathcal{L}^1(\mu)/\overset{\mu}{\sim}$ we are only interested in the equivalence class of f under the equivalence relation (1.35) rather than a specific choice of an element of this equivalence class.

1.7 Exercises

Exercise 1.57. Let X be an uncountable set and let $\mathcal{A} \subset 2^X$ be the set of all subsets $A \subset X$ such either A or A^{c} is countable. Define

$$\mu(A) := \begin{cases} 0 & \text{if } A \text{ is countable,} \\ 1 & \text{if } A^{\mathrm{c}} \text{ is countable,} \end{cases}$$

for $A \in \mathcal{A}$. Show that $(X, \mathcal{A}, \mu)$ is a measure space. Describe the measurable functions and their integrals. (See Examples 1.4 and 1.32.)

Exercise 1.58. Let $(X, \mathcal{A}, \mu)$ be a measure space such that $\mu(X) < \infty$ and let $f_n : X \to [0, \infty)$ be a sequence of bounded measurable functions that converges uniformly to $f : X \to [0, \infty)$. Prove that

$$\int_X f \, d\mu = \lim_{n\to\infty} \int_X f_n \, d\mu. \tag{1.46}$$

Find an example of a measure space $(X, \mathcal{A}, \mu)$ with $\mu(X) = \infty$ and a sequence of bounded measurable functions $f_n : X \to [0, \infty)$ converging uniformly to f such that (1.46) does not hold.

Exercise 1.59. (i) Let $f_n : [0, 1] \to [-1, 1]$ be a sequence of continuous functions that converges uniformly to zero. Show that

$$\lim_{n\to\infty} \int_0^1 f_n(x) \, dx = 0.$$

(ii) Let $f_n : [0, 1] \to [-1, 1]$ be a sequence of continuous functions such that

$$\lim_{n\to\infty} f_n(x) = 0 \quad \text{for all } x \in [0, 1].$$

Prove that

$$\lim_{n\to\infty} \int_0^1 f_n(x) \, dx = 0,$$

without using Theorem 1.45. A good reference is Eberlein [3].

(iii) Construct a sequence of continuous functions $f_n\colon [0,1] \to [-1,1]$ that converges pointwise, but not uniformly, to zero.

(iv) Construct a sequence of continuous functions $f_n\colon [0,1] \to [-1,1]$ such that $\int_0^1 f_n(x)\,dx = 0$ for all n and $f_n(x)$ does not converge for any $x \in [0,1]$.

Exercise 1.60. Let $(X, \mathcal{A}, \mu)$ be a measure space and $f\colon X \to [0,\infty]$ be a measurable function such that

$$0 < c := \int_X f\,d\mu < \infty.$$

Prove that

$$\lim_{n\to\infty} \int_X n \log\left(1 + \frac{f^\alpha}{n^\alpha}\right) d\mu = \begin{cases} \infty & \text{if } \alpha < 1, \\ c & \text{if } \alpha = 1, \\ 0 & \text{if } \alpha > 1, \end{cases} \quad \text{for } 0 < \alpha < \infty.$$

Hint. The integrand can be estimated by αf when $\alpha \geq 1$.

Exercise 1.61. Let $X := \mathbb{N}$ and $\mathcal{A} := 2^{\mathbb{N}}$ and let $\mu\colon 2^{\mathbb{N}} \to [0,\infty]$ be the counting measure (Example 1.30). Prove that a function $f\colon \mathbb{N} \to \mathbb{R}$ is μ-integrable if and only if the sequence $(f(n))_{n\in\mathbb{N}}$ of real numbers is absolutely summable and that in this case

$$\int_{\mathbb{N}} f\,d\mu = \sum_{n=1}^{\infty} f(n).$$

Exercise 1.62. Let $(X, \mathcal{A})$ be a measurable space and let $\mu_n\colon \mathcal{A} \to [0,\infty]$ be a sequence of measures. Show that the formula

$$\mu(A) := \sum_{n=1}^{\infty} \mu_n(A)$$

for $A \in \mathcal{A}$ defines a measure $\mu\colon \mathcal{A} \to [0,\infty]$. Let $f\colon X \to \mathbb{R}$ be a measurable function. Show that f is μ-integrable if and only if

$$\sum_{n=1}^{\infty} \int_X |f|\,d\mu_n < \infty.$$

If f is μ-integrable, prove that

$$\int_X f\,d\mu = \sum_{n=1}^{\infty} \int_X f\,d\mu_n.$$

Exercise 1.63. Let $(X, \mathcal{A}, \mu)$ be a measure space such that $\mu(X) < \infty$ and let $f: X \to \mathbb{R}$ be a measurable function. Show that f is integrable if and only if

$$\sum_{n=1}^{\infty} |\mu(\{x \in X \mid |f(x)| > n\})| < \infty.$$

Exercise 1.64. Let $(X, \mathcal{A}, \mu)$ be a measure space and let $f: X \to \mathbb{R}$ be a μ-integrable function.

(i) Prove that for every $\varepsilon > 0$ there exists a $\delta > 0$ such that, for any $A \in \mathcal{A}$,

$$\mu(A) < \delta \implies \left| \int_A f \, d\mu \right| < \varepsilon.$$

Hint. Argue indirectly. See Lemma 5.21.

(ii) Prove that for every $\varepsilon > 0$ there exists a measurable set $A \in \mathcal{A}$ such that, for any $B \in \mathcal{A}$,

$$B \supset A \implies \left| \int_X f \, d\mu - \int_B f \, d\mu \right| < \varepsilon.$$

Exercise 1.65. Let $(X, \mathcal{A})$ be a measurable space and define

$$\mu(A) := \begin{cases} 0 & \text{if } A = \emptyset, \\ \infty & \text{if } A \in \mathcal{A} \text{ and } A \neq \emptyset. \end{cases}$$

Determine the completion $(X, \mathcal{A}^*, \mu^*)$ and the space $L^1(\mu)$.

Exercise 1.66. Let $(X, \mathcal{A}, \mu)$ be a measure space such that $\mu = \delta_{x_0}$ is the Dirac measure at some point $x_0 \in X$ (Example 1.31). Determine the completion $(X, \mathcal{A}^*, \mu^*)$ and the space $L^1(\mu)$.

Exercise 1.67. Let $(X, \mathcal{A}, \mu)$ be a complete measure space. Prove that $(X, \mathcal{A}, \mu)$ is equal to its own completion.

Exercise 1.68. Let $(X, \mathcal{A}, \mu)$ and $(X, \mathcal{A}', \mu')$ be two measure spaces with $\mathcal{A} \subset \mathcal{A}'$ and $\mu'|_{\mathcal{A}} = \mu$. Prove that $\mathcal{L}^1(\mu) \subset \mathcal{L}^1(\mu')$ and

$$\int_X f \, d\mu = \int_X f \, d\mu'$$

for every $f \in \mathcal{L}^1(\mu)$.

Hint. Prove the following.

(i) Let $f: X \to [0, \infty]$ be $\mathcal{A}$-measurable and define

$$f_\delta(x) := \begin{cases} 0 & \text{if } f(x) \le \delta, \\ f(x) & \text{if } \delta < f(x) \le \delta^{-1}, \\ \delta^{-1} & \text{if } f(x) > \delta^{-1}. \end{cases}$$

Then f_δ is $\mathcal{A}$-measurable for every $\delta > 0$ and

$$\lim_{\delta \to 0} \int_X f_\delta \, d\mu = \int_X f \, d\mu.$$

(ii) Let $0 < c < \infty$, let $f: X \to [0, c]$ be $\mathcal{A}$-measurable, and assume that

$$\mu(\{x \in X \mid f(x) > 0\}) < \infty.$$

Then

$$\int_X f \, d\mu = \int_X f \, d\mu'.$$

(Consider also the function $c - f$.)

Exercise 1.69 (pushforward of a measure). Let $(X, \mathcal{A}, \mu)$ be a measure space, let Y be a set, and let $\phi: X \to Y$ be a map. The *pushforward of* $\mathcal{A}$ is the σ-algebra

$$\phi_*\mathcal{A} := \{B \subset Y \mid \phi^{-1}(B) \in \mathcal{A}\} \subset 2^Y. \tag{1.47}$$

The *pushforward of* μ is the function

$$\phi_*\mu: \phi_*\mathcal{A} \longrightarrow [0, \infty]$$

defined by

$$(\phi_*\mu)(B) := \mu(\phi^{-1}(B)), \quad \text{for } B \in \phi_*\mathcal{A}. \tag{1.48}$$

(i) Prove that $(Y, \phi_*\mathcal{A}, \phi_*\mu)$ is a measure space.

(ii) Let $(X, \mathcal{A}^*, \mu^*)$ be the completion of $(X, \mathcal{A}, \mu)$ and let $(Y, (\phi_*\mathcal{A})^*, (\phi_*\mu)^*)$ be the completion of $(Y, \phi_*\mathcal{A}, \phi_*\mu)$. Prove that

$$(\phi_*\mu)^*(E) = \mu^*(\phi^{-1}(E)) \quad \text{for all } E \in (\phi_*\mathcal{A})^* \subset \phi_*\mathcal{A}^*. \tag{1.49}$$

Deduce that $(Y, \phi_*\mathcal{A}, \phi_*\mu)$ is complete whenever $(X, \mathcal{A}, \mu)$ is complete. Find an example where $(\phi_*\mathcal{A})^* \subsetneq \phi_*\mathcal{A}^*$.

(iii) Fix a function $f: Y \to [0,\infty]$. Prove that f is $\phi_*\mathcal{A}$-measurable if and only if $f \circ \phi$ is $\mathcal{A}$-measurable. If f is $\phi_*\mathcal{A}$-measurable, prove that

$$\int_Y f \, d(\phi_*\mu) = \int_X (f \circ \phi)\, d\mu. \tag{1.50}$$

(iv) Determine the pushforward of $(X, \mathcal{A}, \mu)$ under a constant map.

The following extended remark contains a brief introduction to some of the basic concepts and terminology in probability theory. It will not be used elsewhere in this book and can be skipped at first reading.

Remark 1.70 (probability theory). A *probability space* is a measure space $(\Omega, \mathcal{F}, P)$ such that

$$P(\Omega) = 1.$$

The underlying set Ω is called the *sample space*, the σ-algebra $\mathcal{F} \subset 2^\Omega$ is called the *set of events,* and the measure

$$P: \mathcal{F} \longrightarrow [0,1]$$

is called a *probability measure.* Examples of finite sample spaces are

- the set $\Omega = \{\mathrm{h}, \mathrm{t}\}$ for tossing a coin,
- the set $\Omega = \{1,2,3,4,5,6\}$ for rolling a dice,
- the set $\Omega = \{00, 0, 1, \dots, 36\}$ for spinning a roulette wheel, and
- the set $\Omega = \{2, \dots, 10, \mathrm{j}, \mathrm{q}, \mathrm{k}, \mathrm{a}\} \times \{\diamondsuit, \heartsuit, \spadesuit, \clubsuit\}$ for drawing a card from a deck.

Examples of infinite sample spaces are the set $\Omega = \mathbb{N} \cup \{\infty\}$ for repeatedly tossing a coin until the first tail shows up, a compact interval of real numbers for random arrival times, and a disc in the plane for throwing a dart.

A *random variable* is an integrable function

$$X: \Omega \longrightarrow \mathbb{R}.$$

Its *expectation* $\mathrm{E}(X)$ and *variance* $\mathrm{V}(X)$ are defined by

$$\mathrm{E}(X) := \int_\Omega X \, dP,$$

and

$$\mathrm{V}(X) := \int_{\Omega} (X - \mathrm{E}(X))^2 \, dP = \mathrm{E}(X^2) - \mathrm{E}(X)^2,$$

respectively.

Given a random variable $X\colon \Omega \to \mathbb{R}$ one is interested in the value of the probability measure on the set $X^{-1}(B)$ for a Borel set $B \subset \mathbb{R}$. This value is the *probability* of the *event* that the random variable X takes its value in the set B and is denoted by

$$P(X \in B) := P(X^{-1}(B)) = (X_* P)(B).$$

Here $X_* P$ denotes the pushforward of the probability measure P to the Borel σ-algebra $\mathcal{B} \subset 2^{\mathbb{R}}$ (Exercise 1.69). By (1.50) the expectation and variance of X are given by

$$\mathrm{E}(X) = \int_{\mathbb{R}} x \, d(X_* P)(x)$$

and

$$\mathrm{V}(X) = \int_{\mathbb{R}} (x - \mathrm{E}(X))^2 \, d(X_* P)(x),$$

respectively.

The *(cumulative) distribution function* of a random variable X is the function

$$F_X\colon \mathbb{R} \longrightarrow [0, 1]$$

defined by

$$F_X(x) := P(X \le x) = P(\{\omega \in \Omega \mid X(\omega) \le x\}) = (X_* P)((-\infty, x]).$$

It is nondecreasing and right continuous, satisfies

$$\lim_{x \to -\infty} F_X(x) = 0, \qquad \lim_{x \to \infty} F_X(x) = 1,$$

and the integral of a continuous function on $\mathbb{R}$ with respect to the pushforward measure $X_* P$ agrees with the Riemann–Stieltjes integral (Exercise 6.20) with respect to F_X. Moreover,

$$F_X(x) - \lim_{t \to x^-} F_X(t) = P(X^{-1}(x))$$

by Theorem 1.28. Thus F_X is continuous at x if and only if $P(X^{-1}(x)) = 0$. This leads to the following notions of convergence. Let $X\colon \Omega \to \mathbb{R}$ be a random variable.

A sequence $(X_i)_{i\in\mathbb{N}}$ of random variables is said to

- *converge in probability* to X if

$$\lim_{i\to\infty} P(|X_i - X| \geq \varepsilon) = 0 \quad \text{for all } \varepsilon > 0,$$

- *converge in distribution* to X if

$$F_X(x) = \lim_{i\to\infty} F_{X_i}(x)$$

  for every $x \in \mathbb{R}$ such that F_X is continuous at x.

We prove that that convergence almost everywhere implies convergence in probability. Let $\varepsilon > 0$ and define

$$A_i := \{\omega \in \Omega \mid |X_i(\omega) - X(\omega)| \geq \varepsilon\}$$

Let $E \subset \Omega$ be the set of all $\omega \in \Omega$ such that the sequence $X_i(\omega)$ does not converge to $X(\omega)$. This set is measurable by Example 1.46 and has measure zero by convergence almost everywhere. Moreover,

$$\bigcap_{i\in\mathbb{N}} \bigcup_{j\geq i} A_j \subset E$$

and so

$$\lim_{i\to\infty} P\Big(\bigcup_{j\geq i} A_j\Big) = P(E) = 0$$

by Theorem 1.28. Thus

$$\lim_{i\to\infty} P(A_i) = 0.$$

We prove that convergence in probability implies convergence in distribution. Let $x \in \mathbb{R}$ such that F_X is continuous at x. Let $\varepsilon > 0$ and choose $\delta > 0$ such that

$$F_X(x) - \frac{\varepsilon}{2} < F_X(x-\delta) \leq F_X(x+\delta) < F_X(x) + \frac{\varepsilon}{2}.$$

Now choose $i_0 \in \mathbb{N}$ such that

$$P(|X_i - X| \geq \delta) < \frac{\varepsilon}{2} \quad \text{for all } i \geq i_0.$$

Then

$$F_X(x-\delta) - P(|X_i - X| \geq \delta) \leq F_{X_i}(x) \\ \leq F_X(x+\delta) + P(|X_i - X| \geq \delta)$$

and hence

$$F_X(x) - \varepsilon < F_{X_i}(x) < F_X(x) + \varepsilon \quad \text{for all } i \geq i_0.$$

This shows that

$$\lim_{i\to\infty} F_{X_i}(x) = F_X(x)$$

as claimed.

A finite collection of random variables $X_1, \dots, X_n$ is called *independent* if, for every collection of Borel sets $B_1, \dots, B_n \subset \mathbb{R}$,

$$P\Big(\bigcap_{i=1}^n X_i^{-1}(B_i)\Big) = \prod_{i=1}^n P(X_i^{-1}(B_i)).$$

In Chapter 7 we shall see that this condition asserts that the pushforward of P under the map

$$X := (X_1, \dots, X_n)\colon \Omega \longrightarrow \mathbb{R}^n$$

agrees with the product of the measures $(X_i)_* P$. Two foundational theorems in probability theory are the law of large numbers and the central limit theorem. These are results about sequences of random variables

$$X_k\colon \Omega \longrightarrow \mathbb{R}$$

that satisfy the following:

(a) the random variables $X_1, \dots, X_n$ are independent for all n;

(b) the X_k have expectation $\mathrm{E}(X_k) = 0$;

(c) the X_k are *identically distributed,* i.e., $F_{X_k} = F_{X_\ell}$ for all k and ℓ.

For $n \in \mathbb{N}$ define

$$S_n := X_1 + \cdots + X_n.$$

Kolmogorov's *strong law of large numbers* asserts that, under these assumptions, the sequence S_n/n converges almost everywhere to zero. (This continues to hold when (c) is replaced by the assumption $\sum_{k=1}^\infty \frac{1}{k^2} \mathrm{V}(X_k) < \infty$.) If, in addition, $\mathrm{V}(X_k) = \sigma^2$ for all k and some positive real number σ, then the *central limit theorem* of Lindeberg and Lévy asserts that the sequence

$$T_n := S_n/\sigma\sqrt{n}$$

converges in distribution to a so-called *standard normal random variable* with expectation zero and variance one, i.e.,

$$\lim_{n\to\infty} F_{T_n}(x) = \frac{1}{\sqrt{2\pi}} \int_{-\infty}^{x} e^{-t^2/2}\, dt \quad \text{for all } x \in \mathbb{R}.$$

The hypotheses listed above are quite restrictive and in modern probability theory these theorems are often needed in much greater generality. For proofs, many examples, and comprehensive expositions of probability theory see Ash [1], Fremlin [4, Chapter 27], Malliavin [13].

An important class of random variables are those where the distribution functions $F_X\colon \mathbb{R} \to [0,1]$ are absolutely continuous (Theorem 6.19). This means that the pushforward measures X_*P on the Borel σ-algebra $\mathcal{B} \subset 2^{\mathbb{R}}$ admit densities as in Theorem 1.40 with respect to the Lebesgue measure. The Lebesgue measure is introduced in Chapter 2 and the existence of a density is the subject of Chapter 5 on the Radon–Nikodým theorem.

Chapter 2

The Lebesgue measure

This chapter introduces the most important example, namely the Lebesgue measure on Euclidean space. Let $n \in \mathbb{N}$ and denote by $\mathcal{B} \subset 2^{\mathbb{R}^n}$ the σ-algebra of all Borel sets in $\mathbb{R}^n$, i.e., the smallest σ-algebra on $\mathbb{R}^n$ that contains all open sets in the standard topology (Definition 1.15). Then

$$B + x := \{y + x \mid y \in B\} \in \mathcal{B} \quad \text{for all } B \in \mathcal{B} \text{ and all } x \in \mathbb{R}^n,$$

because the translation $\mathbb{R}^n \to \mathbb{R}^n : y \mapsto y + x$ is a homeomorphism. A measure $\mu : \mathcal{B} \to [0, \infty]$ is called *translation-invariant* if it satisfies

$$\mu(B + x) = \mu(B) \quad \text{for all } B \in \mathcal{B} \text{ and all } x \in \mathbb{R}^n. \tag{2.1}$$

Theorem 2.1. *There exists a unique measure $\mu : \mathcal{B} \to [0, \infty]$ that is translation-invariant and satisfies the normalization condition $\mu([0,1]^n) = 1$.*

Proof. See page 74. □

Definition 2.2. Let $(\mathbb{R}^n, \mathcal{B}, \mu)$ be the measure space in Theorem 2.1 and denote by $(\mathbb{R}^n, \mathcal{A}, m)$ its completion as in Theorem 1.55. Thus

$$\mathcal{A} := \left\{ A \subset \mathbb{R}^n \;\middle|\; \begin{array}{l} \text{there exist Borel sets } B_0, B_1 \in \mathcal{B} \\ \text{such that } B_0 \subset A \subset B_1 \text{ and } \mu(B_1 \setminus B_0) = 0 \end{array} \right\} \tag{2.2}$$

and

$$m(A) := \mu(B_0) \quad \text{for } A \in \mathcal{A},$$

where $B_0, B_1 \in \mathcal{B}$ are chosen such that $B_0 \subset A \subset B_1$ and $\mu(B_1 \setminus B_0) = 0$. The elements of $\mathcal{A}$ are called *Lebesgue measurable subsets of* $\mathbb{R}^n$, the function

$$m : \mathcal{A} \longrightarrow [0, \infty]$$

is called the *Lebesgue measure,* and the triple $(\mathbb{R}^n, \mathcal{A}, m)$ is called the *Lebesgue measure space.* A function

$$f : \mathbb{R}^n \longrightarrow \mathbb{R}$$

is called *Lebesgue measurable* if it is measurable with respect to the Lebesgue σ-algebra $\mathcal{A}$ on $\mathbb{R}^n$ (and the Borel σ-algebra on the target space $\mathbb{R}$).

2.1 Outer measures

In preparation for the proof of Theorem 2.1 we now take up the idea, announced in the beginning of Chapter 1, of assigning a measure to every subset of a given set, but requiring only subadditivity. Here is the basic definition.

Definition 2.3. Let X be a set. A function

$$\nu: 2^X \longrightarrow [0, \infty]$$

is called an *outer measure* if it satisfies the following three axioms:

(a) $\nu(\emptyset) = 0$;

(b) if $A \subset B \subset X$, then

$$\nu(A) \leq \nu(B);$$

(c) if $A_i \subset X$ for $i \in \mathbb{N}$, then

$$\nu\Big(\bigcup_{i=1}^{\infty} A_i\Big) \leq \sum_{i=1}^{\infty} \nu(A_i).$$

Let $\nu: 2^X \to [0, \infty]$ be an outer measure. A subset $A \subset X$ is called ν*-measurable* if it satisfies

$$\nu(D) = \nu(D \cap A) + \nu(D \setminus A) \tag{2.3}$$

for every subset $D \subset X$.

The inequality $\nu(D) \leq \nu(D \cap A) + \nu(D \setminus A)$ holds for every outer measure and any two subsets $A, D \subset X$ by (a) and (c). However, the outer measure of a disjoint union need not be equal to the sum of the outer measures. Carathéodory's theorem (Theorem 2.4) below asserts that the ν-measurable sets form a σ-algebra $\mathcal{A}$ and that the restriction of ν to $\mathcal{A}$ is a measure. Theorem 2.5 (the *Carathéodory criterion*) characterizes outer measures ν on metric spaces such that every Borel set is ν-measurable.

Theorem 2.4 (Carathéodory). *Let X be a set, let $\nu: 2^X \to [0, \infty]$ be an outer measure, and define*

$$\mathcal{A} := \mathcal{A}(\nu) := \{A \subset X \mid A \textit{ is } \nu\textit{-measurable}\} \tag{2.4}$$

Then $\mathcal{A}$ is a σ-algebra, the function

$$\mu := \nu|_{\mathcal{A}}: \mathcal{A} \to [0, \infty]$$

is a measure, and the measure space $(X, \mathcal{A}, \mu)$ is complete.

Proof. The proof is carried out in six steps.

Step 1. $X \in \mathcal{A}$.

For every subset $D \subset X$, we have

$$\nu(D \cap X) + \nu(D \setminus X) = \nu(D) + \nu(\emptyset) = \nu(D)$$

by condition (a) in Definition 2.3. Hence $X \in \mathcal{A}$.

Step 2. *If $A \in \mathcal{A}$, then $A^{c} \in \mathcal{A}$.*

Let $A \in \mathcal{A}$. Since

$$D \cap A^{c} = D \setminus A, \quad D \setminus A^{c} = D \cap A,$$

it follows from (2.3) that

$$\nu(D) = \nu(D \cap A^{c}) + \nu(D \setminus A^{c})$$

for every subset $D \subset X$. Hence $A^{c} \in \mathcal{A}$.

Step 3. *If $A, B \in \mathcal{A}$, then $A \cup B \in \mathcal{A}$.*

Let $A, B \in \mathcal{A}$. Then, for every subset $D \subset X$,

$$\begin{aligned}
\nu(D) &= \nu(D \cap A) + \nu(D \setminus A) \\
&= \nu(D \cap A) + \nu(D \cap A^{c}) \\
&= \nu(D \cap A) + \nu(D \cap A^{c} \cap B) + \nu((D \cap A^{c}) \setminus B) \\
&\geq \nu((D \cap A) \cup (D \cap A^{c} \cap B)) + \nu(D \cap A^{c} \cap B^{c}) \\
&- \nu(D \cap (A \cup B)) + \nu(D \cap (A \cup B)^{c}) \\
&= \nu(D \cap (A \cup B)) + \nu(D \setminus (A \cup B)).
\end{aligned}$$

Here the inequality follows from axioms (a) and (c) in Definition 2.3. Using axioms (a) and (c) again, we obtain

$$\nu(D) = \nu(D \cap (A \cup B)) + \nu(D \setminus (A \cup B))$$

for every subset $D \subset X$ and hence $A \cup B \in \mathcal{A}$.

Step 4. *Let $A_i \in \mathcal{A}$ for $i \in \mathbb{N}$ such that $A_i \cap A_j = \emptyset$ for $i \neq j$. Then*

$$A := \bigcup_{i=1}^{\infty} A_i \in \mathcal{A}, \quad \nu(A) = \sum_{i=1}^{\infty} \nu(A_i).$$

For $k \in \mathbb{N}$ define

$$B_k := A_1 \cup A_2 \cup \cdots \cup A_k.$$

Then $B_k \in \mathcal{A}$ for all $k \in \mathbb{N}$ by Step 3. Now let $D \subset X$. Then, for all $k \geq 2$,

$$\begin{aligned} \nu(D \cap B_k) &= \nu(D \cap B_k \cap A_k) + \nu((D \cap B_k) \setminus A_k) \\ &= \nu(D \cap A_k) + \nu(D \cap B_{k-1}) \end{aligned}$$

and so, by induction on k,

$$\nu(D \cap B_k) = \sum_{i=1}^{k} \nu(D \cap A_i).$$

Since $B_k \in \mathcal{A}$, this implies

$$\begin{aligned} \nu(D) &= \nu(D \cap B_k) + \nu(D \setminus B_k) \\ &= \sum_{i=1}^{k} \nu(D \cap A_i) + \nu(D \setminus B_k) \\ &\geq \sum_{i=1}^{k} \nu(D \cap A_i) + \nu(D \setminus A). \end{aligned}$$

Here the last inequality follows from axiom (b) in Definition 2.3. Since this holds for all $k \in \mathbb{N}$ and $D \cap A = \bigcup_{i=1}^{\infty}(D \cap A_i)$, it follows that

$$\begin{aligned} \nu(D) &\geq \sum_{i=1}^{\infty} \nu(D \cap A_i) + \nu(D \setminus A) \\ &\geq \nu(D \cap A) + \nu(D \setminus A) \\ &\geq \nu(D). \end{aligned}$$

Here the last two inequalities follow from axiom (c). Hence

$$\nu(D) = \sum_{i=1}^{\infty} \nu(D \cap A_i) + \nu(D \setminus A) = \nu(D \cap A) + \nu(D \setminus A) \tag{2.5}$$

for all $D \subset X$. This shows that $A \in \mathcal{A}$. Now take $D = A$ to obtain $D \setminus A = \emptyset$ and $D \cap A_i = A_i$. Then it follows from (2.5) that $\nu(A) = \sum_{i=1}^{\infty} \nu(A_i)$.

Step 5. *Let $A_i \in \mathcal{A}$ for $i \in \mathbb{N}$. Then*

$$A := \bigcup_{i=1}^{\infty} A_i \in \mathcal{A}.$$

Define

$$B_i = \begin{cases} A_1 & \text{for } i = 1, \\ A_i \setminus (A_1 \cup \dots \cup A_{i-1}) & \text{for } i \geq 2. \end{cases}$$

Then $B_i \cap B_j = \emptyset$ for $i \neq j$ and $B_i = (A_1 \cup \dots \cup A_{i-1} \cup A_i^{\mathrm{c}})^{\mathrm{c}} \in \mathcal{A}$ for all i by Steps 2 and 3. Hence $A = \bigcup_{i=1}^{\infty} B_i \in \mathcal{A}$ by Step 4.

Step 6. *$(X, \mathcal{A}, \mu)$ is a complete measure space.*

It follows from Steps 1, 2, 4, and 5 that $(X, \mathcal{A}, \mu = \nu|_{\mathcal{A}})$ is a measure space. We prove that it is complete. To see this, let $A \subset X$ and suppose that $A \subset N$ where $N \in \mathcal{A}$ satisfies $\mu(N) = 0$. Then it follows from axiom (b) in Definition 2.3 that $\nu(A) \leq \nu(N) = \mu(N) = 0$ and therefore $\nu(A) = 0$. Now use axioms (a), (b) and (c) to obtain

$$\nu(D) \leq \nu(D \cap A) + \nu(D \setminus A) \leq \nu(A) + \nu(D) = \nu(D),$$

and so $\nu(D) = \nu(D \cap A) + \nu(D \setminus A)$ for all $D \subset X$, which shows that $A \in \mathcal{A}$. This yields Step 6 and completes the proof of Theorem 2.4. □

Theorem 2.5 (Carathéodory criterion). *Let (X, d) be a metric space. Consider an outer measure $\nu: 2^X \to [0, \infty]$. Let $\mathcal{A}(\nu) \subset 2^X$ be the σ-algebra given by* (2.4) *and $\mathcal{B} \subset 2^X$ be the Borel σ-algebra of (X, d). Then the following statements are equivalent*:

(i) $\mathcal{B} \subset \mathcal{A}(\nu)$;

(ii) *if $A, B \subset X$ satisfy $d(A, B) := \inf_{a \in A, b \in B} d(a, b) > 0$, then*

$$\nu(A \cup B) = \nu(A) + \nu(B).$$

Proof. We prove that (i) implies (ii). Thus assume that ν satisfies (i). Let $A, B \subset X$ such that $\varepsilon := d(A, B) > 0$. Define

$$U := \{x \in X \mid \text{there exists } a \in A \text{ such that } d(a, x) < \varepsilon\} = \bigcup_{a \in A} B_\varepsilon(a).$$

Then U is open, $A \subset U$, and $U \cap B = \emptyset$. Hence $U \in \mathcal{B} \subset \mathcal{A}(\nu)$ by assumption and hence $\nu(A \cup B) = \nu((A \cup B) \cap U) + \nu((A \cup B) \setminus U) = \nu(A) + \nu(B)$. Thus the outer measure ν satisfies (ii).

We prove that (ii) implies (i). Thus assume that ν satisfies (ii). We show that every closed set $A \subset X$ is ν-measurable, i.e., $\nu(D) = \nu(D \cap A) + \nu(D \setminus A)$ for all $D \subset X$. Since $\nu(D) \le \nu(D \cap A) + \nu(D \setminus A)$, by definition of an outer measure, it suffices to prove the following.

Claim 1. *Fix a closed set $A \subset X$ and a set $D \subset X$ such that $\nu(D) < \infty$. Then*

$$\nu(D) \ge \nu(D \cap A) + \nu(D \setminus A).$$

To see this, replace the set $D \setminus A$ by the smaller set $D \setminus U_k$, where

$$U_k := \{x \in X \mid \text{there exists } a \in A \text{ such that } d(a,x) < 1/k\} = \bigcup_{a \in A} B_{1/k}(a).$$

For each $k \in \mathbb{N}$, the set U_k is open and $d(x,y) \ge 1/k$ for all $x \in D \cap A$ and all $y \in D \setminus U_k$. Hence

$$d(D \cap A, D \setminus U_k) \ge \frac{1}{k}.$$

By (ii) and axiom (b) this implies

$$\nu(D \cap A) + \nu(D \setminus U_k) = \nu((D \cap A) \cup (D \setminus U_k)) \le \nu(D) \tag{2.6}$$

for every subset $D \subset X$ and every $k \in \mathbb{N}$. We will prove the following.

Claim 2. $\lim_{k\to\infty} \nu(D \setminus U_k) = \nu(D \setminus A)$.

Claim 1 follows directly from Claim 2 and (2.6). To prove Claim 2 note that

$$A = \bigcap_{i=1}^{\infty} U_i,$$

because A is closed. (If $x \in U_i$ for all $i \in \mathbb{N}$, then there exists a sequence $a_i \in A$ such that $d(a_i, x) < 1/i$ and hence $x = \lim_{i\to\infty} a_i \in A$.) This implies

$$U_k \setminus A = \bigcup_{i=1}^{\infty} (U_k \setminus U_i) = \bigcup_{i=k}^{\infty} (U_i \setminus U_{i+1})$$

and hence

$$\begin{aligned} D \setminus A &= (D \setminus U_k) \cup (D \cap (U_k \setminus A)) \\ &= (D \setminus U_k) \cup \bigcup_{i=k}^{\infty} (D \cap (U_i \setminus U_{i+1})). \end{aligned}$$

Thus

$$D \setminus A = (D \setminus U_k) \cup \bigcup_{i=k}^{\infty} E_i, \quad E_i := (D \cap U_i) \setminus U_{i+1}. \tag{2.7}$$

Claim 3. *The outer measures of the E_i satisfy*

$$\sum_{i=1}^{\infty} \nu(E_i) < \infty.$$

Claim 3 $\Longrightarrow$ *Claim* 2. It follows from Claim 3 that the sequence

$$\varepsilon_k := \sum_{i=k}^{\infty} \nu(E_i)$$

converges to zero. Moreover, it follows from equation (2.7) and axiom (c) in Definition 2.3 that

$$\nu(D \setminus A) \leq \nu(D \setminus U_k) + \sum_{i=k}^{\infty} \nu(E_i) = \nu(D \setminus U_k) + \varepsilon_k.$$

Hence it follows from axiom (b) in Definition 2.3 that

$$\nu(D \setminus A) - \varepsilon_k \leq \nu(D \setminus U_k) \leq \nu(D \setminus A)$$

for every $k \in \mathbb{N}$. Since ε_k converges to zero, this implies Claim 2. The proof of Claim 3 relies on the next assertion.

Claim 4. $d(E_i, E_j) > 0$ *for* $i \geq j + 2$.

Claim 4 $\Longrightarrow$ *Claim* 3. It follows from Claim 4, axiom (b), and (ii) that

$$\sum_{i=1}^{n} \nu(E_{2i}) = \nu\Big(\bigcup_{i=1}^{n} E_{2i}\Big) \leq \nu(D)$$

and

$$\sum_{i=1}^{n} \nu(E_{2i-1}) = \nu\Big(\bigcup_{i=1}^{n} E_{2i-1}\Big) \leq \nu(D)$$

for every $n \in \mathbb{N}$. Hence

$$\sum_{i=1}^{\infty} \nu(E_i) \leq 2\nu(D) < \infty$$

and this shows that Claim 4 implies Claim 3.

Proof of Claim 4. We show that

$$d(E_i, E_j) \geq \frac{1}{(i+1)(i+2)} \quad \text{for } j \geq i+2.$$

To see this, fix indices i, j with $j \geq i+2$. Let $x \in E_i$ and $y \in X$ such that

$$d(x, y) < \frac{1}{(i+1)(i+2)}.$$

Then $x \notin U_{i+1}$ because $E_i \cap U_{i+1} = \emptyset$. (See (2.7).) Hence

$$d(a, x) \geq \frac{1}{i+1} \quad \text{for all } a \in A.$$

This implies

$$\begin{aligned} d(a, y) &\geq d(a, x) - d(x, y) \\ &> \frac{1}{i+1} - \frac{1}{(i+1)(i+2)} \\ &= \frac{1}{i+2} \\ &\geq \frac{1}{j} \end{aligned}$$

for all $a \in A$. Hence $y \notin U_j$ and consequently $y \notin E_j$, because $E_j \subset U_j$. This proves Claim 4 and Theorem 2.5. □

2.2 The Lebesgue outer measure

The purpose of this section is to introduce the Lebesgue outer measure ν on $\mathbb{R}^n$, construct the Lebesgue measure as the restriction of ν to the σ-algebra of all ν-measurable subsets of $\mathbb{R}^n$, and prove Theorem 2.1.

Definition 2.6. A *closed cuboid in* $\mathbb{R}^n$ is a set of the form

$$\begin{aligned} Q &:= Q(a, b) \\ &:= [a_1, b_1] \times [a_2, b_2] \times \cdots \times [a_n, b_n] \\ &= \{x = (x_1, \ldots, x_n) \in \mathbb{R}^n \mid a_j \leq x_j \leq b_j \text{ for } j = 1, \ldots, n\} \end{aligned} \tag{2.8}$$

for $a_1, \ldots, a_n, b_1, \ldots, b_n \in \mathbb{R}$ with $a_j < b_j$ for all j. The (*n-dimensional*) *volume* of the cuboid $Q(a, b)$ is defined by

$$\operatorname{Vol}(Q(a, b)) := \operatorname{Vol}_n(Q(a, b)) := \prod_{j=1}^{n} (b_j - a_j). \tag{2.9}$$

The *volume of the open cuboid*

$$U := \operatorname{int}(Q) = \prod_{i=1}^{n} (a_i, b_i)$$

is defined by

$$\operatorname{Vol}(U) := \operatorname{Vol}(Q).$$

The set of all closed cuboids in $\mathbb{R}^n$ will be denoted by

$$\mathcal{Q}_n := \left\{ Q(a,b) \,\middle|\, \begin{array}{l} a_1, \dots, a_n, b_1, \dots, b_n \in \mathbb{R}, \\ a_j < b_j \text{ for } j = 1, \dots, n \end{array} \right\}.$$

Definition 2.7. A subset $A \subset \mathbb{R}^n$ is called a *Jordan null set* if, for every $\varepsilon > 0$, there exist finitely many closed cuboids $Q_1, \dots, Q_\ell \in \mathcal{Q}_n$ such that

$$A \subset \bigcup_{i=1}^{\ell} Q_i, \quad \sum_{i=1}^{\ell} \operatorname{Vol}(Q_i) < \varepsilon.$$

Definition 2.8. A subset $A \subset \mathbb{R}^n$ is called a *Lebesgue null set* if, for every $\varepsilon > 0$, there is a sequence of closed cuboids $Q_i \in \mathcal{Q}_n$, $i \in \mathbb{N}$, such that

$$A \subset \bigcup_{i=1}^{\infty} Q_i, \quad \sum_{i=1}^{\infty} \operatorname{Vol}(Q_i) < \varepsilon.$$

Definition 2.9. The *Lebesgue outer measure* on $\mathbb{R}^n$ is the function

$$\nu = \nu_n : 2^{\mathbb{R}^n} \longrightarrow [0, \infty]$$

defined by

$$\nu(A) := \inf \left\{ \sum_{i=1}^{\infty} \operatorname{Vol}_n(Q_i) \,\middle|\, Q_i \in \mathcal{Q}_n,\ A \subset \bigcup_{i=1}^{\infty} Q_i \right\} \quad \text{for } A \subset \mathbb{R}^n. \tag{2.10}$$

Theorem 2.10 (the Lebesgue outer measure). *Let $\nu : 2^{\mathbb{R}^n} \to [0, \infty]$ be the function defined by* (2.10). *Then the following holds.*

(i) *ν is an outer measure;*

(ii) *ν is **translation-invariant**, i.e., for all $A \subset \mathbb{R}^n$ and all $x \in \mathbb{R}^n$*

$$\nu(A + x) = \nu(A);$$

(iii) *if $A, B \subset \mathbb{R}^n$ are such that $d(A, B) > 0$, then $\nu(A \cup B) = \nu(A) + \nu(B)$;*

(iv) *$\nu(\operatorname{int}(Q)) = \nu(Q) = \operatorname{Vol}(Q)$ for all $Q \in \mathcal{Q}_n$.*

Proof. We prove (i). The empty set is contained in every cuboid $Q \in \mathcal{Q}_n$. Since there are cuboids with arbitrarily small volume, it follows that $\nu(\emptyset) = 0$. If $A \subset B \subset \mathbb{R}^n$ it follows directly from Definition 2.9 that $\nu(A) \leq \nu(B)$. Now let $A_i \subset \mathbb{R}^n$ for $i \in \mathbb{N}$, define

$$A := \bigcup_{i=1}^{\infty} A_i,$$

and fix a constant $\varepsilon > 0$. Then it follows from Definition 2.9 that, for $i \in \mathbb{N}$, there exists a sequence of cuboids $Q_{ij} \in \mathcal{Q}_n$, $j \in \mathbb{N}$, such that

$$A_i \subset \bigcup_{j=1}^{\infty} Q_{ij}, \quad \sum_{j=1}^{\infty} \mathrm{Vol}(Q_{ij}) < \frac{\varepsilon}{2^i} + \nu(A_i).$$

Hence

$$A \subset \bigcup_{i,j\in\mathbb{N}} Q_{ij}, \quad \sum_{i,j\in\mathbb{N}} \mathrm{Vol}(Q_{ij}) < \sum_{i=1}^{\infty} \Big(\frac{\varepsilon}{2^i} + \nu(A_i)\Big) = \varepsilon + \sum_{i=1}^{\infty} \nu(A_i).$$

This implies

$$\nu(A) < \varepsilon + \sum_{i=1}^{\infty} \nu(A_i)$$

for every $\varepsilon > 0$, and thus $\nu(A) \leq \sum_{i=1}^{\infty} \nu(A_i)$. This proves part (i).

We prove (ii). If $A \subset \bigcup_{i=1}^{\infty} Q_i$ with $Q_i \in \mathcal{Q}_n$, then $A + x \subset \bigcup_{i=1}^{\infty}(Q_i + x)$ for every $x \in \mathbb{R}^n$ and $\mathrm{Vol}(Q_i + x) = \mathrm{Vol}(Q_i)$, by the definition of the volume. Hence part (ii) follows from Definition 2.9.

We prove (iii). Let $A, B \subset \mathbb{R}^n$ such that $d(A, B) > 0$. Choose a sequence of closed cuboids $Q_i \in \mathcal{Q}_n$ such that

$$A \cup B \subset \bigcup_{i=1}^{\infty} Q_i, \quad \sum_{i=1}^{\infty} \mathrm{Vol}(Q_i) < \nu(A \cup B) + \varepsilon.$$

Subdividing each Q_i into finitely many smaller cuboids, if necessary, we may assume without loss of generality that

$$\mathrm{diam}(Q_i) := \sup_{x,y\in Q_i} |x - y| < \frac{d(A, B)}{2}.$$

Here $|\cdot|$ denotes the Euclidean norm on $\mathbb{R}^n$. Then, for every $i \in \mathbb{N}$, we have either $Q_i \cap A = \emptyset$ or $Q_i \cap B = \emptyset$. This implies

$$I \cap J = \emptyset, \quad I := \{i \in \mathbb{N} \mid Q_i \cap A \neq \emptyset\}, \quad J := \{i \in \mathbb{N} \mid Q_i \cap B \neq \emptyset\}.$$

Hence

$$\begin{aligned}\nu(A) + \nu(B) &\le \sum_{i \in I} \mathrm{Vol}(Q_i) + \sum_{i \in J} \mathrm{Vol}(Q_i) \\ &\le \sum_{i=1}^{\infty} \mathrm{Vol}(Q_i) \\ &< \nu(A \cup B) + \varepsilon.\end{aligned}$$

Thus $\nu(A) + \nu(B) < \nu(A \cup B) + \varepsilon$ for all $\varepsilon > 0$, so $\nu(A) + \nu(B) \le \nu(A \cup B)$, and hence $\nu(A) + \nu(B) = \nu(A \cup B)$, by axioms (a) and (c) in Definition 2.3. This proves part (iii).

We prove (iv) by an argument due to von Neumann. Fix a closed cuboid

$$Q = I_1 \times \cdots \times I_n, \quad I_i = [a_i, b_i].$$

We claim that

$$\mathrm{Vol}(Q) \le \nu(Q). \tag{2.11}$$

Equivalently, if $Q_i \in \mathcal{Q}_n$, $i \in \mathbb{N}$, is a sequence of closed cuboids, then

$$Q \subset \bigcup_{i=1}^{\infty} Q_i \implies \mathrm{Vol}(Q) \le \sum_{i=1}^{\infty} \mathrm{Vol}(Q_i). \tag{2.12}$$

For a closed interval $I = [a, b] \subset \mathbb{R}$ with $a < b$ define

$$|I| := b - a.$$

Then

$$|I| - 1 \le \#(I \cap \mathbb{Z}) \le |I| + 1.$$

Hence

$$N|I| - 1 \le \#(NI \cap \mathbb{Z}) \le N|I| + 1,$$

and thus

$$|I| - \frac{1}{N} \le \frac{1}{N}\#\Big(I \cap \frac{1}{N}\mathbb{Z}\Big) \le |I| + \frac{1}{N}$$

for every integer $N \in \mathbb{N}$. Take the limit $N \to \infty$ to obtain

$$|I| = \lim_{N \to \infty} \frac{1}{N}\#\Big(I \cap \frac{1}{N}\mathbb{Z}\Big).$$

Thus

$$\begin{aligned}\mathrm{Vol}(Q) &= \lim_{N\to\infty} \prod_{j=1}^{n} \frac{1}{N} \#\Big(I_j \cap \frac{1}{N}\mathbb{Z}\Big)\\ &= \lim_{N\to\infty} \frac{1}{N^n} \#\Big(Q \cap \frac{1}{N}\mathbb{Z}^n\Big).\end{aligned} \tag{2.13}$$

Now suppose $Q_i \in \mathcal{Q}_n$, $i \in \mathbb{N}$, is a sequence of closed cuboids such that

$$Q \subset \bigcup_{i=1}^{\infty} Q_i.$$

Fix a constant $\varepsilon > 0$ and choose a sequence of open cuboids $U_i \subset \mathbb{R}^n$ such that

$$Q_i \subset U_i, \quad \mathrm{Vol}(U_i) < \mathrm{Vol}(Q_i) + \frac{\varepsilon}{2^i}.$$

Since Q is compact and the U_i form an open cover of Q, there exists a constant $k \in \mathbb{N}$ such that

$$Q \subset \bigcup_{i=1}^{k} U_i.$$

This implies

$$\begin{aligned}\frac{1}{N^n} \#\Big(Q \cap \frac{1}{N}\mathbb{Z}^n\Big) &\leq \sum_{i=1}^{k} \frac{1}{N^n} \#\Big(U_i \cap \frac{1}{N}\mathbb{Z}^n\Big)\\ &\leq \sum_{i=1}^{k} \frac{1}{N^n} \#\Big(\bar{U}_i \cap \frac{1}{N}\mathbb{Z}^n\Big).\end{aligned}$$

Take the limit $N \to \infty$ and use (2.13) to obtain

$$\begin{aligned}\mathrm{Vol}(Q) &\leq \sum_{i=1}^{k} \mathrm{Vol}(U_i)\\ &\leq \sum_{i=1}^{\infty} \mathrm{Vol}(U_i)\\ &\leq \sum_{i=1}^{\infty} \Big(\frac{\varepsilon}{2^i} + \mathrm{Vol}(Q_i)\Big)\\ &= \varepsilon + \sum_{i=1}^{\infty} \mathrm{Vol}(Q_i).\end{aligned}$$

Since $\varepsilon > 0$ can be chosen arbitrarily small, this proves (2.12) and (2.11). Thus we have proved that $\nu(Q) \le \mathrm{Vol}(Q) \le \nu(Q)$, and so $\nu(Q) = \mathrm{Vol}(Q)$. To prove that $\nu(\mathrm{int}(Q)) = \mathrm{Vol}(Q)$, fix a constant $\varepsilon > 0$ and choose a closed cuboid $P \in \mathcal{Q}_n$ such that

$$P \subset \mathrm{int}(Q), \quad \mathrm{Vol}(Q) - \varepsilon < \mathrm{Vol}(P).$$

Then

$$\mathrm{Vol}(Q) - \varepsilon < \mathrm{Vol}(P) = \nu(P) \le \nu(\mathrm{int}(Q)).$$

Thus $\mathrm{Vol}(Q) - \varepsilon < \nu(\mathrm{int}(Q))$ for all $\varepsilon > 0$. Therefore, by axiom (b),

$$\mathrm{Vol}(Q) \le \nu(\mathrm{int}(Q)) \le \nu(Q) = \mathrm{Vol}(Q),$$

and hence $\nu(\mathrm{int}(Q)) = \mathrm{Vol}(Q)$. This proves part (iv) and Theorem 2.10. □

Definition 2.11. Let $\nu\colon 2^{\mathbb{R}^n} \to [0,\infty]$ be the Lebesgue outer measure. A subset $A \subset \mathbb{R}^n$ is called *Lebesgue measurable* if A is ν-measurable, i.e.,

$$\nu(D) = \nu(D \cap A) + \nu(D \setminus A) \quad \text{for all } D \subset \mathbb{R}^n.$$

The set of all Lebesgue measurable subsets of $\mathbb{R}^n$ will be denoted by

$$\mathcal{A} := \{A \subset \mathbb{R}^n \mid A \text{ is Lebesgue measurable}\}.$$

The function

$$m := \nu|_{\mathcal{A}}\colon \mathcal{A} \longrightarrow [0,\infty]$$

is called the *Lebesgue measure* on $\mathbb{R}^n$. A function $f\colon \mathbb{R}^n \to \mathbb{R}$ is called *Lebesgue measurable* if it is measurable with respect to the Lebesgue σ-algebra $\mathcal{A}$ on $\mathbb{R}^n$ (and the Borel σ-algebra on the target space $\mathbb{R}$).

Corollary 2.12. (i) *$(\mathbb{R}^n, \mathcal{A}, m)$ is a complete measure space.*

(ii) *m is translation-invariant, i.e., if $A \in \mathcal{A}$ and $x \in \mathbb{R}^n$, then*

$$A + x \in \mathcal{A} \quad \textit{and} \quad m(A + x) = m(A).$$

(iii) *Every Borel set in $\mathbb{R}^n$ is Lebesgue measurable.*

(iv) *If $Q \in \mathcal{Q}_n$, then*

$$Q, \mathrm{int}(Q) \in \mathcal{A}, \quad \textit{and} \quad m(\mathrm{int}(Q)) = m(Q) = \mathrm{Vol}(Q).$$

Proof. Assertion (i) follows from Theorem 2.4 and part (i) of Theorem 2.10. Assertion (ii) follows from the definitions and part (ii) of Theorem 2.10. Assertion (iii) follows from Theorem 2.5 and part (iii) of Theorem 2.10. Finally, assertion (iv) follows from (iii) and part (iv) of Theorem 2.10. □

The restriction of the measure m in Corollary 2.12 to the Borel σ-algebra of $\mathbb{R}^n$ satisfies the requirements of Theorem 2.1 (translation invariance and normalization) and hence settles the existence problem. The uniqueness proof relies on certain regularity properties of the measure m which are established in the next theorem along with continuity from below for the Lebesgue outer measure ν. Theorem 2.14 shows that m is the completion of its restriction to the Borel σ-algebra of $\mathbb{R}^n$ and, with that at hand, we can then prove uniqueness in Theorem 2.1.

Theorem 2.13 (regularity of the Lebesgue outer measure). *The Lebesgue outer measure* $\nu\colon 2^{\mathbb{R}^n} \to [0,\infty]$ *satisfies the following.*

(i) *For every subset* $A \subset \mathbb{R}^n$,

$$\nu(A) = \inf\{\nu(U) \mid A \subset U \subset \mathbb{R}^n \textit{ and } U \textit{ is open}\}.$$

(ii) *If* $A \subset \mathbb{R}^n$ *is Lebesgue measurable, then*

$$\nu(A) = \sup\{\nu(K) \mid K \subset A \textit{ and } K \textit{ is compact}\}.$$

(iii) *If* A_i *is a sequence of subsets of* $\mathbb{R}^n$ *such that* $A_i \subset A_{i+1}$ *for all* $i \in \mathbb{N}$, *then their union*

$$A := \bigcup_{i=1}^{\infty} A_i$$

has Lebesgue outer measure

$$\nu(A) = \lim_{i\to\infty} \nu(A_i).$$

Proof. (i) Fix a subset $A \subset \mathbb{R}^n$ and a constant $\varepsilon > 0$. The assertion is obvious when $\nu(A) = \infty$. Hence assume $\nu(A) < \infty$ and choose a sequence of closed cuboids $Q_i \in \mathcal{Q}_n$ such that

$$A \subset \bigcup_{i=1}^{\infty} Q_i, \quad \sum_{i=1}^{\infty} \mathrm{Vol}(Q_i) < \nu(A) + \frac{\varepsilon}{2}.$$

Now choose a sequence of open cuboids $U_i \subset \mathbb{R}^n$ such that

$$Q_i \subset U_i, \quad \mathrm{Vol}(U_i) < \mathrm{Vol}(Q_i) + \frac{\varepsilon}{2^{i+1}}.$$

Then $U := \bigcup_{i=1}^{\infty} U_i$ is an open subset of $\mathbb{R}^n$ containing A and

$$\nu(U) \le \sum_{i=1}^{\infty} \nu(U_i) = \sum_{i=1}^{\infty} \mathrm{Vol}(U_i) < \sum_{i=1}^{\infty} \Big(\mathrm{Vol}(Q_i) + \frac{\varepsilon}{2^{i+1}} \Big) < \nu(A) + \varepsilon.$$

(ii) Assume first that $A \subset \mathbb{R}^n$ is Lebesgue measurable and bounded. Choose $r > 0$ so large that $A \subset B_r := \{x \in \mathbb{R}^n \mid |x| < r\}$. Fix a constant $\varepsilon > 0$. By (i), there exists an open set $U \subset \mathbb{R}^n$ such that $\overline{B}_r \setminus A \subset U$ and $\nu(U) \leq \nu(\overline{B}_r \setminus A) + \varepsilon$. Hence $K := \overline{B}_r \setminus U$ is a compact subset of A and

$$\nu(K) = \nu(\overline{B}_r) - \nu(U) \geq \nu(\overline{B}_r) - \nu(\overline{B}_r \setminus A) - \varepsilon = \nu(A) - \varepsilon.$$

Here the first equality uses the fact that K and U are disjoint Lebesgue measurable sets with union $\overline{B}_r$ and the last equality uses the fact that A and $\overline{B}_r \setminus A$ are disjoint Lebesgue measurable sets with union $\overline{B}_r$. This proves (ii) for bounded Lebesgue measurable sets. If $A \in \mathcal{A}$ is unbounded, then

$$\begin{aligned} \nu(A) &= \sup_{r>0} \nu(A \cap \overline{B}_r) \\ &= \sup_{r>0} \sup\{\nu(K) \mid K \subset (A \cap \overline{B}_r) \text{ and } K \text{ is compact}\} \\ &= \sup\{\nu(K) \mid K \subset A \text{ and } K \text{ is compact}\}. \end{aligned}$$

(iii) If $\nu(A_i) = \infty$ for some i, then the assertion is obvious. Hence assume $\nu(A_i) < \infty$ for all i and fix a constant $\varepsilon > 0$. By (i), there is a sequence of open sets $U_i \subset \mathbb{R}^n$ such that $\nu(U_i) < \nu(A_i) + \varepsilon 2^{-i}$ for all i. Since $A_i \subset U_i \cap U_{i+1}$, this yields

$$\begin{aligned} \nu(U_{i+1} \setminus U_i) &= \nu(U_{i+1}) - \nu(U_i \cap U_{i+1}) \\ &< \nu(A_{i+1}) - \nu(A_i) + \varepsilon 2^{-i-1} \end{aligned}$$

for all $i \in \mathbb{N}$. It follows that

$$\nu\Big(\bigcup_{i=1}^{k} U_i\Big) = \nu(U_1) + \sum_{i=1}^{k-1} \nu(U_{i+1} \setminus U_i) < \nu(A_k) + \varepsilon.$$

Take the limit $k \to \infty$ to obtain

$$\nu\Big(\bigcup_{i=1}^{\infty} U_i\Big) \leq \lim_{k\to\infty} \nu(A_k) + \varepsilon.$$

Thus

$$\nu(A) \leq \lim_{k\to\infty} \nu(A_k) + \varepsilon, \qquad \text{for all } \varepsilon > 0,$$

and so

$$\nu(A) \leq \lim_{k\to\infty} \nu(A_k).$$

The converse inequality is obvious. This proves Theorem 2.13. □

Theorem 2.14 (the Lebesgue measure as a completion). *Let* $\nu: 2^{\mathbb{R}^n} \to [0, \infty]$ *be the Lebesgue outer measure in Definition 2.9, let*

$$m = \nu|_{\mathcal{A}}: \mathcal{A} \longrightarrow [0, \infty]$$

be the Lebesgue measure, let $\mathcal{B} \subset \mathcal{A}$ *be the Borel* σ*-algebra of* $\mathbb{R}^n$*, and define*

$$\mu := \nu|_{\mathcal{B}}: \mathcal{B} \longrightarrow [0, \infty].$$

Then $(\mathbb{R}^n, \mathcal{A}, m)$ *is the completion of* $(\mathbb{R}^n, \mathcal{B}, \mu)$.

Proof. Let $(\mathbb{R}^n, \mathcal{B}^*, \mu^*)$ denote the completion of $(\mathbb{R}^n, \mathcal{B}, \mu)$.

Claim. *Let* $A \subset \mathbb{R}^n$*. Then the following are equivalent.*

(I) $A \in \mathcal{A}$, *i.e.,*

$$\nu(D) = \nu(D \cap A) + \nu(D \setminus A) \quad \textit{for all } D \subset \mathbb{R}^n.$$

(II) $A \in \mathcal{B}^*$, *i.e., there exist Borel measurable sets* $B_0, B_1 \in \mathcal{B}$ *such that*

$$B_0 \subset A \subset B_1 \quad \textit{and} \quad \nu(B_1 \setminus B_0) = 0.$$

If the set A satisfies both (I) and (II), then

$$\nu(A) \le \nu(B_1) = \nu(B_0) + \nu(B_1 \setminus B_0) = \nu(B_0) \le \nu(A),$$

and hence

$$m(A) = \nu(A) = \nu(B_0) = \mu^*(A).$$

This shows that $\mathcal{A} = \mathcal{B}^*$ and $m = \mu^*$. Thus it remains to prove the claim. Fix a subset $A \subset \mathbb{R}^n$.

We show that (II) implies (I). Thus assume that $A \in \mathcal{B}^*$ and choose Borel measurable sets $B_0, B_1 \in \mathcal{B}$ such that

$$B_0 \subset A \subset B_1, \quad \nu(B_1 \setminus B_0) = 0.$$

Then $\nu(A \setminus B_0) \le \nu(B_1 \setminus B_0) = 0$ and hence $\nu(A \setminus B_0) = 0$. Since ν is an outer measure, by part (i) of Theorem 2.10, it follows from Theorem 2.4 that $A \setminus B_0 \in \mathcal{A}$ and hence $A = B_0 \cup (A \setminus B_0) \in \mathcal{A}$.

We now show that (I) implies (II). Thus assume that $A \in \mathcal{A}$. Suppose first that $\nu(A) < \infty$. By Theorem 2.13, there exist a sequence of compact sets $K_i \subset \mathbb{R}^n$ and a sequence of open sets $U_i \subset \mathbb{R}^n$ such that

$$K_i \subset A \subset U_i, \quad \nu(A) - \frac{1}{i} \le \nu(K_i) \le \nu(U_i) \le \nu(A) + \frac{1}{i}.$$

Define

$$B_0 := \bigcup_{i=1}^{\infty} K_i, \quad B_1 := \bigcap_{i=1}^{\infty} U_i.$$

These are Borel sets satisfying $B_0 \subset A \subset B_1$ and

$$\nu(A) - \frac{1}{i} \le \nu(K_i) \le \nu(B_0) \le \nu(B_1) \le \nu(U_i) \le \nu(A) + \frac{1}{i}.$$

Take the limit $i \to \infty$ to obtain

$$\nu(A) \le \nu(B_0) \le \nu(B_1) \le \nu(A),$$

hence

$$\nu(B_0) = \nu(B_1) = \nu(A) < \infty,$$

and hence

$$\nu(B_1 \setminus B_0) = \nu(B_1) - \nu(B_0) = 0.$$

This shows that $A \in \mathcal{B}^*$ for every $A \in \mathcal{A}$ with $\nu(A) < \infty$.

Now suppose that our set $A \in \mathcal{A}$ satisfies $\nu(A) = \infty$ and define

$$A_k := \{x \in A \mid |x_i| \le k \text{ for } i = 1, \dots, n\} \quad \text{for } k \in \mathbb{N}.$$

Then $A_k \in \mathcal{A}$ and $\nu(A_k) \le (2k)^n$ for all k. Hence $A_k \in \mathcal{B}^*$ for all k and so there exist sequences of Borel sets $B_k, B'_k \in \mathcal{B}$ such that

$$B_k \subset A_k \subset B'_k \quad \text{and} \quad \nu(B'_k \setminus B_k) = 0.$$

Define

$$B := \bigcup_{k=1}^{\infty} B_k \quad \text{and} \quad B' := \bigcup_{k=1}^{\infty} B'_k.$$

Then $B, B' \in \mathcal{B}$, $B \subset A \subset B'$, and

$$\nu(B' \setminus B) \le \sum_{k=1}^{\infty} \nu(B'_k \setminus B) \le \sum_{k=1}^{\infty} \nu(B'_k \setminus B_k) = 0.$$

This shows that $A \in \mathcal{B}^*$ for every $A \in \mathcal{A}$. Thus we have proved that (I) implies (II) and this completes the proof of Theorem 2.14. □

Proof of Theorem 2.1. The existence of a translation-invariant normalized Borel measure on $\mathbb{R}^n$ follows from Corollary 2.12. We prove uniqueness. Thus assume that $\mu' : \mathcal{B} \to [0, \infty]$ is a translation-invariant measure such that $\mu'([0,1]^n) = 1$. Define $\lambda := \mu'([0,1)^n)$. Then $0 \le \lambda \le 1$. We show in five steps that $\lambda = 1$ and $\mu' = \mu$.

Step 1. *For* $x = (x_1, \dots, x_n)$ *and* $k \in \mathbb{N}_0 := \mathbb{N} \cup \{0\}$ *define*

$$R(x,k) := [x_1, x_1 + 2^{-k}) \times \cdots \times [x_n, x_n + 2^{-k}).$$

Then $\mu'(R(x,k)) = \lambda 2^{-nk} = \lambda \mu(R(x,k))$.

Fix an integer $k \in \mathbb{N}_0$. Since $R(x,k) = R(0,k) + x$ for every $x \in \mathbb{R}^n$, it follows from the translation invariance of μ' that there is a constant $c_k \ge 0$ such that

$$\mu'(R(x,k)) = c_k \quad \text{for all } x \in \mathbb{R}^n.$$

Since $R(x,0)$ can be expressed as the disjoint union

$$R(x,0) = \bigcup_{\ell \in \mathbb{Z}^n,\, 0 \le \ell_j \le 2^k - 1} R(x + 2^{-k}\ell, k),$$

this implies

$$\lambda = \mu'(R(x,0)) = \sum_{\ell \in \mathbb{Z}^n,\, 0 \le \ell_j \le 2^k - 1} \mu'(R(x + 2^{-k}\ell, k)) = 2^{nk} c_k.$$

Hence $c_k = \lambda 2^{-nk} = \lambda \mu(R(x,k))$. Here the last equation follows from the fact that $(0, 2^{-k})^n \subset R(0,k) \subset [0, 2^{-k}]^n$ and hence $\mu(R(x,k)) = \mu(R(0,k)) = 2^{-k}$ by part (iv) of Corollary 2.12.

Step 2. $\mu'(U) = \lambda \mu(U)$ *for every open set* $U \subset \mathbb{R}^n$.

Let $U \subset \mathbb{R}^n$ be open. We prove that U can be expressed as a countable union of sets $R_i = R(x_i, k_i)$ as in Step 1. To see this, define

$$\mathcal{R}_0 := \{R(x,0) \mid x \in \mathbb{Z}^n,\ R(x,0) \subset U\},$$

$$\mathcal{R}_1 := \left\{ R(x,1) \,\middle|\, \begin{array}{l} x \in 2^{-1}\mathbb{Z}^n,\ R(x,1) \subset U, \\ R(x,1) \not\subset R, \text{ for all } R \in \mathcal{R}_0 \end{array} \right\},$$

$$\mathcal{R}_k := \left\{ R(x,k) \,\middle|\, \begin{array}{l} x \in 2^{-k}\mathbb{Z}^n,\ R(x,k) \subset U, \\ R(x,k) \not\subset R, \text{ for all } R \in \mathcal{R}_0 \cup \mathcal{R}_1 \cup \cdots \cup \mathcal{R}_{k-1} \end{array} \right\} \quad \text{for } k \ge 2,$$

and denote

$$\mathcal{R} := \bigcup_{k=0}^{\infty} \mathcal{R}_k.$$

Then U can be expressed as the disjoint union $U = \bigcup_{R\in\mathcal{R}} R$ and $\mu'(R) = \mu(R)$ for all $R \in \mathcal{R}$ by Step 1. Hence

$$\mu'(U) = \sum_{R\in\mathcal{R}} \mu'(R) = \sum_{R\in\mathcal{R}} \lambda\mu(R) = \lambda\mu(U).$$

Step 3. $\mu'(K) = \lambda\mu(K)$ *for every compact set* $K \subset \mathbb{R}^n$.

Let $K \subset \mathbb{R}^n$ be compact. Choose $r > 0$ so large that $K \subset U := (-r, r)^n$. Then U and $U \setminus K$ are open. Hence, by Step 2,

$$\mu'(K) = \mu'(U) - \mu'(U \setminus K) = \lambda\mu(U) - \lambda\mu(U \setminus K) = \lambda\mu(K).$$

Step 4. $\mu'(B) = \lambda\mu(B)$ *for every Borel set* $B \in \mathcal{B}$.

Let $B \in \mathcal{B}$. It follows from Step 2, Step 3, and Theorem 2.13 that

$$\begin{aligned} \mu'(B) &\le \inf\{\mu'(U) \mid B \subset U \subset \mathbb{R}^n \text{ and } U \text{ is open}\} \\ &= \inf\{\lambda\mu(U) \mid B \subset U \subset \mathbb{R}^n \text{ and } U \text{ is open}\} \\ &= \lambda\mu(B) \\ &= \sup\{\lambda\mu(K) \mid K \subset B \text{ and } K \text{ is compact}\} \\ &= \sup\{\mu'(K) \mid K \subset B \text{ and } K \text{ is compact}\} \\ &\le \mu'(B). \end{aligned}$$

Step 5. $\lambda = 1$ *and* $\mu' = \mu$.

By Step 4, we have $\lambda = \lambda\mu([0,1]^n) = \mu'([0,1]^n) = 1$ and hence $\mu' = \lambda\mu = \mu$. This proves Step 5 and Theorem 2.1. □

We have given two definitions of the Lebesgue measure $m\colon \mathcal{A} \to [0,\infty]$. The first one, in Definition 2.2, uses the existence and uniqueness of a normalized translation-invariant Borel measure $\mu\colon \mathcal{B} \to [0,\infty]$, established in Theorem 2.1, and then defines $(\mathbb{R}^n, \mathcal{A}, m)$ as the completion of that measure. The second one, in Definition 2.11, uses the Lebesgue outer measure $\nu\colon 2^{\mathbb{R}^n} \to [0,\infty]$ of Definition 2.9 and Theorem 2.10, and defines the Lebesgue measure as the restriction of ν to the σ-algebra of ν-measurable subsets of $\mathbb{R}^n$ (see Theorem 2.4). Theorem 2.14 asserts that the two definitions agree. The next lemma uses the Axiom of Choice to establish the existence of subsets of $\mathbb{R}^n$ that are not Lebesgue measurable.

Lemma 2.15. *Let $A \subset \mathbb{R}$ be a Lebesgue measurable set such that $m(A) > 0$. Then there exists a set $B \subset A$ that is not Lebesgue measurable.*

Proof. Consider the equivalence relation on $\mathbb{R}$ defined by

$$x \sim y \overset{\text{def}}{\iff} x - y \in \mathbb{Q}$$

for $x, y \in \mathbb{R}$. By the Axiom of Choice, there exists a subset $E \subset \mathbb{R}$ which contains precisely one element of each equivalence class. This means that E satisfies the following two conditions:

(I) for every $x \in \mathbb{R}$ there exists a rational number $q \in \mathbb{Q}$ such that $x - q \in E$;

(II) if $x, y \in E$ and $x \neq y$, then $x - y \notin \mathbb{Q}$.

For $q \in \mathbb{Q}$ define the set

$$B_q := A \cap (E + q) = \{x \in A \mid x - q \in E\}.$$

Then it follows from (I) that $A = \bigcup_{q \in \mathbb{Q}} B_q$.

Fix a rational number $q \in \mathbb{Q}$. We prove that if B_q is Lebesgue measurable, then $m(B_q) = 0$. Indeed, let $n \in \mathbb{N}$, and define

$$B_{q,q',n} := (B_q \cap [-n, n]) + q' = \{x + q' \mid x \in B_q,\ |x| \leq n\} \quad \text{for } q' \in \mathbb{Q}.$$

This set is Lebesgue measurable, its Lebesgue measure is independent of q', and $B_{q,q',n} \cap B_{q,q'',n} = \emptyset$ for all $q', q'' \in \mathbb{Q}$ with $q' \neq q''$ by condition (II). Since $B_{q,q',n} \subset [-n, n+1]$ for $q' \in [0,1] \cap \mathbb{Q}$, we have $\sum_{q' \in [0,1] \cap \mathbb{Q}} m(B_{q,q',n}) \leq 2n + 1$. This sum is infinite and all summands agree, so $m(B_q \cap [-n, n]) = 0$. This holds for all $n \in \mathbb{N}$, and hence $m(B_q) = 0$, as claimed.

If B_q is Lebesgue measurable for all $q \in \mathbb{Q}$, it follows that $A = \bigcup_{q \in \mathbb{Q}} B_q$ is a Lebesgue null set, a contradiction. Thus one of the sets B_q is not Lebesgue measurable and this proves Lemma 2.15. □

Remark 2.16. (i) Using Lemma 2.15 one can construct a continuous function $f : \mathbb{R} \to \mathbb{R}$ and a Lebesgue measurable function $g : \mathbb{R} \to \mathbb{R}$ such that the composition $g \circ f$ is not Lebesgue measurable (see Example 6.24).

(ii) Let $E \subset \mathbb{R}$ be the set constructed in the proof of Lemma 2.15. Then the set $E \times \mathbb{R} \subset \mathbb{R}^2$ is not Lebesgue measurable. This follows from a similar argument as in Lemma 2.15 using the sets $((E \cap [-n, n]) + q) \times [0, 1]$. On the other hand, the set $E \times \{0\} \subset \mathbb{R}^2$ is Lebesgue measurable and has Lebesgue measure zero. However, it is not a Borel set, because its pre-image in $\mathbb{R}$ under the continuous map $\mathbb{R} \to \mathbb{R}^2 : x \mapsto (x, 0)$ is the original set E and hence is not a Borel set.

2.3 The transformation formula

The transformation formula describes how the integral of a Lebesgue measurable function transforms under composition with a C^1 diffeomorphism. Fix a positive integer $n \in \mathbb{N}$ and denote by $(\mathbb{R}^n, \mathcal{A}, m)$ the Lebesgue measure space. For any Lebesgue measurable set $X \subset \mathbb{R}^n$ denote by

$$\mathcal{A}_X := \{A \in \mathcal{A} \mid A \subset X\}$$

the restricted Lebesgue σ-algebra and by

$$m_X := m|_{\mathcal{A}_X} : \mathcal{A}_X \longrightarrow [0, \infty]$$

the restriction of the Lebesgue measure to $\mathcal{A}_X$.

Theorem 2.17 (transformation formula). *Suppose $\phi: U \to V$ is a C^1 diffeomorphism between open subsets of $\mathbb{R}^n$.*

(i) *If $f: V \to [0, \infty]$ is Lebesgue measurable, then $f \circ \phi: U \to [0, \infty]$ is Lebesgue measurable and*

$$\int_U (f \circ \phi)|\det(d\phi)|\, dm = \int_V f\, dm. \tag{2.14}$$

(ii) *If $E \in \mathcal{A}_U$ and $f \in \mathcal{L}^1(m_V)$, then $\phi(E) \in \mathcal{A}_V$, $(f \circ \phi)|\det(d\phi)| \in \mathcal{L}^1(m_U)$, and*

$$\int_E (f \circ \phi)|\det(d\phi)|\, dm = \int_{\phi(E)} f\, dm. \tag{2.15}$$

Proof. See page 82. □

The proof of Theorem 2.17 relies on the next two lemmas. The first lemma is the special case where ϕ is linear and f is the characteristic function of a Lebesgue measurable set. The second lemma is a basic estimate that follows from the linear case and implies formula (2.14) for the characteristic functions of open sets.

Lemma 2.18. *Let $\Phi: \mathbb{R}^n \to \mathbb{R}^n$ be a linear transformation and let $A \subset \mathbb{R}^n$ be a Lebesgue measurable set. Then $\Phi(A)$ is a Lebesgue measurable set and*

$$m(\Phi(A)) = |\det(\Phi)|m(A). \tag{2.16}$$

Proof. If $\det(\Phi) = 0$, then $\Phi(A)$ is contained in a proper linear subspace of $\mathbb{R}^n$ and hence is a Lebesgue null set for every $A \in \mathcal{A}$. In this case both sides of equation (2.16) vanish. Hence it suffices to assume that Φ is a vector space isomorphism. For vector space isomorphisms we prove the assertion in six steps. Denote by $\mathcal{B} \subset 2^{\mathbb{R}^n}$ the Borel σ-algebra and by

$$\mu := m|_{\mathcal{B}}$$

the restriction of the Lebesgue measure to the Borel σ-algebra. Thus μ is the unique translation-invariant Borel measure on $\mathbb{R}^n$ that satisfies the normalization condition $\mu([0,1]^n) = 1$ (Theorem 2.1) and $(\mathbb{R}^n, \mathcal{A}, m)$ is the completion of $(\mathbb{R}^n, \mathcal{B}, \mu)$ (Theorem 2.14).

Step 1. *There exists a unique map*

$$\rho\colon \mathrm{GL}(n,\mathbb{R}) \longrightarrow (0,\infty)$$

such that

$$\mu(\Phi(B)) = \rho(\Phi)\mu(B) \tag{2.17}$$

for every $\Phi \in \mathrm{GL}(n,\mathbb{R})$ *and every Borel set* $B \in \mathcal{B}$.

Fix a vector space isomorphism $\Phi\colon \mathbb{R}^n \to \mathbb{R}^n$. Since Φ is a homeomorphism of $\mathbb{R}^n$ with its standard topology, it follows that $\Phi(B) \in \mathcal{B}$ for every $B \in \mathcal{B}$. Define the number $\rho(\Phi) \in [0,\infty]$ by

$$\rho(\Phi) := \mu(\Phi([0,1)^n)). \tag{2.18}$$

Since $\Phi([0,1)^n)$ has nonempty interior, it follows that $\rho(\Phi) > 0$ and since $\Phi([0,1)^n)$ is contained in the compact set $\Phi([0,1]^n)$, it follows that $\rho(\Phi) < \infty$. Now define the map $\mu_\Phi\colon \mathcal{B} \to [0,\infty]$ by

$$\mu_\Phi(B) := \frac{\mu(\Phi(B))}{\rho(\Phi)} \quad \text{for } B \in \mathcal{B}.$$

Then μ_Φ is a normalized translation-invariant Borel measure. The σ-additivity follows directly from the σ-additivity of μ, the formula $\mu_\Phi(\emptyset) = 0$ is obvious from the definition, that compact sets have finite measure follows from the fact that $\Phi(K)$ is compact if and only if $K \subset \mathbb{R}^n$ is compact, the translation invariance follows immediately from the translation invariance of μ and the fact that $\Phi(B + x) = \Phi(B) + \Phi(x)$ for all $B \in \mathcal{B}$ and all $x \in \mathbb{R}^n$, and the normalization condition $\mu_\Phi([0,1)^n) = 1$ follows directly from the definition of μ_Φ. Hence $\mu_\Phi = \mu$ by Theorem 2.1.

Step 2. *Let ρ be as in Step 1 and let $A \in \mathcal{A}$ and $\Phi \in \mathrm{GL}(n, \mathbb{R})$. Then $\Phi(A) \in \mathcal{A}$ and $m(\Phi(A)) = \rho(\Phi)m(A)$.*

By Theorem 2.14, there exist Borel sets $B_0, B_1 \in \mathcal{B}$ such that $B_0 \subset A \subset B_1$ and $\mu(B_1 \setminus B_0) = 0$. Then $\Phi(B_0) \subset \Phi(A) \subset \Phi(B_1)$ and, by Step 1,

$$\mu(\Phi(B_1) \setminus \Phi(B_0)) = \mu(\Phi(B_1 \setminus B_0)) = \rho(\Phi)\mu(B_1 \setminus B_0) = 0.$$

Hence $\Phi(A)$ is a Lebesgue measurable set and

$$m(\Phi(A)) = \mu(\Phi(B_0)) = \rho(\Phi)\mu(B_0) = \rho(\Phi)m(A)$$

by Theorem 2.14 and Step 1.

Step 3. *Let ρ be as in Step 1 and let $\Phi = \mathrm{diag}(\lambda_1, \dots, \lambda_n)$ be a diagonal matrix with nonzero diagonal entries $\lambda_i \in \mathbb{R} \setminus \{0\}$. Then $\rho(\Phi) = |\lambda_1| \cdots |\lambda_n|$.*

Define $I := [-1, 1]$ and $I_i := [-|\lambda_i|, |\lambda_i|]$ for $i = 1, \dots, n$. Then $Q := I^n$ has Lebesgue measure $m(Q) = 2^n$ and the cuboid $\Phi(Q) = I_1 \times \cdots \times I_n$ has Lebesgue measure $m(\Phi(Q)) = 2^n|\lambda_1| \cdots |\lambda_n|$ by part (iv) of Corollary 2.12. Hence Step 3 follows from Step 2.

Step 4. *The map $\rho\colon \mathrm{GL}(n, \mathbb{R}) \to (0, \infty)$ in Step 1 is a group homomorphism from the general linear group of automorphisms of $\mathbb{R}^n$ to the multiplicative group of positive real numbers.*

Let $\Phi, \Psi \in \mathrm{GL}(n, \mathbb{R})$. Then it follows from (2.17) with $B := \Psi([0, 1)^n)$ and from the definition of $\rho(\Psi)$ in (2.18) that

$$\rho(\Phi\Psi) = \mu(\Phi\Psi([0, 1)^n)) = \rho(\Phi)\mu(\Psi([0, 1)^n)) = \rho(\Phi)\rho(\Psi).$$

Thus ρ is a group homomorphism as claimed and this proves Step 4.

Step 5. *The map $\rho\colon \mathrm{GL}(n, \mathbb{R}) \to (0, \infty)$ in Step 1 is continuous with respect to the standard topologies on $\mathrm{GL}(n, \mathbb{R})$ and $(0, \infty)$.*

It suffices to prove continuity at the identity. Define the norms

$$\|x\|_\infty := \max_{i=1,\dots,n} |x_i|, \qquad \|\Phi\|_\infty := \sup_{0 \neq x \in \mathbb{R}^n} \frac{\|\Phi x\|_\infty}{\|x\|_\infty} \tag{2.19}$$

for $x \in \mathbb{R}^n$ and a linear map $\Phi\colon \mathbb{R}^n \to \mathbb{R}^n$. Denote the closed unit ball in $\mathbb{R}^n$ by $Q := \{x \in \mathbb{R}^n \mid \|x\|_\infty \le 1\} = [-1, 1]^n$. Fix a constant $0 < \delta < 1$ and a linear map $\Phi\colon \mathbb{R}^n \to \mathbb{R}^n$ such that $\|\Phi - \mathbb{1}\|_\infty < \delta$. Then $\Phi \in \mathrm{GL}(n, \mathbb{R})$ and

$$\Phi^{-1} = \sum_{k=0}^{\infty} (\mathbb{1} - \Phi)^k, \qquad \|\Phi^{-1}\|_\infty < \frac{1}{1 - \delta}.$$

Thus $\Phi(Q) \subset (1+\delta)Q$ and $(1-\delta)\Phi^{-1}(Q) \subset Q$. Hence

$$(1-\delta)Q \subset \Phi(Q) \subset (1+\delta)Q.$$

Since $\rho(\Phi) = m(\Phi(Q))/m(Q)$ by Step 2 and $m(rQ) = r^n m(Q)$ for $r > 0$ by Steps 2 and 3, this shows that $(1-\delta)^n \le \rho(\Phi) \le (1+\delta)^n$. Given $\varepsilon > 0$, choose a constant $0 < \delta < 1$ such that $1-\varepsilon < (1-\delta)^n < (1+\delta)^n < 1+\varepsilon$. Then

$$\|\Phi - \mathbb{1}\|_\infty < \delta \implies |\rho(\Phi) - 1|_\infty < \varepsilon$$

for all $\Phi \in \mathrm{GL}(n,\mathbb{R})$.

Step 6. $\rho(\Phi) = |\det(\Phi)|$ *for all* $\Phi \in \mathrm{GL}(n,\mathbb{R})$.

If $\Phi \in \mathrm{GL}(n,\mathbb{R})$ is diagonalizable with real eigenvalues, then $\rho(\Phi) = |\det(\Phi)|$ by Step 3 and Step 4. If $\Phi \in \mathrm{GL}(n,\mathbb{R})$ has only real eigenvalues, then it can be approximated by a sequence of diagonalizable automorphisms with real eigenvalues and hence it follows from Step 5 that $\rho(\Phi) = |\det(\Phi)|$. Since every automorphism of $\mathbb{R}^n$ is a finite composition of automorphisms with real eigenvalues (elementary matrices), this proves Step 6. Lemma 2.18 follows immediately from Step 2 and Step 6. □

Define the metric

$$d_\infty \colon \mathbb{R}^n \times \mathbb{R}^n \longrightarrow [0,\infty)$$

by

$$d_\infty(x,y) := \|x-y\|_\infty \quad \text{for } x, y \in \mathbb{R}^n,$$

where $\|\cdot\|_\infty$ is as in (2.19). The open ball of radius $r > 0$ about a point $a = (a_1,\dots,a_n) \in \mathbb{R}^n$ with respect to this metric is the open cube

$$B_r(a) := (a_1 - r, a_1 + r) \times \cdots \times (a_n - r, a_n + r)$$

and its closure is $\overline{B}_r(a) = [a_1 - r, a_1 + r] \times \cdots \times [a_n - r, a_n + r]$.

Lemma 2.19. *Let $U \subset \mathbb{R}^n$ be an open set and let $K \subset U$ be a compact subset. Let $\phi\colon U \to \mathbb{R}^n$ be a continuously differentiable map such that* $\det(d\phi(x)) \ne 0$ *for all $x \in K$. For every $\varepsilon > 0$ there exists a constant $\delta > 0$ such that the following holds: if $0 < r < \delta$, $a \in \mathbb{R}^n$, and $R \subset \mathbb{R}^n$ satisfy $B_r(a) \subset R \subset \overline{B}_r(a) \subset K$, then*

$$|m(\phi(R)) - |\det(d\phi(a))|\, m(R)| < \varepsilon\, m(R). \tag{2.20}$$

Proof. The maps $K \to \mathbb{R}: x \mapsto \|d\phi(x)^{-1}\|_\infty$ and $K \to \mathbb{R}: x \mapsto |\det(d\phi(x))|$ are continuous by assumption. Since K is compact, these maps are bounded. Hence there is a constant $c > 0$ such that

$$\left\|d\phi(x)^{-1}\right\|_\infty \le c, \quad |\det(d\phi(x))| \le c \quad \text{for all } x \in K. \tag{2.21}$$

Let $\varepsilon > 0$ and choose a constant $0 < \alpha < 1$ so small that

$$1 - \frac{\varepsilon}{c} < (1-\alpha)^n < (1+\alpha)^n < 1 + \frac{\varepsilon}{c}. \tag{2.22}$$

Choose $\delta > 0$ so small that, for all $x, y \in \mathbb{R}^n$,

$$x, y \in K, \ \|x - y\|_\infty < \delta \implies \|d\phi(x) - d\phi(y)\|_\infty < \frac{\alpha}{c}. \tag{2.23}$$

Such a constant exists because the map $d\phi: U \to \mathbb{R}^{n\times n}$ is uniformly continuous on the compact set $K \subset U$. We prove that the assertion of Lemma 2.19 holds with this constant δ.

Choose $a \in \mathbb{R}^n$ and $0 < r < \delta$ such that $\overline{B}_r(a) \subset K$. Then $\|a - x\|_\infty < \delta$ for all $x \in \overline{B}_r(a)$. By (2.23) with $\Phi := d\phi(a)$ this implies

$$\|d\phi(x) - \Phi\|_\infty < \frac{\alpha}{c} \le \frac{\alpha}{\|\Phi^{-1}\|_\infty} \quad \text{for all } x \in \overline{B}_r(a).$$

Here the first step follows from (2.23) and the second step follows from (2.21). Define the map

$$\psi: U \longrightarrow \mathbb{R}^n$$

by

$$\psi(x) := \Phi^{-1}(\phi(x) - \phi(a)).$$

Then $\psi(a) = 0$ and $d\psi(x) = \Phi^{-1} d\phi(x)$ and hence, by (2.23),

$$\|d\psi(x) - \mathbb{1}\|_\infty = \left\|\Phi^{-1}(d\phi(x) - \Phi)\right\|_\infty \le \left\|\Phi^{-1}\right\|_\infty \|d\phi(x) - \Phi\|_\infty \le \alpha$$

for all $x \in \overline{B}_r(a)$. By Theorem C.1, this implies

$$B_{(1-\alpha)s}(0) \subset \psi(B_r(a)) \subset \psi(\overline{B}_r(a)) \subset \overline{B}_{(1+\alpha)s}(0) \tag{2.24}$$

Now fix a subset $R \subset \mathbb{R}^n$ such that $B_r(a) \subset R \subset \overline{B}_r(a)$. Then by (2.24)

$$(1-\alpha)\Phi(B_r(0)) \subset \phi(R) - \phi(a) \subset (1+\alpha)\Phi(\overline{B}_r(0)).$$

Since $m(R) = m(B_r(0)) = m(\overline{B}_r(0))$ by part (iv) of Corollary 2.12, it follows from Lemma 2.18 and the inequalities (2.21) and (2.22) that

$$\begin{aligned}
|\det(\Phi)|\, m(R) - \varepsilon\, m(R) &\le \left(1 - \frac{\varepsilon}{c}\right) |\det(\Phi)|\, m(R) \\
&< (1-\alpha)^n\, |\det(\Phi)|\, m(R) \\
&= m((1-\alpha)\Phi(B_r(0))) \\
&\le m(\phi(R)) \\
&\le m((1+\alpha)\Phi(\overline{B}_r(0))) \\
&= (1+\alpha)^n\, |\det(\Phi)|\, m(R) \\
&< \left(1 + \frac{\varepsilon}{c}\right) |\det(\Phi)|\, m(R) \\
&\le |\det(\Phi)|\, m(R) + \varepsilon\, m(R).
\end{aligned}$$

This proves (2.20) and Lemma 2.19. □

Proof of Theorem 2.17. The proof has seven steps. The first four steps establish equation (2.14) for the characteristic functions of open sets, compact sets, Borel sets, and Lebesgue measurable sets with compact closure in U.

Step 1. *If $W \subset \mathbb{R}^n$ is an open set with compact closure $\overline{W} \subset U$, then*

$$m(\phi(W)) = \int_W |\det(d\phi)|\, dm.$$

Fix a constant $\varepsilon > 0$. Then there exists a constant $\delta > 0$ that satisfies the following two conditions.

(a) If $a \in \mathbb{R}^n$, $0 < r < \delta$, $R \subset \mathbb{R}^n$ satisfy $B_r(a) \subset R \subset \overline{B}_r(a) \subset \overline{W}$, then

$$|m(\phi(R)) - |\det(d\phi(a))|\, m(R)| < \frac{\varepsilon\, m(R)}{2\, m(W)}.$$

(b) For all $x, y \in \overline{W}$

$$\|x - y\|_\infty < \delta \implies |\det(d\phi(x)) - \det(d\phi(y))| < \frac{\varepsilon}{2\, m(W)}.$$

That $\delta > 0$ can be chosen so small that (a) holds follows from Lemma 2.19, and that it can be chosen so small that (b) holds follows from the fact that the function $\det(d\phi)\colon U \to \mathbb{R}$ is uniformly continuous on the compact set $\overline{W}$. Now write W as a countable union of pairwise disjoint half-open cubes $R_i \subset \mathbb{R}^n$ centered at $a_i \in \mathbb{R}^n$ with side lengths $2r_i$ such that $0 < r_i < \delta$. (See page 74.) Then

$$B_{r_i}(a_i) \subset R_i \subset \bar{B}_{r_i}(a_i) \subset \overline{W} \quad \text{for all } i$$

and

$$m(W) = \sum_i m(R_i), \quad m(\phi(W)) = \sum_i m(\phi(R_i)). \tag{2.25}$$

It follows from (2.25) and (a) that

$$\left| m(\phi(W)) - \sum_i |\det(d\phi(a_i))|\, m(R_i) \right| < \frac{\varepsilon}{2}. \tag{2.26}$$

It follows from (b) that

$$\left| |\det(d\phi)| - \sum_i |\det(d\phi(a_i))|\, \chi_{R_i} \right| < \frac{\varepsilon}{2\, m(W)} \quad \text{on } W.$$

Integrate this inequality over W to obtain

$$\left| \int_W |\det(d\phi)|\, dm - \sum_i |\det(d\phi(a_i))|\, m(R_i) \right| < \frac{\varepsilon}{2}. \tag{2.27}$$

By (2.26) and (2.27), we have

$$\left| m(\phi(W)) - \int_W |\det(d\phi)|\, dm \right| < \varepsilon.$$

Since this holds for all $\varepsilon > 0$, Step 1 follows.

Step 2. *$K \subset U$ is compact, then*

$$m(\phi(K)) = \int_K |\det(d\phi)|\, dm.$$

Choose an open set $W \supset K$ with compact closure $\overline{W} \subset U$. Then

$$\begin{aligned} m(\phi(K)) &= m(\phi(W)) - m(\phi(W \setminus K)) \\ &= \int_W |\det(d\phi)|\, dm - \int_{W \setminus K} |\det(d\phi)|\, dm \\ &= \int_K |\det(d\phi)|\, dm. \end{aligned}$$

Here the second equality follows from Step 1.

Step 3. *If $B \in \mathcal{B}$ has compact closure $\overline{B} \subset U$, then $\phi(B) \in \mathcal{B}$ and*

$$m(\phi(B)) = \int_B |\det(d\phi)|\, dm.$$

That $\phi(B)$ is a Borel set follows from the fact that it is the pre-image of the Borel set B under the continuous map $\phi^{-1}\colon V \to U$ (Theorem 1.20). Abbreviate

$$b := m(\phi(B)).$$

Assume first that $b < \infty$ and fix a constant $\varepsilon > 0$. Then it follows from Theorem 2.13 that there exist an open set $W' \subset \mathbb{R}^n$ with compact closure $\overline{W'} \subset V$ such that $\phi(B) \subset W'$ and $m(W') < b + \varepsilon$ and a compact set $K' \subset B$ such that $\mu(K') > b - \varepsilon$. Define $K := \phi^{-1}(K')$ and $W := \phi^{-1}(W')$. Then K is compact, W is open, $\overline{W} \subset U$ is compact, and

$$K \subset B \subset W, \quad b - \varepsilon < m(\phi(K)) \le m(\phi(W)) < b + \varepsilon.$$

Hence it follows from Step 1 and Step 2 that

$$b - \varepsilon < \int_K |\det(d\phi)|\, dm \le \int_B |\det(d\phi)|\, dm \le \int_W |\det(d\phi)|\, dm < b + \varepsilon.$$

Thus

$$b - \varepsilon < \int_B |\det(d\phi)|\, dm < b + \varepsilon \quad \text{for every } \varepsilon > 0,$$

and so

$$\int_B |\det(d\phi)|\, dm = b = m(\phi(B)).$$

If $b = \infty$, then by Theorem 2.13, there exists a sequence of compact sets $K_i' \subset \phi(B)$ such that $\mu(K_i') > i$. Hence

$$K_i := \phi^{-1}(K_i')$$

is compact and

$$\int_{K_i} |\det(d\phi)|\, dm = \mu(\phi(K_i)) > i$$

by Step 2. Since $K_i \subset B$, this implies

$$\int_B |\det(d\phi)|\, dm > i \quad \text{for all } i \in \mathbb{N},$$

and hence

$$\int_B |\det(d\phi)|\, dm = \infty = m(\phi(B)).$$

Step 4. *If $A \in \mathcal{A}$ has compact closure $\overline{A} \subset U$, then $\phi(A) \in \mathcal{A}$ and*

$$m(\phi(A)) = \int_A |\det(d\phi)|\, dm.$$

Let $A \in \mathcal{A}$. By Theorem 2.14, there exist Borel sets $B_0, B_1 \in \mathcal{B}$, with compact closure contained in U, such that $B_0 \subset A \subset B_1$ and $m(B_1 \setminus B_0) = 0$. Then $\phi(B_0) \subset \phi(A) \subset \phi(B_1)$ and it follows from Step 3 that $\phi(B_0)$ and $\phi(B_1)$ are Borel sets and

$$m(\phi(B_1) \setminus \phi(B_0)) = m(\phi(B_1 \setminus B_0)) = \int_{B_1 \setminus B_0} |\det(d\phi)|\, dm = 0.$$

Now Theorem 2.14 shows that $\phi(A)$ is a Lebesgue measurable set and

$$m(\phi(A)) = m(\phi(B_0)) = \int_{B_0} |\det(d\phi)|\, dm = \int_A |\det(d\phi)|\, dm.$$

Here the last equation follows from the fact that the set $A \setminus B_0$ is Lebesgue measurable and has Lebesgue measure zero.

Step 5. *Assertion* (i) *of Theorem* 2.17 *holds for every Lebesgue measurable step function $f = s\colon V \to \mathbb{R}$ whose support is a compact subset of V.*

Write

$$s = \sum_{i=1}^{\ell} \alpha_i \chi_{A_i}$$

with $\alpha_i \in \mathbb{R}$ and $A_i \in \mathcal{A}$ such that $\overline{A_i}$ is a compact subset of V for all i. Then $\phi^{-1}(A_i)$ is a Lebesgue measurable set with compact closure in U by Step 4. Hence $s \circ \phi = \sum_{i=1}^{\ell} \alpha_i \chi_{\phi^{-1}(A_i)}$ is a Lebesgue measurable step function and

$$\begin{aligned} \int_U (s \circ \phi) |\det(d\phi)|\, dm &= \sum_{i=1}^{\ell} \alpha_i \int_{\phi^{-1}(A_i)} |\det(d\phi)|\, dm \\ &= \sum_{i=1}^{\ell} \alpha_i m(A_i) \\ &= \int_V s\, dm. \end{aligned}$$

Here the second equality follows from Step 4.

Step 6. *We prove* (i).

By Theorem 1.26, there is a sequence of Lebesgue measurable step functions

$$s_i : V \longrightarrow [0, \infty)$$

such that

$$0 \le s_1 \le s_2 \le \cdots$$

and

$$f(x) = \lim_{i \to \infty} s_i(x) \quad \text{for every } x \in V.$$

Choose an *exhaustive* sequence of compact sets $K_i \subset V$ such that $K_i \subset K_{i+1}$ for all i and $\bigcup_i K_i = V$, and replace s_i by $s_i \chi_{K_i}$. Then part (i) follows from Step 5 and the Lebesgue monotone convergence theorem (Theorem 1.37).

Step 7. *We prove* (ii).

For $E = U$ part (ii) follows from part (i) and the fact that $(f \circ \phi)^{\pm} = f^{\pm} \circ \phi$. If $F \in \mathcal{A}_V$, then $f := \chi_F|_V$ is Lebesgue measurable, hence $f \circ \phi = \chi_{\phi^{-1}(F)}|_U$ is Lebesgue measurable by part (i), and so $\phi^{-1}(F) \in \mathcal{A}_U$. Replace ϕ by ϕ^{-1} to deduce that if $E \in \mathcal{A}_U$, then $\phi(E) \in \mathcal{A}_V$. Then (ii) follows for all $E \in \mathcal{A}_U$ by replacing f with $f \chi_{\phi(E)}$. This proves Step 7 and Theorem 2.17. □

2.4 Lebesgue equals Riemann

The main theorem of this section asserts that the Lebesgue integral of a function on $\mathbb{R}^n$ agrees with the Riemann integral whenever the latter is defined and the function in question has compact support. The section begins with a recollection of the definition of the Riemann integral. (For more details see [9, 19, 21].)

The Riemann integral

Recall the notation

$$R(x, k) := x + [0, 2^{-k})^n \quad \text{for } x \in \mathbb{R}^n \text{ and } k \in \mathbb{N},$$

which was used in the proof of Theorem 2.1 on page 74. The closure of $R(x, k)$ is the closed cube $\overline{R(x, k)} = x + [0, 2^{-k}]^n$. The sets $R(\ell, k)$, with ℓ ranging over the countable set $2^{-k}\mathbb{Z}^n$, form a partition of the Euclidean space $\mathbb{R}^n$.

Definition 2.20. Let $f: \mathbb{R}^n \to \mathbb{R}$ be a bounded function whose support

$$\mathrm{supp}(f) := \overline{\{x \in \mathbb{R}^n \mid f(x) \neq 0\}}$$

is a bounded subset of $\mathbb{R}^n$. For $k \in \mathbb{N}$ define the *lower sum* $\underline{S}(f,k) \in \mathbb{R}$ and the *upper sum* $\overline{S}(f,k) \in \mathbb{R}$ by

$$\underline{S}(f,k) := \sum_{\ell \in 2^{-k}\mathbb{Z}^n} \inf_{\overline{R(\ell,k)}} f \, 2^{-nk}, \quad \overline{S}(f,k) := \sum_{\ell \in 2^{-k}\mathbb{Z}^n} \sup_{\overline{R(\ell,k)}} f \, 2^{-nk}. \tag{2.28}$$

These are finite sums and satisfy

$$\sup_k \underline{S}(f,k) \leq \inf_k \overline{S}(f,k).$$

The function $f: \mathbb{R}^n \to \mathbb{R}$ is called *Riemann integrable* if

$$\sup_k \underline{S}(f,k) = \inf_k \overline{S}(f,k).$$

The *Riemann integral* of a Riemann integrable function $f: \mathbb{R}^n \to \mathbb{R}$ is the real number

$$\mathcal{R}(f) := \int_{\mathbb{R}^n} f(x)\,dx := \sup_{k \in \mathbb{N}} \underline{S}(f,k) = \inf_{k \in \mathbb{N}} \overline{S}(f,k) = \lim_{k \to \infty} \underline{S}(f,k). \tag{2.29}$$

Remark 2.21. The Riemann integral can also be defined by allowing for arbitrary partitions of $\mathbb{R}^n$ into cuboids (see [19, Definition 2.3]) or in terms of convergence of the so-called Riemann sums (see [21, Definition 7.1.2]). That all three definitions agree is proved in [19, Satz 2.8] and [21, Theorem 7.1.8]).

Definition 2.22. A bounded set $A \subset \mathbb{R}^n$ is called *Jordan measurable* if its characteristic function $\chi_A: \mathbb{R}^n \to \mathbb{R}$ is Riemann integrable. The *Jordan measure* of a Jordan measurable set $A \subset \mathbb{R}^n$ is the real number

$$\begin{aligned} \mu^{\mathrm{J}}(A) &:= \mathcal{R}(\chi_A) \\ &= \int_{\mathbb{R}^n} \chi_A(x)\,dx \\ &= \lim_{k \to \infty} 2^{-nk} \#\{\ell \in 2^{-k}\mathbb{Z}^n \mid \overline{R(\ell,k)} \cap A \neq \emptyset\}. \end{aligned} \tag{2.30}$$

Exercise 2.23. (i) Prove the last equality in (2.30).

(ii) Prove that a bounded set $A \subset \mathbb{R}^n$ is Jordan measurable if and only if its boundary $\partial A = \overline{A} \setminus \mathrm{int}(A)$ is a Jordan null set. (See Definition 2.7.)

(iii) Prove that the closure of a Jordan null set is a Jordan null set.

The Lebesgue and Riemann integrals agree

Theorem 2.24. (i) *If* $f: \mathbb{R}^n \to \mathbb{R}$ *is Riemann integrable, then* $f \in \mathcal{L}^1(m)$ *and its Lebesgue integral agrees with the Riemann integral, i.e.,*

$$\int_{\mathbb{R}^n} f\, dm = \mathcal{R}(f).$$

(ii) *If* $A \subset \mathbb{R}^n$ *is Jordan measurable, then* A *is Lebesgue measurable and*

$$m(A) = \mu^{\mathrm{J}}(A).$$

Proof. Assertion (ii) follows from (i) by taking $f = \chi_A$. Thus it remains to prove (i). Let $f: \mathbb{R}^n \to \mathbb{R}$ be a Riemann integrable function. Then f is bounded and has bounded support. Define the functions

$$\underline{f}_k, \bar{f}_k: \mathbb{R}^n \longrightarrow \mathbb{R}$$

by

$$\underline{f}_k(x) := \inf_{\overline{R(\ell,k)}} f, \quad \bar{f}_k(x) := \sup_{\overline{R(\ell,k)}} f \quad \text{for } x \in R(\ell,k),\ \ell \in 2^{-k}\mathbb{Z}^n. \tag{2.31}$$

These are Lebesgue measurable step functions and

$$\int_{\mathbb{R}^n} \underline{f}_k\, dm = \underline{S}(f,k), \quad \int_{\mathbb{R}^n} \bar{f}_k\, dm = \bar{S}(f,k).$$

They also satisfy

$$\underline{f}_k \le \underline{f}_{k+1} \le f \le \bar{f}_{k+1} \le \bar{f}_k$$

for all $k \in \mathbb{N}$. Define the functions

$$\underline{f}, \bar{f}: \mathbb{R}^n \longrightarrow \mathbb{R}$$

by

$$\underline{f}(x) := \lim_{k\to\infty} \underline{f}_k(x), \quad \bar{f}(x) := \lim_{k\to\infty} \bar{f}_k(x) \quad \text{for } x \in \mathbb{R}^n.$$

Then

$$\underline{f}(x) \le f(x) \le \bar{f}(x)$$

for every $x \in \mathbb{R}^n$. Moreover, $|\bar{f}_k|$ and $|\underline{f}_k|$ are bounded above by the Lebesgue integrable function $c\chi_A$, where

$$c := \sup_{x\in\mathbb{R}^n} |f(x)|$$

and

$$A := [-N, N]^n,$$

with $N \in \mathbb{N}$ chosen such that $\mathrm{supp}(f) \subset [-N, N]^n$. Hence it follows from the Lebesgue dominated convergence theorem (Theorem 1.45) that $\underline{f}$ and $\bar{f}$ are Lebesgue integrable and

$$\begin{aligned}\int_{\mathbb{R}^n} \underline{f}\, dm &= \lim_{k\to\infty} \int_{\mathbb{R}^n} \underline{f}_k\, dm \\ &= \lim_{k\to\infty} \underline{S}(f, k) \\ &= \mathcal{R}(f) \\ &= \lim_{k\to\infty} \bar{S}(f, k) \\ &= \lim_{k\to\infty} \int_{\mathbb{R}^n} \bar{f}_k\, dm \\ &= \int_{\mathbb{R}^n} \bar{f}\, dm.\end{aligned}$$

By Lemma 1.49, with f replaced by $\bar{f} - \underline{f}$, this implies that $\underline{f} = f = \bar{f}$ Lebesgue almost everywhere. Hence $f \in \mathcal{L}^1(m)$ and

$$\int_{\mathbb{R}^n} f\, dm = \int_{\mathbb{R}^n} \underline{f}\, dm = \mathcal{R}(f).$$

This proves Theorem 2.24. □

The discussion in this section is restricted to Riemann integrable functions $f: \mathbb{R}^n \to \mathbb{R}$ with compact support, and Theorem 2.24 asserts that for such functions the Riemann integral agrees with the Lebesgue integral. When f does not have compact support and is locally Riemann integrable, the *improper Riemann integral* is defined by

$$\int_{\mathbb{R}^n} f(x)dx := \lim_{r\to\infty} \int_{B_r} f(x)\, dx, \tag{2.32}$$

provided that the limit exists. Here $B_r \subset \mathbb{R}^n$ denotes the ball of radius r centered at the origin. There are many examples where the limit (2.32) exists even though the Lebesgue integral $\int_{\mathbb{R}^n} |f|\, dm$ is infinite and so the Lebesgue integral of f does not exist. An example is the function $f: \mathbb{R} \to \mathbb{R}$ given by $f(x) := x^{-1}\sin(x)$ for $x \in \mathbb{R} \setminus \{0\}$ and $f(0) := 1$. This function is continuous and is not Lebesgue integrable, but the improper Riemann integral exists and is equal to π (see Example 7.49). Improper integrals play a central role in Fourier analysis, probability theory, and partial differential equations. However, this topic will not be pursued any further in this book.

2.5 Exercises

Exercise 2.25. Show that the Cantor set in $\mathbb{R}$ is a Jordan null set. Show that $\mathbb{Q} \cap [0, 1]$ is a Lebesgue null set, but not a Jordan null set. Show that $A \subset \mathbb{R}^n$ is a Lebesgue null set if and only if $\nu(A) = 0$. Find an open set $U \subset \mathbb{R}$ whose boundary has positive Lebesgue measure.

Exercise 2.26. Prove that every subset of a proper linear subspace of $\mathbb{R}^n$ is Lebesgue measurable and has Lebesgue measure zero. Find a Jordan measurable subset of $\mathbb{R}^n$ that is not a Borel set. Find a bounded Lebesgue measurable subset of $\mathbb{R}^n$ with positive Lebesgue measure that is neither a Borel set, nor Jordan measurable.

Exercise 2.27. Let $A, B \subset \mathbb{R}^n$ be Lebesgue null sets. Prove that their sum

$$A + B := \{x + y \mid x \in A,\ y \in B\}$$

is a Lebesgue null set.

Exercise 2.28. Let $(X, \mathcal{A}, \mu)$ be a measure space and define the function

$$\nu: 2^X \longrightarrow [0, \infty]$$

by

$$\nu(B) := \inf\{\mu(A) \mid A \in \mathcal{A},\ B \subset A\}. \tag{2.33}$$

(i) Prove that ν is an outer measure and that $\mathcal{A} \subset \mathcal{A}(\nu)$.

(ii) Assume $\mu(X) < \infty$. Prove that the measure space $(X, \mathcal{A}(\nu), \nu|_{\mathcal{A}(\nu)})$ is the completion of $(X, \mathcal{A}, \mu)$. *Hint.* Show that for every subset $B \subset X$ there exists a set $A \in \mathcal{A}$ such that $B \subset A$ and $\nu(B) = \mu(A)$.

(iii) Let X be a set and $A \subsetneq X$ be a nonempty subset. Define

$$\mathcal{A} := \{\emptyset, A, A^c, X\}, \quad \mu(\emptyset) := \mu(A) := 0, \quad \mu(A^c) := \mu(X) := \infty.$$

Prove that $(X, \mathcal{A}, \mu)$ is a measure space. Given $B \subset X$, prove that $\nu(B) = 0$ whenever $B \subset A$ and $\nu(B) = \infty$ whenever $B \not\subset A$. Prove that $\mathcal{A}(\nu) = 2^X$ and that the completion of $(X, \mathcal{A}, \mu)$ is the measure space $(X, \mathcal{A}^*, \mu^*)$ with $\mathcal{A}^* = \{B \subset X \mid B \subset A \text{ or } A^c \subset B\}$ and $\mu^* = \nu|_{\mathcal{A}^*}$. (Thus the hypothesis $\mu(X) < \infty$ cannot be removed in part (ii).)

Exercise 2.29. Let $f: \mathbb{R} \to \mathbb{R}$ be continuously differentiable and define

$$A := \{x \in \mathbb{R} \mid f'(x) = 0\}.$$

Prove that $f(A)$ is a Lebesgue null set. *Hint.* Consider the sets

$$A_{n,\varepsilon} := \{x \in \mathbb{R} \mid |x| < n,\ |f'(x)| < 2^{-n}\varepsilon\}.$$

Exercise 2.30. Find a continuous function $f\colon [0,\infty) \to \mathbb{R}$ such that f is not Lebesgue integrable, but the limit $\lim_{T\to\infty} \int_0^T f(t)\,dt$ exists.

Exercise 2.31. Determine the limits of the sequences

$$a_n := \int_0^n \left(1 - \frac{x}{n}\right)^n e^{x/2} dx, \quad b_n := \int_0^n \left(1 + \frac{x}{n}\right)^n e^{-2x}\,dx, \quad n \in \mathbb{N}.$$

Hint. Use the Lebesgue dominated convergence theorem (Theorem 1.45).

Exercise 2.32. Let $(\mathbb{R}, \mathcal{A}, m)$ be the Lebesgue measure space. Construct a Borel set $E \subset \mathbb{R}$ such that

$$0 < \frac{m(E \cap I)}{m(I)} < 1$$

for every nonempty bounded open interval $I \subset \mathbb{R}$. (See also Exercise 6.21.)

Exercise 2.33. Find the smallest constant c such that

$$\log(1 + e^t) \leq c + t \quad \text{for all } t \geq 0.$$

Does the limit

$$\lim_{n\to\infty} \frac{1}{n} \int_0^1 \log(1 + e^{nf(x)})\,dx$$

exist for every Lebesgue integrable function $f\colon [0,1] \to \mathbb{R}$? Determine the limit when it does exist.

Exercise 2.34. Let $(\mathbb{R}^n, \mathcal{A}, m)$ be the Lebesgue measure space and let $\phi\colon \mathbb{R}^n \to \mathbb{R}^n$ be a C^1-diffeomorphism. Prove that $\phi_*\mathcal{A} = \mathcal{A}$ and that

$$(\phi_* m)(A) = \int_A \frac{1}{|\det(d\phi) \circ \phi^{-1}|}\,dm \quad \text{for all } A \in \mathcal{A}.$$

Hint. See Exercise 1.69 and Theorems 1.40 and 2.17.

Exercise 2.35 (Hausdorff measure). Let (X, ρ) be a metric space and fix a real number $d \geq 0$. The *diameter* of a subset $A \subset X$ is defined by

$$\operatorname{diam}(A) := \sup_{x,y \in A} \rho(x, y). \tag{2.34}$$

For $\varepsilon > 0$ define the function

$$\nu_{d,\varepsilon}: 2^X \longrightarrow [0, \infty]$$

by

$$\nu_{d,\varepsilon}(A) := \inf\left\{ \sum_{i \in I} \operatorname{diam}(D_i)^d \;\middle|\; \begin{array}{l} I \text{ is finite or countably infinite,} \\ D_i \subset X, \operatorname{diam}(D_i) < \varepsilon \text{ for } i \in I, \\ \text{and } A \subset \bigcup_{i \in I} D_i \end{array} \right\}. \qquad (2.35)$$

for $A \subset X$. Thus $\nu_{d,\varepsilon}(\emptyset) = 0$ and $\nu_{d,\varepsilon}(A) = \infty$ whenever A does not admit a countable cover by subsets of diameter less than ε. Moreover, the function $\varepsilon \mapsto \nu_{d,\varepsilon}(A)$ is nonincreasing for every subset $A \subset X$. The *d-dimensional Hausdorff outer measure* is the function

$$\nu_d: 2^X \longrightarrow [0, \infty]$$

defined by

$$\nu_d(A) := \sup_{\varepsilon > 0} \nu_{d,\varepsilon}(A) = \lim_{\varepsilon \to 0} \nu_{d,\varepsilon}(A) \quad \text{for } A \subset X. \qquad (2.36)$$

Prove the following.

(i) ν_d is an outer measure.

(ii) If $A, B \subset X$ satisfy

$$\rho(A, B) := \inf\{\rho(x, y) \mid x \in A,\ y \in B\} > 0,$$

then

$$\nu_d(A \cup B) = \nu_d(A) + \nu_d(B).$$

Hence, by Theorems 2.4 and 2.5, the set

$$\mathcal{A}_d := \{A \subset X \mid A \text{ is } \nu_d\text{-measurable}\}$$

is a σ-algebra containing the Borel sets and

$$\mu_d := \nu_d|_{\mathcal{A}_d}: \mathcal{A}_d \longrightarrow [0, \infty]$$

is a measure. It is called the *d-dimensional Hausdorff measure on X*. Hausdorff measures play a central role in *geometric measure theory.*

(iii) If $d = 0$, then $\mathcal{A}_0 = 2^X$ and $\nu_0 = \mu_0$ is the counting measure.

(iv) The n-dimensional Hausdorff measure on $\mathbb{R}^n$ agrees with the Lebesgue measure up to a factor (the Lebesgue measure of the ball of radius $1/2$).

(v) Let $A \subset X$ be nonempty. The *Hausdorff dimension* of A is the number

$$\dim(A) := \sup\{r \geq 0 \mid \nu_r(A) = \infty\} = \inf\{s \geq 0 \mid \nu_s(A) = 0\}. \quad (2.37)$$

The second equality follows from the fact that $\nu_d(A) > 0$ implies $\nu_r(A) = \infty$ for $0 \leq r < d$, and $\nu_d(A) < \infty$ implies $\nu_s(A) = 0$ for $s > d$.

(vi) The Hausdorff dimension of a smooth embedded curve $\Gamma \subset \mathbb{R}^n$ is $d = 1$ and its 1-dimensional Hausdorff measure $\mu_1(\Gamma)$ is the length of the curve.

(vii) The Hausdorff dimension of the Cantor set is $d = \log(2)/\log(3)$.

Chapter 3

Borel measures

The regularity properties established for the Lebesgue (outer) measure in Theorem 2.13 play an important role in much greater generality. The present chapter is devoted to the study of Borel measures on locally compact Hausdorff spaces that satisfy similar regularity properties. The main result is the Riesz representation theorem (Theorem 3.15). We begin with some further recollections about topological spaces.

Let $(X, \mathcal{U})$ be a topological space (see Definition 1.9). A *neighborhood* of a point $x \in X$ is a subset $A \subset X$ that contains x in its interior, i.e., $x \in U \subset A$ for some open set U. X is called a *Hausdorff space* if any two distinct points in X have disjoint neighborhoods, i.e., for all $x, y \in X$ with $x \neq y$ there exist open sets $U, V \subset X$ such that $x \in U$, $y \in V$, and $U \cap V = \emptyset$. X is called *locally compact* if every point in X has a compact neighborhood. It is called σ*-compact* if there exists a sequence of compact sets $K_i \subset X$, $i \in \mathbb{N}$, such that $K_i \subset K_{i+1}$ for all i and $X = \bigcup_{i=1}^{\infty} K_i$.

3.1 Regular Borel measures

Assume throughout that $(X, \mathcal{U})$ is a locally compact Hausdorff space and denote by $\mathcal{B} \subset 2^X$ the Borel σ-algebra. Thus $\mathcal{B}$ is the smallest σ-algebra on X that contains all open sets. In the context of this chapter it is convenient to include *local finiteness* (compact sets have finite measure) in the definition of a *Borel measure.* There are other geometric settings, such as the study of Hausdorff measures (Exercise 2.35), where one allows for compact sets to have infinite measure, but these are not discussed here.

Definition 3.1. A measure $\mu\colon \mathcal{B} \to [0, \infty]$ is called a *Borel measure* if

$$\mu(K) < \infty$$

for every compact set $K \subset X$. A measure $\mu\colon \mathcal{B} \to [0, \infty]$ is called *outer regular* if

$$\mu(B) = \inf\{\mu(U) \mid B \subset U \subset X \text{ and } U \text{ is open}\} \tag{3.1}$$

for every Borel set $B \in \mathcal{B}$, is called *inner regular* if

$$\mu(B) = \sup\{\mu(K) \mid K \subset B \text{ and } K \text{ is compact}\} \tag{3.2}$$

for every Borel set $B \in \mathcal{B}$, and is called *regular* if it is both outer and inner regular. A *Radon measure* is an inner regular Borel measure.

Example 3.2. The restriction of the Lebesgue measure on $X = \mathbb{R}^n$ to the Borel σ-algebra is a regular Borel measure by Theorem 2.13.

Example 3.3. The counting measure on $X = \mathbb{N}$ with the discrete topology $\mathcal{U} = \mathcal{B} = 2^{\mathbb{N}}$ is a regular Borel measure.

Example 3.4. Let $(X, \mathcal{U})$ be any locally compact Hausdorff space and fix a point $x_0 \in X$. Then the Dirac measure $\mu = \delta_{x_0}$ at x_0 is a regular Borel measure (Example 1.31).

Example 3.5. Let X be an uncountable set equipped with the discrete topology $\mathcal{U} = \mathcal{B} = 2^X$. Define

$$\mu : \mathcal{B} \longrightarrow [0, \infty]$$

by

$$\mu(B) := \begin{cases} 0 & \text{if } B \text{ is countable,} \\ \infty & \text{if } B \text{ is uncountable.} \end{cases}$$

This is a Borel measure. Moreover, a subset $K \subset X$ is compact if and only if it is finite. Hence $\mu(X) = \infty$ and $\mu(K) = 0$ for every compact set $K \subset X$. Thus μ is not a Radon measure.

The next example occupies four pages and illustrates the subtlety of the subject (see also Exercise 18 in Rudin [17, page 59]). It constructs a compact Hausdorff space $(X, \mathcal{U})$ and a Borel measure μ on X that is not a Radon measure. More precisely, there is a point $\kappa \in X$ such that the open set

$$U := X \setminus \{\kappa\}$$

is not σ-compact and satisfies $\mu(U) = 1$ and $\mu(K) = 0$ for every compact subset $K \subset U$. This example is a kind of refinement of Example 3.5. It is due to Dieoudonné.

Example 3.6 (Dieudonné's measure). (i) Let $(X, \preccurlyeq)$ be an uncountable well-ordered set with a maximal element $\kappa \in X$ such that every element $x \in X \setminus \{\kappa\}$ has only countably many predecessors. Here a set is called *countable* if it is finite or countably infinite. (Think of this as the *uncountable Mount Everest*: no sequence reaches the mountain peak κ.) Thus the relation $\preccurlyeq$ on X satisfies the following axioms.

(a) If $x, y, z \in X$ satisfy $x \preccurlyeq y$ and $y \preccurlyeq z$, then $x \preccurlyeq z$.

(b) If $x, y \in X$ satisfy $x \preccurlyeq y$ and $y \preccurlyeq x$, then $x = y$.

(c) If $x, y \in X$, then $x \preccurlyeq y$ or $y \preccurlyeq x$.

(d) If $\emptyset \neq A \subset X$, then there is an $a \in A$ such that $a \preccurlyeq x$ for all $x \in A$.

(e) If $x \in X \setminus \{\kappa\}$, then $x \preccurlyeq \kappa$ and the set $\{y \in X \mid y \preccurlyeq x\}$ is countable.

Define the relation $\prec$ on X by $x \prec y$ if $x \preccurlyeq y$ and $x \neq y$. For $\emptyset \neq A \subset X$ denote by $\min(A) \in A$ the unique element of A that satisfies $\min(A) \preccurlyeq x$ for all $x \in A$. (See conditions (b) and (d).) For $x \in X$ define

$$S_x := \{y \in X \mid x \prec y\}, \quad P_x := \{y \in X \mid y \prec x\}.$$

Thus P_x is the set of *predecessors* of x and S_x is the set of *successors* of x. If $x \in X \setminus \{\kappa\}$, then P_x is countable and S_x is uncountable. Define the map

$$s \colon X \setminus \{\kappa\} \longrightarrow X \setminus \{\kappa\}, \quad s(x) := \min(S_x).$$

Then $X \setminus S_x = P_{s(x)} = P_x \cup \{x\}$ for all $x \in X$. Let $\mathcal{U} \subset 2^X$ be the smallest topology that contains the sets P_x and S_x for all $x \in X$. A set $U \subset X$ is open in this topology if it is a union of sets of the form P_b, S_a and $S_a \cap P_b$.

(ii) We prove that $(X, \mathcal{U})$ is a Hausdorff space. Let $x, y \in X$ such that $x \neq y$ and suppose without loss of generality that $x \prec y$. Then $P_{s(x)}$ and S_x are disjoint open sets such that $x \in P_{s(x)}$ and $y \in S_x$.

(iii) We prove that every nonempty compact set $K \subset X$ contains a largest element $\max(K) \in K$ such that $K \cap S_{\max(K)} = \emptyset$. This is obvious when $\kappa \in K$ because $S_\kappa = \emptyset$. Thus assume $\kappa \notin K$ and define

$$V := \{x \in X \mid K \subset P_x\}.$$

Since $\kappa \in V$, this set is nonempty and $\min(X) \prec \min(V) =: v$ because $K \neq \emptyset$. Since $X \setminus K$ is open and $v \in X \setminus K$, there exist elements $a, b \in X$ such that $a \prec v \prec b$ and $S_a \cap P_b \cap K = \emptyset$. This implies

$$K \subset P_v \setminus (S_a \cap P_b) \subset P_b \setminus (S_a \cap P_b) \subset X \setminus S_a = P_{s(a)}.$$

Hence $K \setminus \{a\} \subset P_{s(a)} \setminus \{a\} = P_a$ and $K \not\subset P_a$ because $a \prec v$ and so $a \notin V$. This implies $a \in K \subset P_{s(a)}$, and hence $K \cap S_a = K \setminus P_{s(a)} = \emptyset$.

(iv) We prove that $(X, \mathfrak{U})$ is compact. Let $\{U_i\}_{i\in I}$ be an open cover of X. We prove by induction that there exist finite sequences

$$x_1, \dots, x_\ell \in X \quad \text{and} \quad i_1, \dots, i_\ell \in I$$

such that

$$x_k \in U_{i_k} \setminus U_{i_{k-1}}$$

and

$$S_{x_k} \subset U_{i_1} \cup \dots \cup U_{i_{k-1}}$$

for $k \geq 2$, and $X = \bigcup_{j=1}^{\ell} U_{i_j}$. Define

$$x_1 := \kappa$$

and choose $i_1 \in I$ such that $\kappa \in U_{i_1}$. If $U_{i_1} = X$ the assertion holds with $\ell = 1$. Now suppose, by induction, that $x_1, \dots, x_k$ and $i_1, \dots, i_k$ have been constructed such that $x_j \in U_{i_j}$ for $j = 1, \dots, k$ and $S_{x_k} \subset U_{i_1} \cup \dots \cup U_{i_{k-1}}$. If

$$U_{i_1} \cup \dots \cup U_{i_k} = X,$$

then we are done with $\ell = k$. Otherwise

$$C_k := X \setminus U_{i_1} \cup \dots \cup U_{i_k}$$

is a nonempty compact set and we define

$$x_{k+1} := \max(C_k)$$

by part (iii). Then $x_{k+1} \in C_k$ and $C_k \cap S_{x_{k+1}} = \emptyset$. Hence $S_{x_{k+1}} \subset U_{i_1} \cup \dots \cup U_{i_k}$. Choose $i_{k+1} \in I$ such that $x_{k+1} \in U_{i_{k+1}}$. This completes the induction argument. The induction must stop because $x_{k+1} \prec x_k$ for all k and every strictly decreasing sequence in X is finite by the well ordering axiom (d). This shows that $(X, \mathfrak{U})$ is compact.

(v) Let $K_i \subset X$, $i \in \mathbb{N}$, be a sequence of uncountable compact sets. We prove that the compact set

$$K := \bigcap_{i\in\mathbb{N}} K_i$$

is uncountable. To see this, we first prove that

$$K \setminus \{\kappa\} \neq \emptyset. \tag{3.3}$$

Choose a sequence $x_n \in X \setminus \{\kappa\}$ such that $x_n \prec x_{n+1}$ for all $n \in \mathbb{N}$ and $x_{2^k+i} \in K_i$ for $1 \le i \le 2^k - 1$ and $k \in \mathbb{N}$. That such a sequence exists follows by induction from the fact that the set $X \setminus S_{x_n} = P_{s(x_n)}$ is countable for each n, while the sets K_i are uncountable for all i. Now the set

$$P := \bigcup_{n \in \mathbb{N}} P_{x_n}$$

is countable and hence the set

$$\begin{aligned} S &:= X \setminus P \\ &= X \setminus \bigcup_{n \in \mathbb{N}} P_{x_n} \\ &= X \setminus \bigcup_{n \in \mathbb{N}} P_{s(x_n)} \\ &= \bigcap_{n \in \mathbb{N}} (X \setminus P_{s(x_n)}) \\ &= \bigcap_{n \in \mathbb{N}} S_{x_n} \end{aligned}$$

is uncountable. Hence

$$x := \min(S) \prec \kappa.$$

We prove that $x \in K_i$ for all $i \in \mathbb{N}$. Assume by contradiction that $x \notin K_i$ for some i. Then there are elements $a, b \in X$ such that $a \prec x \prec b$ and

$$U := P_b \cap S_a \subset X \setminus K_i.$$

If $x_n \preccurlyeq a$ for all $n \in \mathbb{N}$, then $P \subset P_a$ and so $a \in X \setminus P = S$, which is impossible because $a \prec x = \min(S)$. Thus there must be an integer $n_0 \in \mathbb{N}$ such that $a \prec x_{n_0}$. This implies $a \prec x_n \prec x \prec b$ and hence $x_n \in U \subset X \setminus K_i$ for all $n \ge n_0$, contradicting the fact that $x_{2^k+i} \in K_i$ for all $k \in \mathbb{N}$. This contradiction shows that our assumption that $x \notin K_i$ for some $i \in \mathbb{N}$ must have been wrong. Thus $x \in K$ and this proves (3.3).

We prove that K is uncountable. Assume by contradiction that K is countable and choose a sequence $x_i \in K$ such that $K \setminus \{\kappa\} = \{x_i \mid i \in \mathbb{N}\}$. Then $s(x_i) \prec \kappa$ and

$$K_i' := K_i \cap S_{x_i} = K_i \setminus P_{s(x_i)}$$

is an uncountable compact set for every $i \in \mathbb{N}$. Moreover,

$$K' := \bigcap_{i \in \mathbb{N}} K_i' \subset K \setminus \{x_i \mid i \in \mathbb{N}\} = \{\kappa\},$$

contradicting the fact that $K' \setminus \{\kappa\} \neq \emptyset$ by (3.3). This contradiction shows that K is uncountable as claimed.

(vi) Define $\mathcal{A} \subset 2^X$ by

$$\mathcal{A} := \left\{ A \subset X \;\middle|\; \begin{array}{l} A \cup \{\kappa\} \text{ contains an uncountable compact set,} \\ \text{or } A^c \cup \{\kappa\} \text{ contains an uncountable compact set.} \end{array} \right\}.$$

We prove that this is a σ-algebra. To see this, note first that $X \in \mathcal{A}$ and that $A \in \mathcal{A}$ implies $A^c \in \mathcal{A}$ by definition. Now choose a sequence $A_i \in \mathcal{A}$ and denote

$$A := \bigcup_{i \in \mathbb{N}} A_i.$$

If one of the sets $A_i \cup \{\kappa\}$ contains an uncountable compact set, then so does the set $A \cup \{\kappa\}$. If none of the sets $A_i \cup \{\kappa\}$ contains an uncountable compact set, then the set $A_i^c \cup \{\kappa\}$ contains an uncountable compact set for all $i \in \mathbb{N}$ and hence so does the set $\bigcap_{i \in \mathbb{N}} (A_i^c \cup \{\kappa\}) = A^c \cup \{\kappa\}$ by part (v). In both cases it follows that $A \in \mathcal{A}$.

(vii) Define the map

$$\mu \colon \mathcal{A} \longrightarrow [0, \infty]$$

by

$$\mu(A) := \begin{cases} 1 & \text{if } A \cup \{\kappa\} \text{ contains an uncountable compact set,} \\ 0 & \text{if } A^c \cup \{\kappa\} \text{ contains an uncountable compact set.} \end{cases}$$

This map is well defined because the sets $A \cup \{\kappa\}$ and $A^c \cup \{\kappa\}$ cannot both contain uncountable compact sets by part (v). It satisfies $\mu(\emptyset) = 0$. Moreover, if $A_i \in \mathcal{A}$ is a sequence of pairwise disjoint measurable sets, then at most one of the sets $A_i \cup \{\kappa\}$ can contain an uncountable compact set and hence $\mu(\bigcup_{i \in \mathbb{N}} A_i) = \sum_{i \in \mathbb{N}} \mu(A_i)$. Therefore, μ is a measure.

(viii) The σ-algebra $\mathcal{B} \subset 2^X$ of all Borel sets in X is contained in $\mathcal{A}$. To see this, let $U \subset X$ be open. If U^c is uncountable, then $U^c \cup \{\kappa\}$ is an uncountable compact set and hence $U \in \mathcal{A}$. If U^c is countable choose a sequence $x_i \in U^c$ such that $U^c \setminus \{\kappa\} = \{x_i \mid i \in \mathbb{N}\}$ and define

$$S := \bigcap_{i \in \mathbb{N}} S_{x_i}.$$

Then $X \setminus S = \bigcup_{i \in \mathbb{N}} (X \setminus S_{x_i}) = \bigcup_{i \in \mathbb{N}} P_{s(x_i)}$ is a countable set and hence

$$s := \min(S) \prec \kappa.$$

Since $x_i \prec s$ for all $i \in \mathbb{N}$, it follows that $U^c \setminus \{\kappa\} \subset P_s$. Hence $X \setminus P_s$ is an uncountable compact subset of $U \cup \{\kappa\}$ and so $U \in \mathcal{A}$.

(ix) The set

$$U := X \setminus \{\kappa\}$$

is uncountable and every compact subset of U is countable by part (v). Hence $\mu(K) = 0$ for every compact subset $K \subset U$ and $\mu(U) = 1$ because $U \cup \{\kappa\} = X$ is an uncountable compact set. Thus $\mu|_{\mathcal{B}}: \mathcal{B} \to [0, \infty]$ is a Borel measure, but not a Radon measure.

The next lemma and theorem are included here in preparation for the Riesz representation theorem (Theorem 3.15). They explain the relation between the various regularity properties of Borel measures.

Lemma 3.7. *Let $\mu: \mathcal{B} \to [0, \infty]$ be an outer regular Borel measure that is* ***inner regular on open sets****, i.e.,*

$$\mu(U) = \sup\{\mu(K) \mid K \subset U \text{ and } K \text{ is compact}\} \tag{3.4}$$

for every open set $U \subset X$. Then the following holds:

(i) *every Borel set $B \subset X$ with $\mu(B) < \infty$ satisfies* (3.2);

(ii) *if X is σ-compact, then μ is regular.*

Proof. (i) Fix a Borel set $B \subset X$ with $\mu(B) < \infty$ and a constant $\varepsilon > 0$. Since μ is outer regular, there exists an open set $U \subset X$ such that

$$B \subset U, \quad \mu(U) < \mu(B) + \frac{\varepsilon}{2}.$$

Thus $U \setminus B$ is a Borel set and $\mu(U \setminus B) = \mu(U) - \mu(B) < \varepsilon/2$. Use the outer regularity of μ again to obtain an open set $V \subset X$ such that

$$U \setminus B \subset V, \quad \mu(V) < \frac{\varepsilon}{2}.$$

Now it follows from (3.4) that there exists a compact set $K \subset X$ such that

$$K \subset U, \quad \mu(K) > \mu(U) - \frac{\varepsilon}{2}.$$

Define $C := K \setminus V$. Since X is a Hausdorff space, K is closed, hence C is a closed subset of K, and so C is compact (see Lemma A.2). Moreover,

$$C \subset U \setminus V \subset B, \quad B \setminus C \subset (B \setminus K) \cup V \subset (U \setminus K) \cup V,$$

and hence $\mu(B \setminus C) \leq \mu(U \setminus K) + \mu(V) < \varepsilon$.

(ii) Choose a sequence of compact sets $K_i \subset X$ such that $K_i \subset K_{i+1}$ for all $i \in \mathbb{N}$ and $X = \bigcup_{i=1}^{\infty} K_i$. Fix a Borel set $B \in \mathcal{B}$. If $\mu(B) < \infty$, then B satisfies (3.2) by (i). Hence assume $\mu(B) = \infty$. Then it follows from part (iv) of Theorem 1.28 that $\lim_{i\to\infty} \mu(B \cap K_i) = \infty$. For each integer $n \in \mathbb{N}$ choose $i_n \in \mathbb{N}$ such that

$$\mu(B \cap K_{i_n}) > n.$$

Since $\mu(B \cap K_{i_n}) \leq \mu(K_{i_n}) < \infty$, it follows from (i) that (3.2) holds with B replaced by $B \cap K_{i_n}$. Hence there exists a compact set $C_n \subset B \cap K_{i_n}$ such that $\mu(C_n) > n$. This proves Lemma 3.7. □

Theorem 3.8. *Let $\mu_1 : \mathcal{B} \to [0, \infty]$ be an outer regular Borel measure that is inner regular on open sets. Define*

$$\mu_0 : \mathcal{B} \longrightarrow [0, \infty]$$

by

$$\mu_0(B) := \sup\{\mu_1(K) \mid K \subset B \text{ and } K \text{ is compact}\} \quad \text{for } B \in \mathcal{B}. \tag{3.5}$$

Then the following holds.

(i) *μ_0 is a Radon measure, it agrees with μ_1 on all compact sets and all open sets, and $\mu_0(B) \leq \mu_1(B)$ for all $B \in \mathcal{B}$.*

(ii) *If X is σ-compact, then $\mu_0 = \mu_1$.*

(iii) *If $f : X \to \mathbb{R}$ is a compactly supported continuous function, then*

$$\int_X f\, d\mu_0 = \int_X f\, d\mu_1. \tag{3.6}$$

(iv) *Let $\mu : \mathcal{B} \to [0, \infty]$ be a Borel measure that is inner regular on open sets. Then*

$$\int_X f\, d\mu = \int_X f\, d\mu_1$$

for every compactly supported continuous function $f : X \to \mathbb{R}$ if and only if

$$\mu_0(B) \leq \mu(B) \leq \mu_1(B) \quad \text{for all } B \in \mathcal{B}.$$

Proof. We prove that μ_0 is a measure. It follows directly from the definition that $\mu_0(\emptyset) = 0$. Now assume that $B_i \in \mathcal{B}$ is a sequence of pairwise disjoint Borel sets and define

$$B := \bigcup_{i=1}^{\infty} B_i.$$

Choose any compact set $K \subset B$. Then $\mu_1(B_i \cap K) < \infty$, and so, from part (i) of Lemma 3.7,

$$\mu_0(B_i \cap K) = \mu_1(B_i \cap K)$$

for all $i \in \mathbb{N}$. Consequently,

$$\mu_1(K) = \sum_{i=1}^{\infty} \mu_1(B_i \cap K) = \sum_{i=1}^{\infty} \mu_0(B_i \cap K) \le \sum_{i=1}^{\infty} \mu_0(B_i).$$

Take the supremum over all compact sets $K \subset B$ to obtain

$$\mu_0(B) \le \sum_{i=1}^{\infty} \mu_0(B_i). \tag{3.7}$$

To prove the converse inequality, it suffices to assume that $\mu_0(B) < \infty$. Then $\mu_0(B_i) \le \mu_0(B) < \infty$ for all $i \in \mathbb{N}$. Fix a constant $\varepsilon > 0$ and choose a sequence of compact sets $K_i \subset B_i$ such that $\mu_1(K_i) > \mu_0(B_i) - 2^{-i}\varepsilon$ for all i. Then, for every $n \in \mathbb{N}$, the union $K_1 \cup \cdots \cup K_n$ is a compact subset of B and

$$\mu_0(B) \ge \mu_1(K_1 \cup \cdots \cup K_n) = \sum_{i=1}^{n} \mu_1(K_i) > \sum_{i=1}^{n} \mu_0(B_i) - \varepsilon.$$

Now take the limit $n \to \infty$ to obtain

$$\mu_0(B) \ge \sum_{i=1}^{\infty} \mu_0(B_i) - \varepsilon.$$

Since this holds for all $\varepsilon > 0$, it follows that $\mu_0(B) \ge \sum_{i=1}^{\infty} \mu_0(B_i)$ and hence $\mu_0(B) = \sum_{i=1}^{\infty} \mu_0(B_i)$ by (3.7). This shows that μ_0 is a measure. Moreover, it follows directly from the definition of μ_0 that $\mu_0(K) = \mu_1(K)$ for every compact set $K \subset X$. Since μ_1 is inner regular on open sets it follows that $\mu_0(U) = \mu_1(U)$ for every open set $U \subset X$. Since $\mu_0(K) = \mu_1(K)$ for every compact set $K \subset X$, the definition of μ_0 in (3.5) shows that μ_0 is inner regular and hence is a Radon measure. The inequality $\mu_0(B) \le \mu_1(B)$ for $B \in \mathcal{B}$ follows directly from the definition of μ_0. This proves part (i). Part (ii) follows directly from the definition of μ_0 and part (ii) of Lemma 3.7.

We prove part (iii). Assume first that $s\colon X \to \mathbb{R}$ is a Borel measurable step function with compact support. Then

$$s = \sum_{i=1}^{\ell} \alpha_i \chi_{B_i},$$

where $\alpha_i \in \mathbb{R}$ and $B_i \in \mathcal{B}$ with $\mu_1(B_i) < \infty$. Hence $\mu_0(B_i) = \mu_1(B_i)$ by part (i) of Lemma 3.7 and hence

$$\int_X s\, d\mu_0 = \sum_{i=1}^{\ell} \alpha_i \mu_0(B_i) = \int_X s\, d\mu_1.$$

Now let $f\colon X \to [0,\infty]$ be a Borel measurable function with compact support. By Theorem 1.26, there exists a sequence of Borel measurable step functions $s_n\colon X \to [0,\infty)$ such that

$$0 \le s_1(x) \le s_2(x) \le \cdots \quad \text{and} \quad f(x) = \lim_{n\to\infty} s_n(x)$$

for all $x \in X$. Thus s_n has compact support for each n. By the Lebesgue monotone convergence theorem (Theorem 1.37) this implies

$$\int_X f\, d\mu_0 = \lim_{n\to\infty} \int_X s_n\, d\mu_0 = \lim_{n\to\infty} \int_X s_n\, d\mu_1 = \int_X f\, d\mu_1.$$

If $f\colon X \to \mathbb{R}$ is a μ_1-integrable function with compact support, then by what we have just proved,

$$\int_X f^{\pm}\, d\mu_0 = \int_X f^{\pm}\, d\mu_1 < \infty,$$

so f is μ_0-integrable and satisfies (3.6). This proves part (iii).

We prove part (iv) in four steps.

Step 1. *Let $\mu\colon \mathcal{B} \to [0,\infty]$ be a Borel measure such that*

$$\int_X f\, d\mu = \int_X f\, d\mu_1 \tag{3.8}$$

for every compactly supported continuous function $f\colon X \to \mathbb{R}$. Then

$$\mu(K) \le \mu_1(K), \quad \mu_1(U) \le \mu(U)$$

for every compact set $K \subset X$ and every open set $U \subset X$.

Fix an open set $U \subset X$ and a compact set $K \subset U$. Then Urysohn's lemma (Theorem A.1) asserts that there exists a compactly supported continuous function $f : X \to \mathbb{R}$ such that

$$f|_K \equiv 1, \quad \mathrm{supp}(f) \subset U, \quad 0 \le f \le 1.$$

Hence it follows from (3.8) that

$$\mu(K) \le \int_X f \, d\mu = \int_X f \, d\mu_1 \le \mu_1(U)$$

and likewise

$$\mu_1(K) \le \int_X f \, d\mu_1 = \int_X f \, d\mu \le \mu(U).$$

Since $\mu(K) \le \mu_1(U)$ for every open set $U \subset X$ containing K and μ_1 is outer regular, we obtain

$$\mu(K) \le \inf\{\mu_1(U) \mid K \subset U \subset X \text{ and } U \text{ is open}\} = \mu_1(K).$$

Since $\mu_1(K) \le \mu(U)$ for every compact set $K \subset U$ and μ_1 is inner regular on open sets, we obtain

$$\mu_1(U) = \sup\{\mu_1(K) \mid K \subset U \text{ and } K \text{ is compact}\} \le \mu(U).$$

Step 2. *Let μ be as in Step* 1 *and assume in addition that μ is inner regular on open sets. Then $\mu(K) = \mu_1(K)$ for every compact set $K \subset X$ and $\mu(U) = \mu_1(U)$ for every open set $U \subset X$.*

If $U \subset X$ is an open set, then

$$\begin{aligned}
\mu(U) &= \sup\{\mu(K) \mid K \subset U \text{ and } K \text{ is compact}\} \\
&\le \sup\{\mu_1(K) \mid K \subset U \text{ and } K \text{ is compact}\} \\
&= \mu_1(U) \le \mu(U).
\end{aligned}$$

Here the two inequalities follow from Step 1. It follows that $\mu(U) = \mu_1(U)$. Now let K be a compact set. Then $\mu_1(K) < \infty$. Since μ_1 is outer regular, there exists an open set $U \subset X$ such that $K \subset U$ and $\mu_1(U) < \infty$. Since μ and μ_1 agree on open sets it follows that

$$\mu(K) = \mu(U) - \mu(U \setminus K) = \mu_1(U) - \mu_1(U \setminus K) = \mu_1(K).$$

Step 3. *Let μ be as in Step 2. Then*

$$\mu_0(B) \le \mu(B) \le \mu_1(B) \quad \text{for all } B \in \mathcal{B}. \tag{3.9}$$

Fix a Borel set $B \in \mathcal{B}$. Then, by Step 2,

$$\begin{aligned}
\mu_0(B) &= \sup\{\mu_1(K) \mid K \subset B \text{ and } K \text{ is compact}\} \\
&= \sup\{\mu(K) \mid K \subset B \text{ and } K \text{ is compact}\} \\
&\le \mu(B) \\
&\le \inf\{\mu(U) \mid B \subset U \subset X \text{ and } U \text{ is open}\} \\
&= \inf\{\mu_1(U) \mid B \subset U \subset X \text{ and } U \text{ is open}\} \\
&= \mu_1(B).
\end{aligned}$$

Step 4. *Let $\mu\colon \mathcal{B} \to [0, \infty]$ be a Borel measure that satisfies* (3.9). *Then*

$$\int_X f\,d\mu = \int_X f\,d\mu_0 = \int_X f\,d\mu_1$$

for every continuous function $f\colon X \to \mathbb{R}$ with compact support.

By the definition of the integral and part (iii),

$$\int_X f\,d\mu_0 \le \int_X f\,d\mu \le \int_X f\,d\mu_1 = \int_X f\,d\mu_0$$

for every compactly supported continuous function $f\colon X \to [0, \infty)$. Hence

$$\int_X f\,d\mu = \int_X f\,d\mu_0 = \int_X f\,d\mu_1$$

for every compactly supported continuous function $f\colon X \to [0, \infty)$, and hence also for every compactly supported continuous function $f\colon X \to \mathbb{R}$. This proves Step 4 and Theorem 3.8. □

Example 3.9. Let $(X, \mathcal{U})$ be the compact Hausdorff space in Example 3.6 and let $\mu\colon \mathcal{B} \to [0, \infty]$ be Dieudonné's measure.

(i) Take $\mu_1 := \mu$ and define the function $\mu_0\colon \mathcal{B} \to [0, \infty]$ by (3.5). Then

$$\mu_0(X) = 1, \quad \mu_0(\{\kappa\}) = 0, \quad \mu_0(X \setminus \{\kappa\}) = 0,$$

and so μ_0 is not a measure. Hence the assumptions on μ_1 cannot be removed in part (i) of Theorem 3.8.

(ii) Take $\mu_1 := \delta_\kappa$ to be the Dirac measure at the point $\kappa \in X$. This is a regular Borel measure and so the measure μ_0 in (3.5) agrees with μ_1. It is an easy exercise to show that the integral of a continuous function $f\colon X \to \mathbb{R}$ with respect to the Dieudonné measure μ is given by

$$\int_X f\,d\mu = f(\kappa) = \int_X f\,d\mu_0 = \int_X f\,d\mu_1.$$

Moreover, the compact set $K = \{\kappa\}$ satisfies $\mu(K) = 0 < 1 = \mu_1(K)$ and the open set $U := X \setminus \{\kappa\}$ satisfies $\mu_1(U) = 0 < 1 = \mu(U)$. This shows that the inequalities in Step 1 in the proof of Theorem 3.8 can be strict and that the hypothesis that μ is inner regular on open sets cannot be removed in part (iv) of Theorem 3.8.

Remark 3.10. As Example 3.6 shows, it may sometimes be convenient to define a Borel measure first on a σ-algebra that contains the σ-algebra of all Borel measurable sets and then restrict it to $\mathcal{B}$. Thus let $\mathcal{A} \subset 2^X$ be a σ-algebra containing $\mathcal{B}$ and let $\mu\colon \mathcal{A} \to [0,\infty]$ be a measure. Call μ *outer regular* if it satisfies (3.1) for all $B \in \mathcal{A}$, call it *inner regular* if it satisfies (3.2) for all $B \in \mathcal{A}$, and call it *regular* if it is both outer and inner regular. If μ is regular and $(X, \mathcal{B}^*, \mu^*)$ denotes the completion of $(X, \mathcal{B}, \mu|_{\mathcal{B}})$, it turns out that the completion is also regular (exercise). If in addition $(X, \mathcal{A}, \mu)$ is σ-finite (see Definition 4.29 below), then

$$\mathcal{A} \subset \mathcal{B}^*, \qquad \mu = \mu^*|_{\mathcal{A}}. \tag{3.10}$$

To see this, let $A \in \mathcal{A}$ such that $\mu(A) < \infty$. Choose a sequence of compact sets $K_i \subset X$ and a sequence of open sets $U_i \subset X$ such that $K_i \subset A \subset U_i$ and $\mu(A) - 2^{-i} \le \mu(K_i) \le \mu(U_i) \le \mu(A) + 2^{-i}$ for all $i \in \mathbb{N}$. Then $B_0 := \bigcup_{i=1}^\infty K_i$ and $B_1 := \bigcap_{i=1}^\infty U_i$ are Borel sets such that $B_0 \subset A \subset B_1$ and $\mu(B_1 \setminus B_0) = 0$. Thus every set $A \in \mathcal{A}$ with $\mu(A) < \infty$ belongs to $\mathcal{B}^*$ and $\mu^*(A) = \mu(A)$. This proves (3.10) because every $\mathcal{A}$-measurable set is a countable union of $\mathcal{A}$-measurable sets with finite measure. Note that if X is σ-compact and $\mu(K) < \infty$ for every compact set $K \subset X$, then $(X, \mathcal{A}, \mu)$ is σ-finite.

3.2 Borel outer measures

This section is of preparatory nature. It discusses outer measures on a locally compact Hausdorff space that satisfy suitable regularity properties and shows that the resulting measures on the Borel σ-algebra are outer/inner regular. The result will play a central role in the proof of the Riesz representation theorem. As in Section 3.1, we assume that $(X, \mathcal{U})$ is a locally compact Hausdorff space and denote by $\mathcal{B}$ the Borel σ-algebra of $(X, \mathcal{U})$.

Definition 3.11. A *Borel outer measure on* X is an outer measure

$$\nu: 2^X \longrightarrow [0, \infty]$$

that satisfies the following axioms:

(a) if $K \subset X$ is compact, then $\nu(K) < \infty$;

(b) if $K_0, K_1 \subset X$ are disjoint compact sets, then $\nu(K_0 \cup K_1) = \nu(K_0) + \nu(K_1)$;

(c) $\nu(A) = \inf\{\nu(U) \mid A \subset U \subset X,\ U \text{ is open}\}$ for every subset $A \subset X$;

(d) $\nu(U) = \sup\{\nu(K) \mid K \subset U,\ K \text{ is compact}\}$ for every open set $U \subset X$.

Theorem 3.12. *Let* $\nu: 2^X \to [0, \infty]$ *be a Borel outer measure. Then* $\nu|_{\mathcal{B}}$ *is an outer regular Borel measure and is inner regular on open sets.*

One can deduce Theorem 3.12 from Carathéodory's theorem (Theorem 2.4) and use axioms (a) and (b) (instead of the Carathéodory criterion in Theorem 2.5) to show that the σ-algebra of ν-measurable sets contains the Borel σ-algebra. That the resulting Borel measure has the required regularity properties is then obvious from axioms (c) and (d). We choose a different route, following Rudin [17], and give a direct proof of Theorem 3.12 which does not rely on Theorem 2.4. The former approach is left to the reader as well as the verification that both proofs give rise to the the same σ-algebra, i.e., the σ-algebra $\mathcal{A}$ in (3.11) agrees with the σ-algebra of ν-measurable subsets of X.

Proof of Theorem 3.12. Define

$$\mathcal{A}_e := \{E \subset X \mid \nu(E) = \sup\{\nu(K) \mid K \subset E,\ K \text{ is compact}\} < \infty\} \quad (3.11a)$$

and

$$\mathcal{A} := \{A \subset X \mid A \cap K \in \mathcal{A}_e \text{ for every compact set } K \subset X\}. \quad (3.11b)$$

Here the subscript "e" stands for "endlich" and indicates that the elements of $\mathcal{A}_e$ are sets of finite measure. We prove in seven steps that $\mathcal{A}$ is a σ-algebra containing $\mathcal{B}$, that

$$\mu := \nu|_{\mathcal{A}}: \mathcal{A} \longrightarrow [0, \infty]$$

is a measure, and that $(X, \mathcal{A}, \mu)$ is a complete measure space. That μ is outer regular and is inner regular on open sets follows immediately from conditions (c) and (d) in Definition 3.11.

Step 1. *Let $E_1, E_2, E_3, \dots$ be a sequence of pairwise disjoint sets in $\mathcal{A}_e$ and define*

$$E := \bigcup_{i=1}^{\infty} E_i.$$

Then the following holds:

(i) $\nu(E) = \sum_{i=1}^{\infty} \nu(E_i)$;

(ii) *if $\nu(E) < \infty$, then $E \in \mathcal{A}_e$.*

The assertions are obvious when $\nu(E) = \infty$ because $\nu(E) \le \sum_{i=1}^{\infty} \nu(E_i)$. Hence assume $\nu(E) < \infty$. We argue as in the proof of Theorem 3.8. Fix a constant $\varepsilon > 0$. Since $E_i \in \mathcal{A}_e$ for all i, there is a sequence of compact sets $K_i \subset E_i$ such that $\nu(K_i) > \nu(E_i) - 2^{-i}\varepsilon$ for all i. Then for all $n \in \mathbb{N}$

$$\begin{aligned} \nu(E) &\ge \nu(K_1 \cup \dots \cup K_n) \\ &= \nu(K_1) + \dots + \nu(K_n) \\ &\ge \nu(E_1) + \dots + \nu(E_n) - \varepsilon. \end{aligned} \tag{3.12}$$

Here the equality follows from condition (b) in Definition 3.11. Take the limit $n \to \infty$ to obtain

$$\sum_{i=1}^{\infty} \nu(E_i) \le \nu(E) + \varepsilon.$$

Since this holds for all $\varepsilon > 0$, it follows that

$$\sum_{i=1}^{\infty} \nu(E_i) \le \nu(E) \le \sum_{i=1}^{\infty} \nu(E_i)$$

and hence

$$\sum_{i=1}^{\infty} \nu(E_i) = \nu(E). \tag{3.13}$$

Now (3.12) and (3.13) yield

$$\begin{aligned} \nu(E) &\ge \nu(K_1 \cup \dots \cup K_n) \\ &\ge \sum_{i=1}^{n} \nu(E_i) - \varepsilon \\ &= \nu(E) - \sum_{i=n+1}^{\infty} \nu(E_i) - \varepsilon \end{aligned}$$

for all $n \in \mathbb{N}$.

By (3.13), there exists an $n_\varepsilon \in \mathbb{N}$ such that

$$\sum_{i=n_\varepsilon+1}^{\infty} \nu(E_i) < \varepsilon.$$

Hence the compact set

$$K_\varepsilon := K_1 \cup \cdots \cup K_{n_\varepsilon} \subset E$$

satisfies

$$\nu(E) \geq \nu(K_\varepsilon) \geq \nu(E) - 2\varepsilon.$$

Since this holds for all $\varepsilon > 0$, we obtain

$$\nu(E) = \sup\{\nu(K) \mid K \subset E,\ K \text{ is compact}\}$$

and hence $E \in \mathcal{A}_e$. This proves Step 1.

Step 2. *If* $E_0, E_1 \in \mathcal{A}_e$, *then*

$$E_0 \cup E_1 \in \mathcal{A}_e, \quad E_0 \cap E_1 \in \mathcal{A}_e, \quad E_0 \setminus E_1 \in \mathcal{A}_e.$$

We first prove that $E_0 \setminus E_1 \in \mathcal{A}_e$. Fix a constant $\varepsilon > 0$. Since $E_0, E_1 \in \mathcal{A}_e$, and by condition (c) in Definition 3.11, there exist compact sets $K_0, K_1 \subset X$ and open sets $U_0, U_1 \subset X$ such that

$$K_i \subset E_i \subset U_i, \quad \nu(E_i) - \varepsilon < \nu(K_i) \leq \nu(U_i) < \nu(E_i) + \varepsilon, \quad i = 0, 1.$$

Moreover, every compact set with finite outer measure is an element of $\mathcal{A}_e$, by definition, and every open set with finite outer measure is an element of $\mathcal{A}_e$ by condition (d) in Definition 3.11. Hence

$$K_i, U_i, U_i \setminus K_i \in \mathcal{A}_e$$

for $i = 0, 1$ and it follows from Step 1 that

$$\nu(E_i \setminus K_i) \leq \nu(U_i \setminus K_i) = \nu(U_i) - \nu(K_i) \leq 2\varepsilon, \tag{3.14a}$$

$$\nu(U_i \setminus E_i) \leq \nu(U_i \setminus K_i) = \nu(U_i) - \nu(K_i) \leq 2\varepsilon, \tag{3.14b}$$

for $i = 0, 1$. Define

$$K := K_0 \setminus U_1 \subset E_0 \setminus E_1. \tag{3.15}$$

Then K is a compact set and

$$E_0 \setminus E_1 \subset (E_0 \setminus K_0) \cup (K_0 \setminus U_1) \cup (U_1 \setminus E_1).$$

By the definition of an outer measure, this implies

$$\nu(E_0 \setminus E_1) \le \nu(E_0 \setminus K_0) + \nu(K_0 \setminus U_1) + \nu(U_1 \setminus E_1) \le \nu(K) + 4\varepsilon.$$

Here the last inequality follows from the definition of K in (3.15) and the inequalities in (3.14). Since $\varepsilon > 0$ was chosen arbitrarily, it follows that

$$\nu(E_0 \setminus E_1) = \sup\{\nu(K) \mid K \subset E_0 \setminus E_1,\ K \text{ is compact}\}$$

and hence $E_0 \setminus E_1 \in \mathcal{A}_e$. With this understood, it follows from Step 1 that

$$E_0 \cup E_1 = (E_0 \setminus E_1) \cup E_1 \in \mathcal{A}_e, \quad E_0 \cap E_1 = E_0 \setminus (E_0 \setminus E_1) \in \mathcal{A}_e.$$

This proves Step 2.

Step 3. *$\mathcal{A}$ is a σ-algebra.*

First, $X \in \mathcal{A}$ because $K \in \mathcal{A}_e$ for every compact set $K \subset X$.

Second, assume $A \in \mathcal{A}$ and let $K \subset X$ be a compact set. Then by definition $A \cap K \in \mathcal{A}_e$. Moreover $K \in \mathcal{A}_e$ and hence, by Step 2,

$$A^c \cap K = K \setminus (A \cap K) \in \mathcal{A}_e.$$

Since this holds for every compact set $K \subset X$, we have $A^c \in \mathcal{A}_e$.

Third, let $A_i \in \mathcal{A}$ for $i \in \mathbb{N}$ and denote

$$A := \bigcup_{i=1}^{\infty} A_i.$$

Fix a compact set $K \subset X$. Then

$$A_i \cap K \in \mathcal{A}_e$$

for all i by the definition of $\mathcal{A}$. Hence, by Step 2,

$$B_i := A_i \cap K \in \mathcal{A}_e$$

for all i and then, again by Step 2,

$$E_i := B_i \setminus (B_1 \cup \cdots \cup B_{i-1}) \in \mathcal{A}_e$$

for all i.

The sets E_i are pairwise disjoint and

$$\bigcup_{i=1}^{\infty} E_i = \bigcup_{i=1}^{\infty} B_i = A \cap K.$$

Since $\nu(A \cap K) \leq \nu(K) < \infty$ by condition (a) in Definition 3.11, it follows from Step 1 that $A \cap K \in \mathcal{A}_e$. This holds for every compact set $K \subset X$ and hence $A \in \mathcal{A}$. This proves Step 3.

Step 4. $\mathcal{B} \subset \mathcal{A}$.

Let $F \subset X$ be closed. If $K \subset X$ is compact, then $F \cap K$ is a closed subset of a compact set and hence is compact (see Lemma A.2). Thus $F \cap K \in \mathcal{A}_e$ for every compact subset $K \subset X$ and so $F \in \mathcal{A}$. Therefore we have proved that $\mathcal{A}$ contains all closed subsets of X. Since $\mathcal{A}$ is a σ-algebra by Step 3, it also contains all open subsets of X and thus $\mathcal{B} \subset \mathcal{A}$. This proves Step 4.

Step 5. *Let $A \subset X$. Then $A \in \mathcal{A}_e$ if and only if $A \in \mathcal{A}$ and $\nu(A) < \infty$.*

If $A \in \mathcal{A}_e$, then $A \cap K \in \mathcal{A}_e$ for every compact set $K \subset X$ by Step 2 and hence $A \in \mathcal{A}$. Conversely, let $A \in \mathcal{A}$ such that $\nu(A) < \infty$. Fix a constant $\varepsilon > 0$. By condition (c) in Definition 3.11, there exists an open set $U \subset X$ such that $A \subset U$ and $\nu(U) < \infty$. By condition (d) in Definition 3.11, there exists a compact set $K \subset X$ such that

$$K \subset U, \quad \nu(K) > \nu(U) - \varepsilon.$$

Since $K, U \in \mathcal{A}_e$ and $U = (U \setminus K) \cup K$, it follows from Step 1 that

$$\nu(U \setminus K) = \nu(U) - \nu(K) < \varepsilon.$$

Moreover, $A \cap K \in \mathcal{A}_e$ because $A \in \mathcal{A}$. Hence it follows from the definition of $\mathcal{A}_e$ that there exists a compact set $H \subset A \cap K$ such that

$$\begin{aligned}
\nu(H) &\geq \nu(A \cap K) - \varepsilon \\
&= \nu(A \setminus (A \setminus K)) - \varepsilon \\
&\geq \nu(A) - \nu(A \setminus K) - \varepsilon \\
&\geq \nu(A) - \nu(U \setminus K) - \varepsilon \\
&\geq \nu(A) - 2\varepsilon.
\end{aligned}$$

Since $\varepsilon > 0$ was chosen arbitrarily, we conclude that

$$\nu(A) = \sup\{\nu(K) \mid K \subset A,\ K \text{ is compact}\}$$

and hence $A \in \mathcal{A}_e$. This proves Step 5.

Step 6. *$\mu := \nu|_{\mathcal{A}}$ is an outer regular extended Borel measure and μ is inner regular on open sets.*

We prove that μ is a measure. By definition, $\mu(\emptyset) = 0$. Now let $A_i \in \mathcal{A}$ be a sequence of pairwise disjoint measurable sets and define

$$A := \bigcup_{i=1}^{\infty} A_i.$$

If $\mu(A_i) < \infty$ for all i, then $A_i \in \mathcal{A}_e$ by Step 5 and hence $\mu(A) = \sum_{i=1}^{\infty} \mu(A_i)$ by Step 1. If $\nu(A_i) = \infty$ for some i, then $\mu(A) \geq \mu(A_i)$ and so $\mu(A) = \infty$. Thus μ is a measure. Moreover, $\mathcal{B} \subset \mathcal{A}$ by Step 4, $\mu(K) < \infty$ for every compact set $K \subset X$ by condition (a) in Definition 3.11, μ is outer regular by condition (c) in Definition 3.11, and μ is inner regular on open sets by condition (d) in Definition 3.11. This proves Step 6.

Step 7. *$(X, \mathcal{A}, \mu)$ is a complete measure space.*

If $E \subset X$ satisfies $\nu(E) = 0$, then $E \in \mathcal{A}_e$ by the definition of $\mathcal{A}_e$, and hence $E \in \mathcal{A}$ by Step 5. This proves Step 7 and Theorem 3.12. □

3.3 The Riesz representation theorem

Let $(X, \mathcal{U})$ be a locally compact Hausdorff space and $\mathcal{B}$ be its Borel σ-algebra. A function $f: X \to \mathbb{R}$ is called *compactly supported* if its support

$$\mathrm{supp}(f) := \overline{\{x \in X \mid f(x) \neq 0\}}$$

is a compact subset of X. The set of compactly supported continuous functions on X will be denoted by

$$C_c(X) := \left\{ f: X \to \mathbb{R} \,\middle|\, \begin{array}{l} f \text{ is continuous and} \\ \mathrm{supp}(f) \text{ is a compact subset of } X \end{array} \right\}.$$

Thus a continuous function $f: X \to \mathbb{R}$ belongs to $C_c(X)$ if and only if there exists a compact set $K \subset X$ such that $f(x) = 0$ for all $x \in X \setminus K$. The set $C_c(X)$ is a real vector space.

Definition 3.13. A linear functional $\Lambda: C_c(X) \to \mathbb{R}$ is called *positive* if

$$f \geq 0 \implies \Lambda(f) \geq 0$$

for all $f \in C_c(X)$.

The next lemma shows that every positive linear functional on $C_c(X)$ is continuous with respect to the topology of uniform convergence when restricted to the subspace of functions with support contained in a fixed compact subset of X.

Lemma 3.14. *Let* $\Lambda: C_c(X) \to \mathbb{R}$ *be a positive linear functional and let* $f_i \in C_c(X)$ *be a sequence of compactly supported continuous functions that converges uniformly to* $f \in C_c(X)$. *If there exists a compact set* $K \subset X$ *such that*

$$\mathrm{supp}(f_i) \subset K \quad \textit{for all } i \in \mathbb{N},$$

then

$$\Lambda(f) = \lim_{i \to \infty} \Lambda(f_i).$$

Proof. Since f_i converges uniformly to f, the sequence

$$\varepsilon_i := \sup_{x \in X} |f_i(x) - f(x)|$$

converges to zero. By Urysohn's lemma (Theorem A.1), there exists a compactly supported continuous function $\phi: X \to [0,1]$ such that $\phi(x) = 1$ for all $x \in K$. This function satisfies

$$-\varepsilon_i \phi \le f_i - f \le \varepsilon_i \phi \quad \text{for all } i.$$

Hence

$$-\varepsilon_i \Lambda(\phi) \le \Lambda(f_i) - \Lambda(f) \le \varepsilon_i \Lambda(\phi),$$

because Λ is positive, and hence $|\Lambda(f_i) - \Lambda(f)| \le \varepsilon_i \Lambda(\phi)$ for all i. Since ε_i converges to zero, so does $|\Lambda(f_i) - \Lambda(f)|$ and this proves Lemma 3.14. □

Let $\mu: \mathcal{B} \to [0, \infty]$ be a Borel measure. Then every continuous function $f: X \to \mathbb{R}$ with compact support is integrable with respect to μ. Define the map

$$\Lambda_\mu: C_c(X) \longrightarrow \mathbb{R}$$

by

$$\Lambda_\mu(f) := \int_X f \, d\mu. \tag{3.16}$$

Then Λ_μ is a positive linear functional. The Riesz representation theorem asserts that every positive linear functional on $C_c(X)$ has this form. It also asserts uniqueness under certain regularity hypotheses on the Borel measure. The following theorem includes two versions of the uniqueness statement.

Theorem 3.15 (Riesz representation theorem). *Let $\Lambda\colon C_c(X) \to \mathbb{R}$ be a positive linear functional. Then the following holds.*

(i) *There exists a unique Radon measure*

$$\mu_0\colon \mathcal{B} \longrightarrow [0, \infty]$$

such that

$$\Lambda_{\mu_0} = \Lambda.$$

(ii) *There exists a unique outer regular Borel measure*

$$\mu_1\colon \mathcal{B} \longrightarrow [0, \infty]$$

such that μ_1 is inner regular on open sets and

$$\Lambda_{\mu_1} = \Lambda.$$

(iii) *The Borel measures μ_0 and μ_1 in* (i) *and* (ii) *agree on all compact sets and on all open sets. Moreover,*

$$\mu_0(B) \le \mu_1(B) \quad \textit{for all } B \in \mathcal{B}.$$

(iv) *Let $\mu\colon \mathcal{B} \to [0, \infty]$ be a Borel measure that is inner regular on open sets. Then*

$$\Lambda_\mu = \Lambda \iff \mu_0(B) \le \mu(B) \le \mu_1(B) \textit{ for all } B \in \mathcal{B}.$$

Proof. The proof has nine steps. Step 1 defines a function $\nu\colon 2^X \to [0, \infty]$, Step 2 shows that it is an outer measure, and Steps 3, 4, and 5 show that it satisfies the axioms of Definition 3.11. Step 6 defines μ_1 and Step 7 shows that $\Lambda_{\mu_1} = \Lambda$. Step 8 defines μ_0 and Step 9 proves uniqueness.

Step 1. *Define the function*

$$\nu_{\mathcal{U}}\colon \mathcal{U} \longrightarrow [0, \infty]$$

by

$$\nu_{\mathcal{U}}(U) := \sup\{\Lambda(f) \mid f \in C_c(X),\ 0 \le f \le 1,\ \mathrm{supp}(f) \subset U\} \tag{3.17}$$

for every open set $U \subset X$ and define

$$\nu\colon 2^X \longrightarrow [0, \infty]$$

$$\nu(A) := \inf\{\nu_{\mathcal{U}}(U) \mid A \subset U \subset X,\ U \textit{ is open}\} \tag{3.18}$$

for every subset $A \subset X$. Then

$$\nu(U) = \nu_{\mathcal{U}}(U)$$

for every open set $U \subset X$.

If $U, V \subset X$ are open sets such that $U \subset V$, then $\nu_{\mathcal{U}}(U) \leq \nu_{\mathcal{U}}(V)$ by definition. Hence $\nu(U) = \inf\{\nu_{\mathcal{U}}(V) \mid U \subset V \subset X,\ V \text{ is open}\} = \nu_{\mathcal{U}}(U)$ for every open set $U \subset X$ and this proves Step 1.

Step 2. *The function $\nu\colon 2^X \to [0, \infty]$ in Step 1 is an outer measure.*

By definition, $\nu(\emptyset) = \nu_{\mathcal{U}}(\emptyset) = 0$. Since $\nu_{\mathcal{U}}(U) \leq \nu_{\mathcal{U}}(V)$ for all open sets $U, V \subset X$ with $U \subset V$, it follows also from the definition that $\nu(A) \leq \nu(B)$ whenever $A \subset B \subset X$. Next we prove that for all open sets $U, V \subset X$

$$\nu_{\mathcal{U}}(U \cup V) \leq \nu_{\mathcal{U}}(U) + \nu_{\mathcal{U}}(V). \tag{3.19}$$

To see this, let $f \in C_c(X)$ such that $0 \leq f \leq 1$ and

$$K := \operatorname{supp}(f) \subset U \cup V.$$

By Theorem A.4, there exist functions $\phi, \psi \in C_c(X)$ such that

$$\operatorname{supp}(\phi) \subset U, \quad \operatorname{supp}(\psi) \subset V, \quad \phi, \psi \geq 0, \quad \phi + \psi \leq 1, \quad (\phi + \psi)|_K \equiv 1.$$

Hence $f = \phi f + \psi f$ and so

$$\Lambda(f) = \Lambda(\phi f + \psi f) = \Lambda(\phi f) + \Lambda(\psi f) \leq \nu_{\mathcal{U}}(U) + \nu_{\mathcal{U}}(V).$$

This proves (3.19).

Now choose a sequence of subsets $A_i \subset X$ and define

$$A := \bigcup_{i=1}^{\infty} A_i.$$

We must prove that

$$\nu(A) \leq \sum_{i=1}^{\infty} \nu(A_i). \tag{3.20}$$

If there exists an $i \in \mathbb{N}$ such that $\nu(A_i) = \infty$, then $\nu(A) = \infty$ because $A_i \subset A$ and hence $\sum_{i=1}^{\infty} \nu(A_i) = \infty = \nu(A)$. Hence assume $\nu(A_i) < \infty$ for all i. Fix a constant $\varepsilon > 0$. By the definition of ν in (3.18), there exists a sequence of open sets $U_i \subset X$ such that

$$A_i \subset U_i, \quad \nu_{\mathcal{U}}(U_i) < \nu(A_i) + 2^{-i}\varepsilon.$$

Define

$$U := \bigcup_{i=1}^{\infty} U_i.$$

Let $f \in C_c(X)$ such that $0 \le f \le 1$ and $\mathrm{supp}(f) \subset U$. Since f has compact support, there exists an integer $k \in \mathbb{N}$ such that

$$\mathrm{supp}(f) \subset \bigcup_{i=1}^{k} U_i.$$

By the definition of $\nu_{\mathcal{U}}$ and (3.19), this implies

$$\begin{aligned}\Lambda(f) &\le \nu_{\mathcal{U}}(U_1 \cup \cdots \cup U_k)\\ &\le \nu_{\mathcal{U}}(U_1) + \cdots + \nu_{\mathcal{U}}(U_k)\\ &< \nu(A_1) + \cdots + \nu(A_k) + \varepsilon.\end{aligned}$$

Hence

$$\Lambda(f) \le \sum_{i=1}^{\infty} \nu(A_i) + \varepsilon$$

for every $f \in C_c(X)$ such that $0 \le f \le 1$ and $\mathrm{supp}(f) \subset U$. Consequently

$$\nu(A) \le \nu_{\mathcal{U}}(U) \le \sum_{i=1}^{\infty} \nu(A_i) + \varepsilon$$

by the definition of $\nu_{\mathcal{U}}(U)$ in (3.17). Thus

$$\nu(A) \le \sum_{i=1}^{\infty} \nu(A_i) + \varepsilon \quad \text{for every } \varepsilon > 0,$$

and hence

$$\nu(A) \le \sum_{i=1}^{\infty} \nu(A_i).$$

This proves (3.20) and Step 2.

Step 3. *Let $U \subset X$ be an open set. Then*

$$\nu_{\mathcal{U}}(U) = \sup\{\nu(K) \mid K \subset U,\ K \text{ is compact}\}. \tag{3.21}$$

Let $f \in C_c(X)$ such that

$$0 \le f \le 1, \quad K := \mathrm{supp}(f) \subset U.$$

Then it follows from the definition of $\nu_{\mathcal{U}}$ in (3.17) that $\Lambda(f) \le \nu_{\mathcal{U}}(V)$ for every open set $V \subset X$ with $K \subset V$. Hence it follows from the definition of ν in (3.18) that

$$\Lambda(f) \le \nu(K).$$

Hence

$$\begin{aligned}\nu_{\mathcal{U}}(U) &= \sup\{\Lambda(f) \mid f \in C_c(X),\ 0 \le f \le 1,\ \mathrm{supp}(f) \subset U\} \\ &\le \sup\{\nu(K) \mid K \subset U,\ K \text{ is compact}\} \\ &\le \nu(U) \\ &= \nu_{\mathcal{U}}(U).\end{aligned}$$

Hence $\nu_{\mathcal{U}}(U) = \sup\{\nu(K) \mid K \subset U,\ K \text{ is compact}\}$ and this proves Step 3.

Step 4. *Let $K \subset X$ be an compact set. Then*

$$\nu(K) = \inf\{\Lambda(f) \mid f \in C_c(X),\ f \ge 0,\ f|_K \equiv 1\}. \tag{3.22}$$

In particular, $\nu(K) < \infty$.

Define

$$a := \inf\{\Lambda(f) \mid f \in C_c(X),\ f \ge 0,\ f|_K \equiv 1\}.$$

We prove that $a \le \nu(K)$. Let $U \subset X$ be any open set containing K. By Urysohn's lemma (Theorem A.1), there exists a function $f \in C_c(X)$ such that

$$0 \le f \le 1, \quad \mathrm{supp}(f) \subset U, \quad f|_K \equiv 1.$$

Hence,

$$a \le \Lambda(f) \le \nu_{\mathcal{U}}(U).$$

This shows that $a \le \nu_{\mathcal{U}}(U)$ for every open set $U \subset X$ containing K. Take the infimum over all open sets containing K and use the definition of ν in equation (3.18) to obtain $a \le \nu(K)$.

We prove that $\nu(K) \le a$. Choose a function $f \in C_c(X)$ such that $f \ge 0$ and $f(x) = 1$ for all $x \in K$. Fix a constant $0 < \alpha < 1$ and define

$$U_\alpha := \{x \in X \mid f(x) > \alpha\}.$$

Then U_α is open and $K \subset U_\alpha$. Hence

$$\nu(K) \le \nu_{\mathcal{U}}(U_\alpha).$$

Moreover, every function $g \in C_c(X)$ with $0 \le g \le 1$ and $\mathrm{supp}(g) \subset U_\alpha$ satisfies $\alpha g(x) \le \alpha \le f(x)$ for $x \in U_\alpha$, hence $\alpha g \le f$, and so $\alpha\Lambda(g) \le \Lambda(f)$. Take the supremum over all such g to obtain $\alpha\nu_{\mathcal{U}}(U_\alpha) \le \Lambda(f)$ and hence

$$\nu(K) \le \nu_{\mathcal{U}}(U_\alpha) \le \frac{1}{\alpha}\Lambda(f).$$

This shows that $\nu(K) \leq \frac{1}{\alpha}\Lambda(f)$ for all $\alpha \in (0,1)$, and hence

$$\nu(K) \leq \Lambda(f).$$

Since this holds for every function $f \in C_c(X)$ with $f \geq 0$ and $f|_K \equiv 1$, it follows that $\nu(K) \leq a$. This proves Step 4.

Step 5. *Let $K_0, K_1 \subset X$ be compact sets such that $K_0 \cap K_1 = \emptyset$. Then*

$$\nu(K_0 \cup K_1) = \nu(K_0) + \nu(K_1).$$

The inequality $\nu(K_0 \cup K_1) \leq \nu(K_0) + \nu(K_1)$ holds because ν is an outer measure by Step 2. To prove the converse inequality, choose $f \in C_c(X)$ such that

$$0 \leq f \leq 1, \quad f|_{K_0} \equiv 0, \quad f|_{K_1} \equiv 1.$$

That such a function exists follows from Urysohn's lemma (Theorem A.1) with

$$K := K_1 \quad \text{and} \quad U := X \setminus K_0.$$

Now fix a constant $\varepsilon > 0$. Then it follows from Step 4 that there exists a function $g \in C_c(X)$ such that

$$g \geq 0, \quad g|_{K_0 \cup K_1} \equiv 1, \quad \Lambda(g) < \nu(K_0 \cup K_1) + \varepsilon.$$

It follows also from Step 4 that

$$\nu(K_0) + \nu(K_1) \leq \Lambda((1-f)g) + \Lambda(fg) = \Lambda(g) < \nu(K_0 + K_1) + \varepsilon.$$

Hence

$$\nu(K_0) + \nu(K_1) < \nu(K_0 + K_1) + \varepsilon \quad \text{for every } \varepsilon > 0,$$

and therefore

$$\nu(K_0) + \nu(K_1) \leq \nu(K_0 + K_1).$$

This proves Step 5.

Step 6. *The function*

$$\mu_1 := \nu|_{\mathcal{B}} : \mathcal{B} \to [0, \infty]$$

is an outer regular Borel measure that is inner regular on open sets.

The function ν is an outer measure by Step 2. It satisfies condition (a) in Definition 3.11 by Step 4, it satisfies condition (b) by Step 5, it satisfies condition (c) by Step 1, and it satisfies condition (d) by Step 3. Hence ν is a Borel outer measure. Hence Step 6 follows from Theorem 3.12.

Step 7. *Let μ_1 be as in Step 6. Then*

$$\Lambda_{\mu_1} = \Lambda.$$

We will prove that

$$\Lambda(f) \le \int_X f \, d\mu_1 \tag{3.23}$$

for all $f \in C_c(X)$. Once this is understood, it follows that

$$-\Lambda(f) = \Lambda(-f) \le \int_X (-f) \, d\mu_1 = -\int_X f \, d\mu_1$$

and hence

$$\int_X f \, d\mu_1 \le \Lambda(f) \quad \text{for all } f \in C_c(X).$$

Thus

$$\Lambda(f) = \int_X f \, d\mu_1 \quad \text{for all } f \in C_c(X),$$

and this proves Step 7.

Thus it remains to prove the inequality (3.23). Consider a continuous function $f\colon X \to \mathbb{R}$ with compact support and denote

$$K := \operatorname{supp}(f), \quad a := \inf_{x \in X} f(x), \quad b := \sup_{x \in X} f(x).$$

Fix a constant $\varepsilon > 0$ and choose real numbers

$$y_0 < a < y_1 < y_2 < \cdots < y_{n-1} < y_n = b$$

such that

$$y_i - y_{i-1} < \varepsilon, \quad i = 1, \dots, n.$$

For $i = 1, \dots, n$ define

$$E_i := \{x \in K \mid y_{i-1} < f(x) \le y_i\}.$$

Then E_i is the intersection of the open set $f^{-1}((y_{i-1}, \infty))$ with the closed set $f^{-1}((-\infty, y_i])$ and hence is a Borel set. Moreover, $E_i \cap E_j = \emptyset$ for $i \ne j$ and

$$K = \bigcup_{i=1}^{n} E_i.$$

Since μ_1 is outer regular, there exist open sets $U_1, \dots, U_n \subset X$ such that

$$E_i \subset U_i, \quad \mu_1(U_i) < \mu_1(E_i) + \frac{\varepsilon}{n}, \quad \sup_{U_i} f < y_i + \varepsilon \tag{3.24}$$

for all i. (For each i, choose first an open set that satisfies the first two conditions in (3.24) and then intersect it with the open set $f^{-1}((-\infty, y_i + \varepsilon))$.) By Theorem A.4, there exist functions $\phi_1, \dots, \phi_n \in C_c(X)$ such that

$$\phi_i \geq 0, \quad \mathrm{supp}(\phi_i) \subset U_i, \quad \sum_{i=1}^n \phi_i \leq 1, \quad \sum_{i=1}^n \phi_i|_K \equiv 1. \tag{3.25}$$

It follows from (3.24), (3.25), and Step 4 that

$$f = \sum_{i=1}^n \phi_i f, \qquad \phi_i f \leq (y_i + \varepsilon)\phi_i,$$

and

$$\mu_1(K) \leq \sum_{i=1}^n \Lambda(\phi_i), \quad \Lambda(\phi_i) \leq \mu_1(U_i) < \mu_1(E_i) + \frac{\varepsilon}{n}.$$

Hence,

$$\begin{aligned}
\Lambda(f) &= \sum_{i=1}^n \Lambda(\phi_i f) \\
&\leq \sum_{i=1}^n (y_i + \varepsilon)\Lambda(\phi_i) \\
&= \sum_{i=1}^n (y_i + |a| + \varepsilon)\Lambda(\phi_i) - |a| \sum_{i=1}^n \Lambda(\phi_i) \\
&\leq \sum_{i=1}^n (y_i + |a| + \varepsilon)\Big(\mu_1(E_i) + \frac{\varepsilon}{n}\Big) - |a|\mu_1(K) \\
&= \sum_{i=1}^n (y_i + \varepsilon)\mu_1(E_i) + \frac{\varepsilon}{n}\sum_{i=1}^n (y_i + |a| + \varepsilon) \\
&\leq \sum_{i=1}^n (y_i - \varepsilon)\mu_1(E_i) + \varepsilon(2\mu_1(K) + b + |a| + \varepsilon) \\
&\leq \int_X f \, d\mu_1 + \varepsilon(2\mu_1(K) + b + |a| + \varepsilon).
\end{aligned}$$

Here we have used the inequality $y_i + |a| + \varepsilon \geq 0$. Since $\varepsilon > 0$ can be chosen arbitrarily small it follows that $\Lambda(f) \leq \int_X f \, d\mu_1$. This proves (3.23).

Step 8. *Define*

$$\mu_0 : \mathcal{B} \longrightarrow [0, \infty]$$

by

$$\mu_0(B) := \sup\{\nu(K) \mid K \subset B,\ K \text{ is compact}\}$$

Then μ_0 is a Radon measure, $\Lambda_{\mu_0} = \Lambda$, and μ_0 and μ_1 satisfy (iii) *and* (iv).

It follows from Step 6 and part (i) of Theorem 3.8 that μ_0 is a Radon measure and it follows from Step 7 and part (iii) of Theorem 3.8 that $\Lambda_{\mu_0} = \Lambda_{\mu_1} = \Lambda$. That the measures μ_0 and μ_1 satisfy assertions (iii) and (iv) follows from parts (i) and (iv) of Theorem 3.8.

Step 9. *We prove uniqueness in* (i) *and* (ii).

By definition, $\mu_0(K) = \nu(K) = \mu_1(K)$ for every compact set $K \subset X$. Second, it follows from and Steps 1 and 3 that $\mu_0(U) = \nu_{\mathcal{U}}(U) = \nu(U) = \mu_1(U)$ for every open set $U \subset X$. Third, Steps 7 and 8 assert that $\Lambda_{\mu_0} = \Lambda_{\mu_1} = \Lambda$. Hence it follows from part (iv) of Theorem 3.8 that every Borel measure $\mu : \mathcal{B} \to [0, \infty]$ that is inner regular on open sets and satisfies $\Lambda_\mu = \Lambda$ agrees with ν on all compact sets and on all open sets. Hence, every Radon measure $\mu : \mathcal{B} \to [0, \infty]$ that satisfies $\Lambda_\mu = \Lambda$ is given by

$$\mu(B) = \sup\{\nu(K) \mid K \subset B,\ K \text{ is compact}\} = \mu_0(B)$$

for every $B \in \mathcal{B}$. Likewise, every outer regular Borel measure $\mu : \mathcal{B} \to [0, \infty]$ that is inner regular on open sets and satisfies $\Lambda_\mu = \Lambda$ is given by

$$\mu(B) = \inf\{\nu(U) \mid B \subset U \subset X,\ U \text{ is open}\} = \nu(B) = \mu_1(B)$$

for every $B \in \mathcal{B}$. This proves Step 9 and Theorem 3.15. □

The following corollary is the converse of Theorem 3.8.

Corollary 3.16. *Let $\mu_0 : \mathcal{B} \to [0, \infty]$ be a Radon measure and define*

$$\mu_1(B) := \inf\{\mu_0(U) \mid B \subset U \subset X,\ U \text{ is open}\} \quad \text{for all } B \in \mathcal{B}. \tag{3.26}$$

Then μ_1 is an outer regular Borel measure, is inner regular on open sets, and

$$\mu_0(B) = \sup\{\mu_1(K) \mid K \subset B,\ K \text{ is compact}\} \quad \text{for all } B \in \mathcal{B}. \tag{3.27}$$

Proof. Let μ_1 be the unique outer regular Borel measure on X that is inner regular on open sets and satisfies $\Lambda_{\mu_1} = \Lambda_{\mu_0}$. Then Theorem 3.15 asserts that μ_0 and μ_1 agree on all compact sets and all open sets. Since μ_1 is outer regular, it follows that μ_1 is given by (3.26). Since μ_0 is inner regular it follows that μ_0 satisfies (3.27). This proves Corollary 3.16. □

Corollary 3.17. *Every Radon measure is outer regular on compact sets.*

Proof. Equation (3.26) with $B = K$ compact and $\mu_0(K) = \mu_1(K)$. □

The next theorem formulates a condition on a locally compact Hausdorff space which guarantees that all Borel measures are regular. The condition (*every open subset is σ-compact*) is shown below to be strictly weaker than *second countability*.

Theorem 3.18. *Let X be a locally compact Hausdorff space.*

(i) *Assume X is σ-compact. Then every Borel measure on X that is inner regular on open sets is regular.*

(ii) *Assume every open subset of X is σ-compact. Then every Borel measure on X is regular.*

Proof. (i) Let $\mu\colon \mathcal{B} \to [0,\infty]$ be a Borel measure that is inner regular on open sets and let $\mu_0, \mu_1\colon \mathcal{B} \to [0,\infty]$ be the Borel measures associated to $\Lambda := \Lambda_\mu$ in parts (i) and (ii) of the Riesz representation theorem (Theorem 3.15). Since μ is inner regular on open sets, it follows from part (iii) of Theorem 3.15 that $\mu_0(B) \le \mu(B) \le \mu_1(B)$ for all $B \in \mathcal{B}$. Since X is σ-compact, it follows from part (ii) of Theorem 3.8 that $\mu_0 = \mu = \mu_1$. Hence μ is regular.

(ii) Let $\mu\colon \mathcal{B} \to [0,\infty]$ be a Borel measure. We prove that μ is inner regular on open sets. Fix an open set $U \subset X$. Since U is σ-compact, there exists a sequence of compact sets $K_i \subset U$ such that $K_i \subset K_{i+1}$ for all $i \in \mathbb{N}$ and $U = \bigcup_{i=1}^\infty K_i$. Hence

$$\mu(U) = \lim_{i\to\infty} \mu(K_i)$$

by Theorem 1.28, so

$$\mu(U) = \sup\{\mu(K) \mid K \subset U \text{ and } K \text{ is compact}\}.$$

This shows that μ is inner regular on open sets and hence it follows from (i) that μ is regular. This proves Theorem 3.18. □

Example 3.9 shows that the assumption that every open set is σ-compact cannot be removed in part (ii) of Theorem 3.18 even if X is compact. Note also that Theorem 3.18 provides another proof of regularity for the Lebesgue measure, which was established in Theorem 2.13.

Corollary 3.19. *Let X be a locally compact Hausdorff space such that every open subset of X is σ-compact. Then for every positive linear functional $\Lambda\colon C_c(X) \to \mathbb{R}$ there exists a unique Borel measure μ such that*

$$\Lambda_\mu = \Lambda.$$

Proof. This follows from Theorem 3.15 and part (ii) of Theorem 3.18. □

Remark 3.20. Let X be a compact Hausdorff space and let $C(X) = C_c(X)$ be the space of continuous real valued functions on X. From a functional analytic viewpoint it is interesting to understand the dual space of $C(X)$, i.e., the space of all bounded linear functionals on $C(X)$ (Definition 4.23). Exercise 5.35 below shows that every bounded linear functional on $C(X)$ is the difference of two positive linear functionals. If every open subset of X is σ-compact, it then follows from Corollary 3.19 that every bounded linear functional on $C(X)$ can be represented uniquely by a *signed Borel measure*. (See Definition 5.10 in Section 5.3 below.)

An important class of locally compact Hausdorff spaces that satisfy the hypotheses of Theorem 3.18 and Corollary 3.19 are the second countable ones. Here are the definitions. A *basis* of a topological space $(X, \mathcal{U})$ is a collection $\mathcal{V} \subset \mathcal{U}$ of open sets such that every open set $U \subset X$ is a union of elements of $\mathcal{V}$. A topological space $(X, \mathcal{U})$ is called *second countable* if it admits a countable basis. It is called *first countable* if, for every $x \in X$, there is a sequence of open sets W_i, $i \in \mathbb{N}$, such that $x \in W_i$ for all i and every open set that contains x contains one of the sets W_i.

Lemma 3.21. *Let X be a locally compact Hausdorff space*:

(i) *if X is second countable, then every open subset of X is σ-compact*;

(ii) *if every open subset of X is σ-compact, then X is first countable.*

Proof. (i) Let $\mathcal{V}$ be a countable basis of the topology and let $U \subset X$ be an open set. Denote by $\mathcal{V}(U)$ the collection of all sets $V \in \mathcal{V}$ such that $\overline{V} \subset U$ and $\overline{V}$ is compact. Let $x \in U$. By Lemma A.3 there is an open set $W \subset X$ with compact closure such that $x \in W \subset \overline{W} \subset U$. Since $\mathcal{V}$ is a basis of the topology, there is an element $V \in \mathcal{V}$ such that $x \in V \subset W$. Hence $\overline{V}$ is a closed subset of the compact set $\overline{W}$ and so is compact by Lemma A.2. Thus $V \in \mathcal{V}(U)$ and $x \in V$. This shows that

$$U = \bigcup_{V \in \mathcal{V}(U)} V.$$

Since $\mathcal{V}$ is countable, so is $\mathcal{V}(U)$. Choose a bijection $\mathbb{N} \to \mathcal{V}(U) : i \mapsto V_i$ and define

$$K_i := \overline{V}_1 \cup \cdots \cup \overline{V}_i$$

for $i \in \mathbb{N}$. Then $K_i \subset K_{i+1}$ for all i and $U = \bigcup_{i=1}^{\infty} K_i$. Hence U is σ-compact.

(ii) Fix an element $x \in X$. Since X is a Hausdorff space, the set $X \setminus \{x\}$ is open and hence is σ-compact by assumption. Choose a sequence of compact sets

$$K_i \subset X \setminus \{x\}$$

such that

$$K_i \subset K_{i+1} \quad \text{for all } i \in \mathbb{N}$$

and

$$\bigcup_{i=1}^{\infty} K_i = X \setminus \{x\}.$$

Then each set

$$U_i := X \setminus K_i$$

is open and contains x. By Lemma A.3, there exists a sequence of open sets $V_i \subset X$ with compact closure such that

$$x \in V_i \subset \overline{V}_i \subset U_i = X \setminus K_i \quad \text{for all } i \in \mathbb{N}.$$

Define

$$W_i := V_1 \cap \cdots \cap V_i \quad \text{for } i \in \mathbb{N}.$$

Then

$$\overline{W}_i \subset \bigcap_{j=1}^{i} (X \setminus K_j) = X \setminus K_i$$

and hence

$$\bigcap_{i=1}^{\infty} \overline{W}_i = \{x\}.$$

This implies that each open set $U \subset X$ that contains x also contains one of the sets $\overline{W}_i$. Namely, if $x \in U$ and U is open, then $\overline{W}_1 \setminus U$ is a compact set contained in

$$X \setminus \{x\} = \bigcup_{i=1}^{\infty} (X \setminus \overline{W}_i),$$

hence there exists a $j \in \mathbb{N}$ such that

$$\overline{W}_1 \setminus U \subset \bigcup_{i=1}^{j} (X \setminus \overline{W}_i) = X \setminus \overline{W}_j,$$

and so $\overline{W}_j \subset U$. Thus the sets W_j form a countable neighborhood basis of x and this proves Lemma 3.21. □

Example 3.22. The *Alexandrov double arrow space* is an example of a compact Hausdorff space in which every open subset is σ-compact and which is not second countable. It is defined as the ordered space $(X, \prec)$, where

$$X := [0, 1] \times \{0, 1\}$$

and $\prec$ denotes the *lexicographic ordering*

$$(s, i) \prec (t, j) \iff \begin{cases} s < t \text{ or} \\ s = t \text{ and } i = 0 \text{ and } j = 1. \end{cases}$$

The topology $\mathcal{U} \subset 2^X$ is defined as the smallest topology containing the sets

$$S_a := \{x \in X \mid a \prec x\}, \quad P_b := \{x \in X \mid x \prec b\}, \quad a, b \in X.$$

It has a basis consisting of the sets S_a, P_b, $S_a \cap P_b$ for all $a, b \in X$.

This topological space $(X, \mathcal{U})$ is a compact Hausdorff space and is *perfectly normal,* i.e., for any two disjoint closed subsets $F_0, F_1 \subset X$ there exists a continuous function $f: X \to [0, 1]$ such that

$$F_0 = f^{-1}(0), \qquad F_1 = f^{-1}(1).$$

(For a proof see Dan Ma's topology blog [12].) This implies that every open subset of X is σ-compact. Moreover, the subsets

$$Y_0 := (0, 1) \times \{0\}, \quad Y_1 := (0, 1) \times \{1\}$$

are both homeomorphic to the *Sorgenfrey line*, defined as the real axis with the (nonstandard) topology in which the open sets are the unions of half open intervals $[a, b)$. Since the Sorgenfrey line is not second countable, neither is the double arrow space $(X, \mathcal{U})$. (The Sorgenfrey line is Hausdorff and perfectly normal, but is not locally compact because every compact subset of it is countable.)

3.4 Exercises

Exercise 3.23. This exercise shows that the measures μ_0, μ_1 in Theorem 3.15 need not agree. Let (X, d) be the metric space given by $X := \mathbb{R}^2$ and

$$d((x_1, y_1), (x_2, y_2)) := |y_1 - y_2| + \begin{cases} 0 & \text{if } x_1 = x_2, \\ 1 & \text{if } x_1 \neq x_2. \end{cases}$$

Let $\mathcal{B} \subset 2^X$ be the Borel σ-algebra of (X, d).

(i) Show that (X, d) is locally compact.

(ii) Show that for every compactly supported continuous function $f: X \to \mathbb{R}$ there exists a finite set $S_f \subset \mathbb{R}$ such that $\mathrm{supp}(f) \subset S_f \times \mathbb{R}$.

(iii) Define the positive linear functional $\Lambda: C_c(X) \to \mathbb{R}$ by

$$\Lambda(f) := \sum_{x \in S_f} \int_{-\infty}^{\infty} f(x, y)\, dy.$$

(Here the integrals on the right are understood as the Riemann integrals or, equivalently by Theorem 2.24, as the Lebesgue integrals.) Let $\mu: \mathcal{B} \to [0, \infty]$ be a Borel measure such that

$$\int_X f\, d\mu = \Lambda(f) \quad \text{for all } f \in C_c(X).$$

Prove that every one-element subset of X has measure zero.

(iv) Let μ be as in (iii) and let $E := \mathbb{R} \times \{0\}$. This set is closed. If μ is inner regular, prove that $\mu(E) = 0$. If μ is outer regular, prove that $\mu(E) = \infty$.

Exercise 3.24. This exercise shows that the *Borel* assumption cannot be removed in Theorem 3.18. (The measure μ in part (ii) is not a Borel measure.) Let $(X, \mathcal{U})$ be the topological space defined by $X := \mathbb{N} \cup \{\infty\}$ and

$$\mathcal{U} := \{U \subset X \mid U \subset \mathbb{N} \text{ or } \#U^c < \infty\}.$$

Thus $(X, \mathcal{U})$ is the *(Alexandrov) one-point compactification* of the set $\mathbb{N}$ of natural numbers with the discrete topology. (If $\infty \in U$, then the condition $\#U^c < \infty$ is equivalent to the assertion that U^c is compact.)

(i) Prove that $(X, \mathcal{U})$ is a compact Hausdorff space and that every subset of X is σ-compact. Prove that the Borel σ-algebra of X is $\mathcal{B} = 2^X$.

(ii) Let $\mu: 2^X \to [0, \infty]$ be the counting measure. Prove that μ is inner regular, but not outer regular.

Exercise 3.25. Let $(X, \mathcal{U}_X)$ and $(Y, \mathcal{U}_Y)$ be locally compact Hausdorff spaces and denote their Borel σ-algebras by $\mathcal{B}_X \subset 2^X$ and $\mathcal{B}_Y \subset 2^Y$. Let $\phi: X \to Y$ be a continuous map and let $\mu_X: \mathcal{B}_X \to [0, \infty]$ be a measure.

(i) Prove that $\mathcal{B}_Y \subset \phi_* \mathcal{B}_X$ (see Exercise 1.69).

(ii) If μ_X is inner regular, show that $\phi_* \mu_X|_{\mathcal{B}_Y}$ is inner regular.

(iii) Find an example where μ_X is outer regular and $\phi_*\mu_X|_{\mathcal{B}_Y}$ is not outer regular. *Hint.* Consider the inclusion of $\mathbb{N}$ into its one-point compactification and use Exercise 3.24. (In this example μ_X is a Borel measure, however, $\phi_*\mu_X$ is not a Borel measure.)

Exercise 3.26. Let (X, d) be a metric space. Prove that (X, d) is perfectly normal, i.e., if $F_0, F_1 \subset X$ are disjoint closed subsets, then there is a continuous function $f\colon X \to [0,1]$ such that $F_0 = f^{-1}(0)$ and $F_1 = f^{-1}(1)$. Compare this with Urysohn's lemma (Theorem A.1). *Hint.* An explicit formula for f is given by

$$f(x) := \frac{d(x, F_0)}{d(x, F_0) + d(x, F_1)},$$

where

$$d(x, F) := \inf_{y \in F} d(x, y)$$

for $x \in X$ and $F \subset X$.

Exercise 3.27. Recall that the *Sorgenfrey line* is the topological space $(\mathbb{R}, \mathcal{U})$, where $\mathcal{U} \subset 2^{\mathbb{R}}$ is the smallest topology that contains all half open intervals $[a, b)$ with $a < b$. Prove that the Borel σ-algebra of $(\mathbb{R}, \mathcal{U})$ agrees with the Borel σ-algebra of the standard topology on $\mathbb{R}$.

Exercise 3.28. Recall from Example 3.22 that the double arrow space is

$$X := [0,1] \times \{0,1\}$$

with the topology induced by the lexicographic ordering. Prove that $B \subset X$ is a Borel set for this topology if and only if there exist a Borel set $E \subset [0,1]$ and two countable sets $F, G \subset X$ such that

$$B = ((E \times \{0,1\}) \cup F) \setminus G. \tag{3.28}$$

Hint 1. Show that the projection $f\colon X \to [0,1]$ onto the first factor is continuous with respect to the standard topology on the unit interval.

Hint 2. Denote by $\mathcal{B} \subset 2^X$ the set of all sets of the form (3.28) with $E \subset [0,1]$ a Borel set and $F, G \subset X$ countable. Prove that $\mathcal{B}$ is a σ-algebra.

Exercise 3.29 (the Baire σ-algebra). Let $(X, \mathcal{U})$ be a locally compact Hausdorff space and define

$$\mathcal{K}_{\mathrm{a}} := \left\{ K \subset X \;\middle|\; \begin{array}{l} K \text{ is compact and there is a sequence of open sets} \\ U_i \text{ such that } U_{i+1} \subset U_i \text{ for all } i \text{ and } K = \bigcap_{i=1}^{\infty} U_i \end{array} \right\}.$$

Let

$$\mathcal{B}_a \subset 2^X$$

be the smallest σ-algebra that contains $\mathcal{K}_a$. It is contained in the Borel σ-algebra $\mathcal{B} \subset 2^X$ and is called the *Baire σ-algebra* of $(X, \mathcal{U})$. The elements of $\mathcal{B}_a$ are called *Baire sets.* A function $f: X \to \mathbb{R}$ is called *Baire measurable* if $f^{-1}(U) \in \mathcal{B}_a$ for every open set $U \subset \mathbb{R}$. A *Baire measure* is a measure $\mu: \mathcal{B}_a \to [0, \infty]$ such that $\mu(K) < \infty$ for all $K \in \mathcal{K}_a$.

(i) Let $f: X \to \mathbb{R}$ be a continuous function with compact support. Prove that $f^{-1}(c) \in \mathcal{K}_a$ for every nonzero real number c.

(ii) Prove that $\mathcal{B}_a$ is the smallest σ-algebra such that every continuous function $f: X \to \mathbb{R}$ with compact support is $\mathcal{B}_a$-measurable.

(iii) If every open subset of X is σ-compact, prove that $\mathcal{B}_a = \mathcal{B}$.

Hint. Show first that every compact set belongs to $\mathcal{K}_a$ and then that every open set belongs to $\mathcal{B}_a$.

Exercise 3.30. (i) Let X be an uncountable set and let $\mathcal{U} := 2^X$ be the discrete topology. Prove that $B \subset X$ is a Baire set if and only if B is countable or has a countable complement. Define

$$\mu: \mathcal{B}_a \longrightarrow [0, 1]$$

by

$$\mu(B) := \begin{cases} 0 & \text{if } B \text{ is countable,} \\ 1 & \text{if } B^c \text{ is countable.} \end{cases}$$

Show that $\int_X f \, d\mu = 0$ for every $f \in C_c(X)$. Thus positive linear functionals $\Lambda: C_c(X) \to \mathbb{R}$ need not be *uniquely* represented by Baire measures.

(ii) Let X be the compact Hausdorff space of Example 3.6. Prove that the Baire sets in X are the countable subsets of $X \setminus \{\kappa\}$ and their complements.

(iii) Let X be the Stone–Čech compactification of $\mathbb{N}$ in Example 4.60 below. Prove that the Baire sets in X are the subsets of $\mathbb{N}$ and their complements.

(iv) Let $X = \mathbb{R}^2$ be the locally compact Hausdorff space in Exercise 3.23 (with a nonstandard topology). Show that $B \subset X$ is a Baire set if and only if the set $B_x := \{y \in \mathbb{R} \mid (x, y) \in B\}$ is a Borel set in $\mathbb{R}$ for every $x \in \mathbb{R}$ and one of the sets $S_0 := \{x \in \mathbb{R} \mid B_x \neq \emptyset\}$ and $S_1 := \{x \in \mathbb{R} \mid B_x \neq \mathbb{R}\}$ is countable.

Exercise 3.31. Let $(X, \mathcal{U})$ be a locally compact Hausdorff space and let

$$\mathcal{B}_{\mathrm{a}} \subset \mathcal{B} \subset 2^X$$

be the Baire and Borel σ-algebras. Let $\mathcal{F}(X)$ denote the real vector space of all functions $f: X \to \mathbb{R}$. For $\mathcal{F} \subset \mathcal{F}(X)$ consider the following conditions:

(a) $C_{\mathrm{c}}(X) \subset \mathcal{F}$;

(b) If $f_i \in \mathcal{F}$ is a sequence converging pointwise to $f \in \mathcal{F}(X)$, then $f \in \mathcal{F}$.

Let $\mathcal{F}_{\mathrm{a}} \subset \mathcal{F}(X)$ be the intersection of all subsets $\mathcal{F} \subset \mathcal{F}(X)$ that satisfy conditions (a) and (b). Prove the following.

(i) $\mathcal{F}_{\mathrm{a}}$ satisfies (a) and (b).

(ii) Every element of $\mathcal{F}_{\mathrm{a}}$ is Baire measurable. *Hint.* The set of Baire measurable functions on X satisfies (a) and (b).

(iii) If $f \in \mathcal{F}_{\mathrm{a}}$ and $g \in C_{\mathrm{c}}(X)$, then $f + g \in \mathcal{F}_{\mathrm{a}}$. *Hint.* Let $g \in C_{\mathrm{c}}(X)$. Then the set $\mathcal{F}_{\mathrm{a}} - g$ satisfy (a) and (b) and hence contains $\mathcal{F}_{\mathrm{a}}$.

(iv) If $f, g \in \mathcal{F}_{\mathrm{a}}$, then $f + g \in \mathcal{F}_{\mathrm{a}}$. *Hint.* Let $g \in \mathcal{F}_{\mathrm{a}}$. Then the set $\mathcal{F}_{\mathrm{a}} - g$ satisfy (a) and (b) and hence contains $\mathcal{F}_{\mathrm{a}}$.

(v) If $f \in \mathcal{F}_{\mathrm{a}}$ and $c \in \mathbb{R}$, then $cf \in \mathcal{F}_{\mathrm{a}}$. *Hint.* Fix a real number $c \neq 0$. Then the set $c^{-1}\mathcal{F}_{\mathrm{a}}$ satisfy (a) and (b) and hence contains $\mathcal{F}_{\mathrm{a}}$.

(vi) If $f \in \mathcal{F}_{\mathrm{a}}$ and $g \in C_{\mathrm{c}}(X)$, then $fg \in \mathcal{F}_{\mathrm{a}}$. *Hint.* Fix a real number c such that $c + g(x) > 0$ for all $x \in \mathbb{R}$. Then the set $(c + g)^{-1}\mathcal{F}_{\mathrm{a}}$ satisfy (a) and (b) and hence contains $\mathcal{F}_{\mathrm{a}}$. Now use (iv) and (v).

(vii) If $A \subset X$ such that $\chi_A \in \mathcal{F}_{\mathrm{a}}$ and $f \in \mathcal{F}_{\mathrm{a}}$, then $f\chi_A \in \mathcal{F}_{\mathrm{a}}$. *Hint.* The set $(1 + \chi_A)^{-1}\mathcal{F}_{\mathrm{a}}$ satisfy (a) and (b) and hence contains $\mathcal{F}_{\mathrm{a}}$.

(viii) The set

$$\mathcal{A} := \{A \subset X \mid \chi_A \in \mathcal{F}_{\mathrm{a}} \text{ or } \chi_{X \setminus A} \in \mathcal{F}_{\mathrm{a}}\}$$

is a σ-algebra. *Hint.* If $\chi_A, \chi_B \in \mathcal{F}_A$, then $\chi_{A \cup B} = \chi_A + \chi_B - \chi_A\chi_B \in \mathcal{F}_{\mathrm{a}}$. If $\chi_{X\setminus A}, \chi_{X\setminus B} \in \mathcal{F}_A$, then $\chi_{X\setminus(A\cup B)} = \chi_{X\setminus A}\chi_{X\setminus B} \in \mathcal{F}_{\mathrm{a}}$. If $\chi_A, \chi_{X\setminus B} \in \mathcal{F}_A$, then $\chi_{X\setminus(A\cup B)} = \chi_{(X\setminus A)\cap(X\setminus B)} = \chi_{X\setminus B} - \chi_A\chi_{X\setminus B} \in \mathcal{F}_{\mathrm{a}}$. Thus

$$A, B \in \mathcal{A} \implies A \cup B \in \mathcal{A}.$$

(ix) $\mathcal{A} = \mathcal{B}_a$. *Hint.* Let $K \in \mathcal{K}_a$. Use Urysohn's lemma (Theorem A.1) to construct a sequence $g_i \in C_c(X)$ that converges pointwise to χ_K.

(x) For every $f \in \mathcal{F}_a$ there exists a sequence of compact sets $K_i \in \mathcal{K}_a$ such that $K_i \subset K_{i+1}$ for all i and $\mathrm{supp}(f) \subset \bigcup_{i\in\mathbb{N}} K_i$. *Hint.* The set of functions $f: X \to \mathbb{R}$ with this property satisfies conditions (a) and (b).

Exercise 3.32. Show that, for every locally compact Hausdorff space X and any two Borel measures μ_0, μ_1 as in Theorem 3.8, there is a Baire set $N \subset X$ such that $\mu_0(N) = 0$ and $\mu_0(B) = \mu_1(B)$ for every Baire set $B \subset X \setminus N$.

Hint 1. Show first that

$$\mu_0(B) = \sup\{\mu_0(K) : K \in \mathcal{K}_a,\ K \subset B\}, \tag{3.29}$$

where $\mathcal{K}_a$ is as in Exercise 3.29. To see this, prove that the right-hand side of equation (3.29) defines a Borel measure μ on X that is inner regular on open sets and satisfies $\mu \le \mu_0$ and $\Lambda_\mu = \Lambda_{\mu_0}$.

Hint 2. Suppose there exists a Baire set $N \subset X$ such that $\mu_0(N) < \mu_1(N)$. Show that $\mu_1(N) = \infty$ and that N can be chosen such that $\mu_0(N) = 0$. Next show that $\chi_{X\setminus N} \in \mathcal{F}_a$, where $\mathcal{F}_a$ is as in Exercise 3.31, and deduce that $X \setminus N$ is contained in a countable union of compact sets.

Example 3.33. Let X be the Stone–Čech compactification of $\mathbb{N}$ discussed in Example 4.60 below and denote by $\mathcal{B}_a \subset \mathcal{B} \subset 2^X$ the Baire and Borel σ-algebras. Thus $B \subset X$ is a Baire set if and only if either $B \subset \mathbb{N}$ or $X \setminus B \subset \mathbb{N}$. (See part (iii) of Exercise 3.30.) For a Borel set $B \subset X$, define

$$\mu_0(B) := \sum_{n\in B} \frac{1}{n},$$

$$\mu_1(B) := \inf\left\{\mu_0(U) \;\middle|\; \begin{array}{l} B \subset U \subset X, \\ U \text{ is open} \end{array}\right\}.$$

As in Example 4.60, denote by $X_0 \subset X$ the union of all open sets $U \subset X$ with $\mu_0(U) < \infty$. Then the restriction of μ_0 to X_0 is a Radon measure, the restriction of μ_1 to X_0 is outer regular and is inner regular on open sets, and μ_0 is given by (3.5) as in Theorem 3.8. Moreover, $X_0 \setminus \mathbb{N}$ is a Baire set in X_0 and $\mu_0(X_0 \setminus \mathbb{N}) = 0$ while $\mu_1(X_0 \setminus \mathbb{N}) = \infty$. Thus we can choose $N := X_0 \setminus \mathbb{N}$ in Exercise 3.32 and μ_0 and μ_1 do not agree on the Baire σ-algebra.

Example 3.34. Let $X = \mathbb{R}^2$ be the locally compact Hausdorff space in Example 3.23 and let μ_0, μ_1 be the Borel measures of Theorem 3.15 associated to the linear functional $\Lambda: C_c(X) \to \mathbb{R}$ in that example. Then it follows from part (iv) of Exercise 3.30 that $\mu_0(B) = \mu_1(B)$ for every Baire set $B \subset X$. Thus we can choose $N = \emptyset$ in Exercise 3.32. However, there does not exist any Borel set $N \subset X$ such that $\mu_0(N) = 0$ and μ_0 agrees with μ_1 on all Borel subsets of $X \setminus N$. (A set $N \subset X$ is a Borel set with $\mu_0(N) = 0$ if and only if $N_x := \{y \in \mathbb{R} \mid (x, y) \in N\}$ is a Borel set and $m(N_x) = 0$ for all $x \in \mathbb{R}$.)

Exercise 3.35. Let Z be the disjoint union of the locally compact Hausdorff spaces X_0 in Example 3.33 and $X = \mathbb{R}^2$ in Example 3.34. Find Baire sets $B_0 \subset X_0$ and $B \subset X$ whose (disjoint) union is not a Baire set in Z.

Chapter 4

L^p spaces

This chapter discusses the Banach space $L^p(\mu)$ associated to a measure space $(X, \mathcal{A}, \mu)$ and a number $1 \le p \le \infty$. Section 4.1 introduces the inequalities of Hölder and Minkowski and Section 4.2 shows that $L^p(\mu)$ is complete. In Section 4.3 we prove that, when X is a locally compact Hausdorff space, μ is a Radon measure, and $1 \le p < \infty$, the subspace of continuous functions with compact support is dense in $L^p(\mu)$. If in addition X is second countable, it follows that $L^p(\mu)$ is separable. When $1 < p < \infty$ (or $p = 1$ and the measure space $(X, \mathcal{A}, \mu)$ is localizable) the dual space of $L^p(\mu)$ is isomorphic to $L^q(\mu)$, where $1/p + 1/q = 1$. For $p = 2$ this follows from elementary Hilbert space theory and is proved in Section 4.4. For general p the proof requires the Radon–Nikodým theorem and is deferred to Chapter 5. Some preparatory results are proved in Section 4.5.

4.1 Hölder and Minkowski

Assume throughout that $(X, \mathcal{A}, \mu)$ is a measure space and that p, q are real numbers such that

$$\frac{1}{p} + \frac{1}{q} = 1, \quad 1 < p < \infty, \quad 1 < q < \infty. \tag{4.1}$$

Then any two nonnegative real numbers a and b satisfy *Young's inequality*

$$ab \le \frac{1}{p}a^p + \frac{1}{q}b^q, \tag{4.2}$$

and equality holds in (4.2) if and only if $a^p = b^q$.

Exercise. Prove this by examining the critical points of the function

$$(0, \infty) \longrightarrow \mathbb{R}: x \longmapsto \frac{1}{p}x^p - xb.$$

Theorem 4.1. *Let $f, g\colon X \to [0,\infty]$ be measurable functions. Then f and g satisfy the* ***Hölder inequality***

$$\int_X fg\,d\mu \le \left(\int_X f^p\,d\mu\right)^{1/p}\left(\int_X g^q\,d\mu\right)^{1/q} \tag{4.3}$$

and the ***Minkowski inequality***

$$\left(\int_X (f+g)^p\,d\mu\right)^{1/p} \le \left(\int_X f^p\,d\mu\right)^{1/p} + \left(\int_X g^p\,d\mu\right)^{1/p}. \tag{4.4}$$

Proof. Define

$$A := \left(\int_X f^p\,d\mu\right)^{1/p}, \qquad B := \left(\int_X g^q\,d\mu\right)^{1/q}.$$

If $A = 0$, then $f = 0$ almost everywhere by Lemma 1.49, hence $fg = 0$ almost everywhere, and hence $\int_X fg\,d\mu = 0$ by Lemma 1.48. This proves the Hölder inequality (4.3) in the case $A = 0$. If $A = \infty$ and $B > 0$, then $AB = \infty$ and so (4.3) holds trivially. Interchanging A and B if necessary, we find that (4.3) holds whenever one of the numbers A, B is zero or infinity. Hence assume $0 < A < \infty$ and $0 < B < \infty$. Then it follows from (4.2) that

$$\begin{aligned}
\frac{\int_X fg\,d\mu}{AB} &= \int_X \frac{f}{A}\frac{g}{B}\,d\mu \\
&\le \int_X \left(\frac{1}{p}\left(\frac{f}{A}\right)^p + \frac{1}{q}\left(\frac{g}{B}\right)^q\right)d\mu \\
&= \frac{1}{p}\frac{\int_X f^p\,d\mu}{A^p} + \frac{1}{q}\frac{\int_X g^q\,d\mu}{B^q} \\
&= \frac{1}{p} + \frac{1}{q} \\
&= 1.
\end{aligned}$$

This proves the Hölder inequality. To prove the Minkowski inequality, put

$$\begin{aligned}
a &:= \left(\int_X f^p\,d\mu\right)^{1/p}, \\
b &:= \left(\int_X g^p\,d\mu\right)^{1/p}, \\
c &:= \left(\int_X (f+g)^p\,d\mu\right)^{1/p}.
\end{aligned}$$

We must prove that $c \le a + b$. This is obvious when $a = \infty$ or $b = \infty$. Hence assume $a, b < \infty$. We first show that $c < \infty$. This holds because

$$f \le (f^p + g^p)^{1/p} \quad \text{and} \quad g \le (f^p + g^p)^{1/p},$$

hence

$$f + g \le 2(f^p + g^p)^{1/p},$$

therefore

$$(f + g)^p \le 2^p (f^p + g^p),$$

and integrating this inequality and raising the integral to the power $1/p$ we obtain

$$c \le 2(a^p + b^p)^{1/p} < \infty.$$

With this understood, it follows from the Hölder inequality that

$$\begin{aligned} c^p &= \int_X f(f+g)^{p-1}\, d\mu + \int_X g(f+g)^{p-1}\, d\mu \\ &\le \left(\int_X f^p\, d\mu\right)^{1/p} \left(\int_X (f+g)^{(p-1)q}\, d\mu\right)^{1/q} \\ &\quad + \left(\int_X g^p\, d\mu\right)^{1/p} \left(\int_X (f+g)^{(p-1)q}\, d\mu\right)^{1/q} \\ &= (a+b) \left(\int_X (f+g)^p\, d\mu\right)^{1-1/p} \\ &= (a+b)c^{p-1}. \end{aligned}$$

Here we have used the identity $pq - q = p$. It follows that $c \le a + b$ and this proves Theorem 4.1. □

Exercise 4.2. (i) Assume $0 < \int_X f^p\, d\mu < \infty$ and $0 < \int_X g^q\, d\mu < \infty$. Prove that equality holds in (4.3) if and only if there exists a constant $\alpha > 0$ such that $g^q = \alpha f^p$ almost everywhere. *Hint.* Use the proof of the Hölder inequality and the fact that equality holds in (4.2) if and only $a^p = b^q$.

(ii) Assume $0 < \int_X f^p\, d\mu < \infty$ and $0 < \int_X g^p\, d\mu < \infty$. Prove that equality holds in (4.4) if and only if there is a real number $\lambda > 0$ such that $g = \lambda f$ almost everywhere. *Hint.* Use part (i) and the proof of the Minkowski inequality.

4.2 The Banach space $L^p(\mu)$

Definition 4.3. Let $(X, \mathcal{A}, \mu)$ be a measure space and let $1 \le p < \infty$. Consider a measurable function $f : X \to \overline{\mathbb{R}}$. The *$L^p$-norm of f* is the number

$$\|f\|_p := \left(\int_X |f|^p \, d\mu \right)^{1/p}. \tag{4.5}$$

A function $f : X \to \mathbb{R}$ is called *p-integrable* or an *L^p-function* if it is measurable and $\|f\|_p < \infty$. The space of L^p-functions is denoted by

$$\mathcal{L}^p(\mu) := \{ f : X \to \mathbb{R} \mid f \text{ is } \mathcal{A}\text{-measurable and } \|f\|_p < \infty \}. \tag{4.6}$$

It follows from the Minkowski inequality (4.4) that the sum of two L^p-functions is again an L^p-function, and hence $\mathcal{L}^p(\mu)$ is a real vector space. Moreover, the function

$$\mathcal{L}^p(\mu) \longrightarrow [0, \infty) : f \longmapsto \|f\|_p$$

satisfies the triangle inequality

$$\|f + g\|_p \le \|f\|_p + \|g\|_p$$

for all $f, g \in \mathcal{L}^p(\mu)$ by (4.4), and

$$\|\lambda f\|_p = |\lambda| \, \|f\|_p ,$$

for all $\lambda \in \mathbb{R}$ and $f \in \mathcal{L}^p(\mu)$ by definition. However, in general $\|\cdot\|_p$ is not a norm on $\mathcal{L}^p(\mu)$ because $\|f\|_p = 0$ if and only if $f = 0$ almost everywhere by Lemma 1.49. We can turn the space $\mathcal{L}^p(\mu)$ into a normed vector space by identifying two functions $f, g \in \mathcal{L}^p(\mu)$ whenever they agree almost everywhere. Thus we introduce the equivalence relation

$$f \overset{\mu}{\sim} g \iff f = g \quad \mu\text{-almost everywhere.} \tag{4.7}$$

Denote the equivalence class of a function $f \in \mathcal{L}^p(\mu)$ under this equivalence relation by $[f]_\mu$ and the quotient space by

$$L^p(\mu) := \mathcal{L}^p(\mu) / \overset{\mu}{\sim} . \tag{4.8}$$

This is again a real vector space. (For $p = 1$ see Definition 1.51.) The L^p-norm in (4.5) depends only on the equivalence class of f and so the map

$$L^p(\mu) \longrightarrow [0,\infty): [f]_\mu \longmapsto \|f\|_p$$

is well defined. It is a norm on $L^p(\mu)$ by Lemma 1.49. Thus we have defined the normed vector space $L^p(\mu)$ for $1 \le p < \infty$. It is often convenient to abuse notation and write $f \in L^p(\mu)$ instead of $[f]_\mu \in L^p(\mu)$, always bearing in mind that then f denotes an equivalence class of p-integrable functions. If $(X, \mathcal{A}^*, \mu^*)$ denotes the completion of $(X, \mathcal{A}, \mu)$, it follows as in Corollary 1.56 that $L^p(\mu)$ is naturally isomorphic to $L^p(\mu^*)$.

Remark 4.4. Assume $1 < p < \infty$ and let $f, g \in \mathcal{L}^p(\mu)$ such that

$$\|f + g\|_p = \|f\|_p + \|g\|_p, \quad \|f\|_p \neq 0.$$

Then it follows from part (ii) of Exercise 4.2 that there exists a real number $\lambda \ge 0$ such that $g = \lambda f$ almost everywhere.

Example 4.5. If $(\mathbb{R}^n, \mathcal{A}, m)$ is the Lebesgue measure space we write

$$L^p(\mathbb{R}^n) := L^p(m).$$

(See Definition 2.2 and Definition 2.11.)

Example 4.6. If $\mu: 2^{\mathbb{N}} \to [0,\infty]$ is the counting measure we write

$$\ell^p := L^p(\mu).$$

Thus the elements of ℓ^p are sequences $(x_n)_{n\in\mathbb{N}}$ of real numbers such that

$$\|(x_n)\|_p := \Big(\sum_{p=1}^{\infty} |x_n|^p\Big)^{1/p} < \infty.$$

If we define

$$f: \mathbb{N} \longrightarrow \mathbb{R}$$

by

$$f(n) := x_n \quad \text{for } n \in \mathbb{N},$$

then

$$\int_{\mathbb{N}} |f|^p \, d\mu = \sum_{p=1}^{\infty} |x_n|^p.$$

For $p = \infty$ there is a similar normed vector space $L^\infty(\mu)$, defined next.

Definition 4.7. Let $(X, \mathcal{A}, \mu)$ be a measure space and let $f: X \to [0, \infty]$ be a measurable function. The *essential supremum* of f is the number $\operatorname{ess\,sup} f \in [0, \infty]$ defined by

$$\operatorname{ess\,sup} f := \inf\{c \in [0, \infty] \mid f \leq c \text{ almost everywhere}\} \tag{4.9}$$

A function $f: X \to \mathbb{R}$ is called an L^∞*-function* if it is measurable and

$$\|f\|_\infty := \operatorname{ess\,sup} |f| < \infty \tag{4.10}$$

The set of L^∞-functions on X will be denoted by

$$\mathcal{L}^\infty(\mu) := \{f: X \to \mathbb{R} \mid f \text{ is measurable and } \operatorname{ess\,sup}|f| < \infty\}$$

and the quotient space by the equivalence relation (4.7) by

$$L^\infty(\mu) := \mathcal{L}^\infty(\mu)/\overset{\mu}{\sim}. \tag{4.11}$$

This is a normed vector space with the norm defined by (4.10), which depends only on the equivalence class of f.

Lemma 4.8. *For every $f \in \mathcal{L}^\infty(\mu)$ there exists a measurable set $E \in \mathcal{A}$ such that*

$$\mu(E) = 0 \quad \textit{and} \quad \sup_{X \setminus E} |f| = \|f\|_\infty.$$

Proof. The set

$$E_n := \{x \in X \mid |f(x)| > \|f\|_\infty + 1/n\}$$

has measure zero for all n. Hence

$$E := \bigcup_{n \in \mathbb{N}} E_n$$

is also a set of measure zero and $|f(x)| \leq \|f\|_\infty$ for all $x \in X \setminus E$. Therefore $\sup_{X \setminus E} |f| = \|f\|_\infty$. This proves Lemma 4.8. □

Theorem 4.9. *$L^p(\mu)$ is a Banach space for $1 \leq p \leq \infty$.*

Proof. Assume first that $1 \leq p < \infty$. In this case the argument is a refinement of the proof of Theorem 1.52 and Theorem 1.53 for the case $p = 1$. Let $f_n \in \mathcal{L}^p(\mu)$ be a Cauchy sequence with respect to the norm (4.5). Choose a sequence of positive integers $n_1 < n_2 < n_3 < \cdots$ such that

$$\|f_{n_i} - f_{n_{i+1}}\|_p < 2^{-i}$$

for all $i \in \mathbb{N}$.

Define

$$g_k := \sum_{i=1}^{k} |f_{n_{i+1}} - f_{n_i}|, \quad g := \sum_{i=1}^{\infty} |f_{n_{i+1}} - f_{n_i}| = \lim_{k\to\infty} g_k.$$

Then it follows from Minkowski's inequality (4.4) that

$$\|g_k\|_p \le \sum_{i=1}^{k} \|f_{n_i} - f_{n_{i+1}}\|_p < \sum_{i=1}^{k} 2^{-i} \le 1$$

for all $k \in \mathbb{N}$. Moreover,

$$g_k^p \le g_{k+1}^p \quad \text{for all } k \in \mathbb{N}$$

and the sequence of functions $g_k^p : X \to [0, \infty]$ converges pointwise to the integrable function g^p. Hence it follows from the Lebesgue monotone convergence theorem (Theorem 1.37) that

$$\|g\|_p = \lim_{k\to\infty} \|g_k\|_p \le 1.$$

Hence, by Lemma 1.47, there is a measurable set $E \in \mathcal{A}$ such that

$$\mu(E) = 0, \quad g(x) < \infty \text{ for all } x \in X \setminus E.$$

Hence the series $\sum_{i=1}^{\infty}(f_{n_{i+1}}(x) - f_{n_i}(x))$ converges absolutely for $x \in X \setminus E$. Define the function

$$f : X \longrightarrow \mathbb{R}$$

by

$$f(x) := f_{n_1}(x) + \sum_{i=1}^{\infty}(f_{n_{i+1}}(x) - f_{n_i}(x))$$

for $x \in X \setminus E$, and by $f(x) := 0$ for $x \in E$. Then the sequence

$$f_{n_k}\chi_{X\setminus E} = f_{n_1}\chi_{X\setminus E} + \sum_{i=1}^{k\ 1}(f_{n_{i+1}} - f_{n_i})\chi_{X\setminus E}$$

converges pointwise to f. Hence f is $\mathcal{A}$-measurable by Theorem 1.24.

We must prove that $f \in \mathcal{L}^p(\mu)$ and that

$$\lim_{n\to\infty} \|f - f_n\|_p = 0.$$

To see this, fix a constant $\varepsilon > 0$. Then there exists an integer $n_0 \in \mathbb{N}$ such that

$$\|f_n - f_m\|_p < \varepsilon \quad \text{for all } n, m \ge n_0.$$

By Fatou's lemma (Theorem 1.41), this implies

$$\begin{aligned}\int_X |f_n - f|^p \, d\mu &= \int_X \liminf_{k\to\infty} |f_n - f_{n_k}\chi_{X\setminus E}|^p \, d\mu \\ &\le \liminf_{k\to\infty} \int_X |f_n - f_{n_k}\chi_{X\setminus E}|^p \, d\mu \\ &= \liminf_{k\to\infty} \int_X |f_n - f_{n_k}|^p \, d\mu \\ &\le \varepsilon^p\end{aligned}$$

for all $n \ge n_0$. Hence $\|f_n - f\|_p \le \varepsilon$ for all $n \ge n_0$, and so

$$\|f\|_p \le \|f_{n_0}\|_p + \|f - f_{n_0}\|_p \le \|f_{n_0}\|_p + \varepsilon < \infty.$$

Thus $f \in \mathcal{L}^p(\mu)$ and $\lim_{n\to\infty} \|f - f_n\|_p = 0$ as claimed. This shows that $L^p(\mu)$ is a Banach space for $p < \infty$.

The proof for $p = \infty$ is simpler. Let $f_n \in \mathcal{L}^\infty(\mu)$ such that the $[f_n]_\mu$ form a Cauchy sequence in $L^\infty(\mu)$. Then there is a set $E \in \mathcal{A}$ such that

$$\mu(E) = 0, \quad \|f_n\|_\infty = \sup_{X\setminus E} |f_n|, \quad \|f_m - f_n\|_\infty = \sup_{X\setminus E} |f_m - f_n| \tag{4.12}$$

for all $m, n \in \mathbb{N}$. To see this, use Lemma 4.8 to find null sets $E_n, E_{m,n} \in \mathcal{A}$ such that

$$\sup_{X\setminus E_n} |f_n| = \|f_n\|_\infty \quad \text{and} \quad \sup_{X\setminus E_{m,n}} |f_m - f_n| = \|f_m - f_n\|_\infty \quad \text{for all } m, n \in \mathbb{N}.$$

Then the union E of the sets E_n and $E_{m,n}$ is measurable and satisfies (4.12). Since $[f_n]_\mu$ is a Cauchy sequence in $L^\infty(\mu)$, we have

$$\lim_{n\to\infty} \varepsilon_n = 0, \quad \varepsilon_n := \sup_{m\ge n} \|f_m - f_n\|_\infty .$$

Since $|f_m(x) - f_n(x)| \le \varepsilon_n$ for all $m \ge n$ and all $x \in X \setminus E$, it follows that $(f_n(x))_{n\in\mathbb{N}}$ is Cauchy sequence in $\mathbb{R}$ and hence converges for every $x \in X \setminus E$. Define

$$f : X \to \mathbb{R}$$

by

$$f(x) := \begin{cases} \lim_{n\to\infty} f_n(x) & \text{for } x \in X \setminus E, \\ 0 & \text{for } x \in E. \end{cases}$$

Then

$$\|f - f_n\|_\infty \leq \sup_{x\in X\setminus E} |f(x) - f_n(x)| = \sup_{x\in X\setminus E} \lim_{m\to\infty} |f_m(x) - f_n(x)| \leq \varepsilon_n$$

for all $n \in \mathbb{N}$. Hence

$$\|f\|_\infty \leq \|f_1\|_\infty + \varepsilon_1 < \infty \quad \text{and} \quad \lim_{n\to\infty} \|f - f_n\|_\infty = 0.$$

This proves Theorem 4.9. □

Corollary 4.10. *Let $(X, \mathcal{A}, \mu)$ be a measure space and, given $1 \leq p \leq \infty$, let $f \in \mathcal{L}^p(\mu)$ and $f_n \in \mathcal{L}^p(\mu)$ be a sequence such that*

$$\lim_{n\to\infty} \|f_n - f\|_p = 0.$$

- *If $p = \infty$, then f_n converges almost everywhere to f.*
- *If $p < \infty$, then there exists a subsequence f_{n_i} that converges almost everywhere to f.*

Proof. For $p = \infty$ this follows directly from the definitions. For $p < \infty$ choose a sequence of integers $0 < n_1 < n_2 < n_3 < \cdots$ such that $\|f_{n_i} - f_{n_{i+1}}\|_p < 2^{-i}$ for all $i \in \mathbb{N}$. Then the proof of Theorem 4.9 shows that f_{n_i} converges almost everywhere to an L^p-function g such that $\lim_{n\to\infty} \|f_n - g\|_p = 0$. Since the limit is unique in $L^p(\mu)$, it follows that $g = f$ almost everywhere. □

4.3 Separability

Definition 4.11. Let X be a topological space. A subset $S \subset X$ is said to be *dense* (*in* X) if its closure is equal to X or, equivalently, $U \cap S \neq \emptyset$ for every nonempty open set $U \subset X$. A subset $S \subset X$ of a metric space is dense if and only if every element of X is the limit of a sequence in S. The topological space X is called *separable* if it admits a countable dense subset.

Every second countable topological space is separable and first countable (see Lemma 3.21). The Sorgenfrey line is separable and first countable, but is not second countable (see Example 3.22). A metric space is separable if and only if it is second countable. (If S is a countable dense subset, then the balls with rational radii centered at the points of S form a basis of the topology.) The Euclidean space $X = \mathbb{R}^n$ with its standard topology is separable ($\mathbb{Q}^n$ is a countable dense subset) and hence is second countable. The next lemma gives a condition for a linear subspace to be dense in $L^p(\mu)$.

Lemma 4.12. *Let $(X, \mathcal{A}, \mu)$ be a measure space and let $1 \le p < \infty$. Let $\mathcal{X}$ be a linear subspace of $L^p(\mu)$ such that $[\chi_A]_\mu \in \mathcal{X}$ for every measurable set $A \in \mathcal{A}$ with $\mu(A) < \infty$. Then $\mathcal{X}$ is dense in $L^p(\mu)$.*

Proof. Let $\mathcal{Y}$ denote the closure of $\mathcal{X}$ in $L^p(\mu)$. Then $\mathcal{Y}$ is a closed linear subspace of $L^p(\mu)$. We prove in three steps that $\mathcal{Y} = L^p(\mu)$.

Step 1. *If $s \in \mathcal{L}^p(\mu)$ is a measurable step function, then $[s]_\mu \in \mathcal{Y}$.*

Write $s = \sum_{i=1}^{\ell} \alpha_i \chi_{A_i}$ where $\alpha_i \in \mathbb{R} \setminus \{0\}$ and $A_i = s^{-1}(\alpha_i) \in \mathcal{A}$. Then

$$|\alpha_i|^p \mu(A_i) = \int_X |\alpha_i \chi_{A_i}|^p \, d\mu \le \int_X |s|^p \, d\mu < \infty$$

and hence $\mu(A_i) < \infty$ for all i. This implies $[\chi_{A_i}]_\mu \in \mathcal{Y}$ for all i. Since $\mathcal{Y}$ is a linear subspace of $L^p(\mu)$, it follows that $[s]_\mu \in \mathcal{Y}$.

Step 2. *If $f \in \mathcal{L}^p(\mu)$ and $f \ge 0$, then $[f]_\mu \in \mathcal{Y}$.*

By Theorem 1.26, there is a sequence of measurable step functions $s_i : X \to \mathbb{R}$ such that $0 \le s_1 \le s_2 \le \cdots$ and s_i converges pointwise to f. Then $s_i \in \mathcal{L}^p(\mu)$ and hence $[s_i]_\mu \in \mathcal{Y}$ for all i by Step 1. Moreover, $|f - s_i|^p \le f^p$, f^p is integrable, and $|f - s_i|^p$ converges pointwise to zero. Hence it follows from the Lebesgue dominated convergence theorem (Theorem 1.45) that $\lim_{i\to\infty} \|f - s_i\|_p = 0$. Since $[s_i]_\mu \in \mathcal{Y}$ for all i and $\mathcal{Y}$ is a closed subspace of $L^p(\mu)$, it follows that $[f]_\mu \in \mathcal{Y}$.

Step 3. $\mathcal{Y} = L^p(\mu)$.

Let $f \in \mathcal{L}^p(\mu)$. Then $f^\pm \in \mathcal{L}^p(\mu)$, hence $[f^\pm]_\mu \in \mathcal{Y}$ by Step 2, and hence $[f]_\mu = [f^+]_\mu - [f^-]_\mu \in \mathcal{Y}$. This proves Step 3 and Lemma 4.12. □

Standing assumption. Assume throughout the remainder of this section that $(X, \mathcal{U})$ is a locally compact Hausdorff space, $\mathcal{B} \subset 2^X$ is its Borel σ-algebra, $\mu : \mathcal{B} \to [0, \infty]$ is a Borel measure, and fix a constant $1 \le p < \infty$.

Theorem 4.13. *If X is second countable, then $L^p(\mu)$ is separable.*

Proof. See page 144. □

Example 4.14. If X is an uncountable set with the discrete topology $\mathcal{U} = 2^X$ and $\mu : 2^X \to [0, \infty]$ is the counting measure, then X is not second countable and $\mathcal{L}^p(\mu) = L^p(\mu)$ is not separable.

Theorem 4.15. *Assume μ is outer regular and is inner regular on open sets. Define*

$$\mathcal{S}_c(X) := \left\{ s\colon X \to \mathbb{R} \;\middle|\; \begin{array}{l} s \text{ is a Borel measurable step function} \\ \text{and } \operatorname{supp}(s) \text{ is a compact subset of } X \end{array} \right\}. \tag{4.13}$$

Then the linear subspaces $\mathcal{S}_c(X)/\overset{\mu}{\sim}$ and $C_c(X)/\overset{\mu}{\sim}$ are dense in $L^p(\mu)$. This continues to hold when μ is a Radon measure.

Proof. See page 145. □

Example 4.16. Let $(X, \mathcal{U})$ be the compact Hausdorff space constructed in Example 3.6, let $\mu\colon \mathcal{A} \to [0,1]$ be the Dieudonné measure constructed in that example, let $\delta\colon 2^X \to [0,1]$ be the Dirac measure at the point $\kappa \in X$, and define $\mu' := \mu|_{\mathcal{B}} + \delta|_{\mathcal{B}}\colon \mathcal{B} \to [0,2]$. Then $L^p(\mu')$ is a 2-dimensional vector space and $C_c(X)/\overset{\mu'}{\sim}$ is a 1-dimensional subspace of $L^p(\mu')$ and hence is not dense. Thus the regularity assumption on μ cannot be removed in Theorem 4.15.

Lemma 4.17. *Assume $\mu = \mu_1$ is outer regular and is inner regular on open sets. Let $\mu_0\colon \mathcal{B} \to [0,\infty]$ be the unique Radon measure such that $\Lambda_{\mu_1} = \Lambda_{\mu_0}$. Then $\mathcal{L}^p(\mu_1) \subset \mathcal{L}^p(\mu_0)$ and the linear map*

$$L^p(\mu_1) \longrightarrow L^p(\mu_0)\colon [f]_{\mu_1} \longmapsto [f]_{\mu_0} \tag{4.14}$$

is a Banach space isometry.

Proof. Since $\mu_0(B) \le \mu_1(B)$ for all $B \in \mathcal{B}$ by Theorem 3.15, it follows that

$$\int_X |f|^p \, d\mu_0 \le \int_X |f|^p \, d\mu_1$$

for every Borel measurable function $f\colon X \to \mathbb{R}$. Hence $\mathcal{L}^p(\mu_1) \subset \mathcal{L}^p(\mu_0)$. We prove that

$$\int_X |f|^p \, d\mu_0 = \int_X |f|^p \, d\mu_1 \quad \text{for all } f \in \mathcal{L}^p(\mu_1). \tag{4.15}$$

Thus the map (4.14) is injective and has a closed image. To prove (4.15), define

$$E_\varepsilon := \{x \in X \mid |f(x)| > \varepsilon\} \quad \text{for } \varepsilon > 0.$$

Then $\mu_1(E_\varepsilon) < \infty$ and hence μ_1 and μ_0 agree on all Borel subsets of E_ε by Lemma 3.7. This implies

$$\int_{E_\varepsilon} |f|^p \, d\mu_0 = \int_{E_\varepsilon} |f|^p \, d\mu_1,$$

and (4.15) follows by taking the limit $\varepsilon \to 0$.

We prove that the map (4.14) is surjective. Denote its image by $\mathcal{X}$. This is a closed linear subspace of $L^p(\mu_0)$, by what we have just proved. Let $B \in \mathcal{B}$ such that $\mu_0(B) < \infty$. By (3.5), there is a sequence of compact sets $K_i \subset B$ such that $K_i \subset K_{i+1}$ and $\mu_1(K_i) = \mu_0(K_i) > \mu_0(B) - 2^{-i}$ for all i. Define

$$A := \bigcup_{i \in \mathbb{N}} K_i \subset B.$$

Then

$$\mu_1(A) = \mu_0(A) = \lim_{i \to \infty} \mu_0(K_i) = \mu_0(B).$$

This implies $\chi_A \in \mathcal{L}^p(\mu_1)$ and $[\chi_B]_{\mu_0} = [\chi_A]_{\mu_0} \in \mathcal{X}$. By Lemma 4.12, it follows that $\mathcal{X} = L^p(\mu_0)$ and this proves Lemma 4.17. □

Proof of Theorem 4.13. Let $\mathcal{V} \subset \mathcal{U}$ be a countable basis for the topology. Assume without loss of generality that $\overline{V}$ is compact for all $V \in \mathcal{V}$. (If $\mathcal{W} \subset \mathcal{U}$ is any countable basis for the topology, then the set $\mathcal{V} := \{V \in \mathcal{W} \mid \overline{V} \text{ is compact}\}$ is also a countable basis for the topology by Lemma A.3.) Choose a bijection

$$\mathbb{N} \longrightarrow \mathcal{V} : i \longmapsto V_i$$

and let

$$\mathcal{I} := \{I \subset \mathbb{N} \mid \#I < \infty\}$$

be the set of finite subsets of $\mathbb{N}$. Then the map

$$\mathcal{I} \longrightarrow \mathbb{N} : I \longmapsto \sum_{i \in I} 2^{i-1}$$

is a bijection, so the set $\mathcal{I}$ is countable. For $I \in \mathcal{I}$ define

$$V_I := \bigcup_{i \in I} V_i.$$

Define the set $\mathcal{V} \subset \mathcal{L}^p(\mu)$ by

$$\mathcal{V} := \Big\{ s = \sum_{j=1}^{\ell} \alpha_j \chi_{V_{I_j}} \,\Big|\, \ell \in \mathbb{N} \text{ and } \alpha_j \in \mathbb{Q},\ I_j \in \mathcal{I} \text{ for } j = 1, \dots, \ell \Big\}.$$

This set is contained in $\mathcal{L}^p(\mu)$ because $\overline{V}$ is compact for all $V \in \mathcal{V}$. It is countable and its closure

$$\mathcal{X} := \overline{\mathcal{V}}$$

in $L^p(\mu)$ is a closed linear subspace. By Lemma 4.12, it suffices to prove that $[\chi_B]_\mu \in \mathcal{X}$ for every $B \in \mathcal{B}$ with $\mu(B) < \infty$. To see this, fix a Borel set $B \in \mathcal{B}$ with $\mu(B) < \infty$ and a constant $\varepsilon > 0$. Since X is second countable, every open subset of X is σ-compact (Lemma 3.21). Hence μ is regular by Theorem 3.18. Therefore, there exist a compact set $K \subset X$ and an open set $U \subset X$ such that

$$K \subset B \subset U, \quad \mu(U \setminus K) < \varepsilon^p.$$

Define

$$\mathcal{I} := \{i \in \mathbb{N} \mid V_i \subset U\}.$$

Since $\mathcal{V}$ is a basis of the topology, we have $K \subset U = \bigcup_{i \in \mathcal{I}} V_i$. Since K is compact, there is a finite set $I \subset \mathcal{I}$ such that

$$K \subset V_I \subset U.$$

Next, since $\chi_B - \chi_{V_I}$ vanishes on $X \setminus (U \setminus K)$ and $|\chi_B - \chi_{V_I}| \le 1$,

$$\|\chi_B - \chi_{V_I}\|_p \le \mu(U \setminus K)^{1/p} < \varepsilon.$$

Since $\chi_{V_I} \in \mathcal{V}$ and $\varepsilon > 0$ was chosen arbitrary, it follows that $[\chi_B]_\mu \in \mathcal{X} = \overline{\mathcal{V}}$. This proves Theorem 4.13. □

Proof of Theorem 4.15. By Corollary 3.16 and Lemma 4.17, it suffices to consider the case where μ is outer regular and is inner regular on open sets. Define

$$\begin{aligned} \mathcal{S} := \{[f]_\mu \in L^p(\mu) \ &| \text{ for all } \varepsilon > 0, \text{ there exists } s \in \mathcal{S}_c(X) \\ &\text{ such that } \|f - s\|_p < \varepsilon\}, \end{aligned}$$

and

$$\begin{aligned} \mathcal{C} := \{[f]_\mu \in L^p(\mu_1) \ &| \text{ for all } \varepsilon > 0, \text{ there exists } g \in C_c(X) \\ &\text{ such that } \|f - g\|_p < \varepsilon\}. \end{aligned}$$

We must prove that $L^p(\mu) = \mathcal{S} = \mathcal{C}$. Since $\mathcal{S}$ and $\mathcal{C}$ are closed linear subspaces of $L^p(\mu)$, it suffices to prove that $[\chi_B]_\mu \in \mathcal{S} \cap \mathcal{C}$ for every Borel set $B \in \mathcal{B}$ with $\mu(B) < \infty$ by Lemma 4.12. Let $B \in \mathcal{B}$ with $\mu(B) < \infty$ and let $\varepsilon > 0$. By Lemma 3.7, there is a compact set $K \subset X$ and an open set $U \subset X$ such that $K \subset B \subset U$ and $\mu(U \setminus K) < \varepsilon^p$. By Urysohn's lemma (Theorem A.1), there is a function $f \in C_c(X)$ such that $0 \le f \le 1$, $f|_K \equiv 1$, and $\mathrm{supp}(f) \subset U$. This implies $0 \le f - \chi_K \le \chi_{U \setminus K}$ and $0 \le \chi_B - \chi_K \le \chi_{U \setminus K}$. Hence,

$$\|\chi_B - \chi_K\|_p \le \|\chi_{U\setminus K}\|_p = \mu(U \setminus K)^{1/p} < \varepsilon$$

and likewise $\|f - \chi_K\|_p < \varepsilon$. By Minkowski's inequality (4.4), this implies

$$\|\chi_B - f\|_p \le \|\chi_B - \chi_K\|_p + \|\chi_K - f\|_p < 2\varepsilon.$$

This shows that $[\chi_B]_\mu \in \mathcal{S} \cap \mathcal{C}$. This proves Theorem 4.15. □

Remark 4.18. The reader may wonder whether Theorem 4.15 continues to hold for all Borel measures $\mu\colon \mathcal{B} \to [0,\infty]$ that are inner regular on open sets. To answer this question one can try to proceed as follows. Let μ_0, μ_1 be the Borel measures on X in Theorem 3.15 that satisfy $\Lambda_{\mu_0} = \Lambda_{\mu_1} = \Lambda_\mu$. Then μ_0 is a Radon measure, μ_1 is outer regular and is inner regular on open sets, and $\mu_0(B) \le \mu(B) \le \mu_1(B)$ for all $B \in \mathcal{B}$. Thus $\mathcal{L}^p(\mu_1) \subset \mathcal{L}^p(\mu) \subset \mathcal{L}^p(\mu_0)$ and one can consider the maps

$$L^p(\mu_1) \longrightarrow L^p(\mu) \longrightarrow L^p(\mu_0).$$

Their composition is a Banach space isometry by Lemma 4.17. The question is now whether or not the first map $L^p(\mu_1) \to L^p(\mu)$ is surjective or, equivalently, whether the second map $L^p(\mu) \to L^p(\mu_0)$ is injective. If this holds, then the subspace $C_c(X)/\overset{\mu}{\sim}$ is dense in $L^p(\mu)$, otherwise it is not. The proof of Lemma 4.17 shows that the answer is affirmative if and only if every Borel set $B \subset X$ with $\mu_0(B) < \mu(B)$ satisfies $\mu(B) = \infty$. Thus the quest for a counterexample can be rephrased as follows.

Question. *Do there exist a locally compact Hausdorff space* $(X, \mathcal{U})$ *and Borel measures*

$$\mu_0, \mu_1, \mu\colon \mathcal{B} \longrightarrow [0,\infty]$$

on its Borel σ*-algebra* $\mathcal{B} \subset 2^X$ *such that all three measures are inner regular on open sets,* μ_1 *is outer regular,* μ_0 *is given by* (3.5),

$$\mu_0(B) \le \mu(B) \le \mu_1(B)$$

for all Borel sets $B \in \mathcal{B}$, *and*

$$0 = \mu_0(B) < \mu(B) < \mu_1(B) = \infty$$

for some Borel set $B \in \mathcal{B}$?

This leads to deep problems in set theory. A *probability measure* on a measurable space $(X, \mathcal{A})$ is a measure $\mu: \mathcal{A} \to [0, 1]$ such that $\mu(X) = 1$. A measure $\mu: \mathcal{A} \to [0, \infty]$ is called *nonatomic* if countable sets have measure zero. Now consider the measure on $X = \mathbb{R}^2$ in Exercise 3.23 with $\mu_0(\mathbb{R} \times \{0\}) = 0$ and $\mu_1(\mathbb{R} \times \{0\}) = \infty$, and define $\iota: \mathbb{R} \to \mathbb{R}^2$ by $\iota(x) := (x, 0)$. If there is a nonatomic probability measure $\mu: 2^{\mathbb{R}} \to [0, 1]$, then the measure $\mu_0 + \iota_* \mu$ provides a positive answer to the above question, and thus Theorem 4.15 would not extend to all Borel measures that are inner regular on open sets. The question of the existence of a nonatomic probability measure is related to the continuum hypothesis. The *generalized continuum hypothesis* asserts that, if X is any set, then each subset of 2^X whose cardinality is strictly larger than that of X admits a bijection to 2^X. It is independent of the other axioms of set theory and implies that nonatomic probability measures $\mu: 2^X \to [0, 1]$ do not exist on any set X. This is closely related to the theory of *measure-free cardinals.* (See Fremlin [4, Section 4.3.7].)

4.4 Hilbert spaces

This section introduces some elementary Hilbert space theory. It serves two purposes. First, it shows that the Hilbert space $L^2(\mu)$ is isomorphic to its own dual space. Second, this result in turn will be used in the proof of the Radon–Nikodým theorem for σ-finite measure spaces in the next chapter.

Definition 4.19. Let H be a real vector space. A bilinear map

$$H \times H \longrightarrow \mathbb{R}: (x, y) \longmapsto \langle x, y \rangle \tag{4.16}$$

is called an *inner product* if it is *symmetric,* i.e., $\langle x, y \rangle = \langle y, x \rangle$ for all $x, y \in H$ and *positive definite,* i.e., $\langle x, x \rangle > 0$ for all $x \in H \setminus \{0\}$. The *norm* associated to an inner product (4.16) is the function

$$H \longrightarrow \mathbb{R}: x \longmapsto \|x\| := \sqrt{\langle x, x \rangle}. \tag{4.17}$$

Lemma 4.20. *Let H be a real vector space equipped with an inner product* (4.16) *and the associated norm* (4.17). *The inner product and norm satisfy the* ***Cauchy–Schwarz inequality***

$$|\langle x, y \rangle| \leq \|x\| \, \|y\| \tag{4.18}$$

and the ***triangle inequality***

$$\|x + y\| \leq \|x\| + \|y\| \tag{4.19}$$

for all $x, y \in H$. Thus (4.17) *is a norm on H.*

Proof. The Cauchy–Schwarz inequality is obvious when $x = 0$ or $y = 0$. Hence assume $x \neq 0$ and $y \neq 0$, and put $\xi := \|x\|^{-1} x$ and $\eta := \|y\|^{-1} y$. Then $\|\xi\| = \|\eta\| = 1$. Hence

$$0 \le \|\eta - \langle \xi, \eta\rangle \xi\|^2 = \langle \eta, \eta - \langle \xi, \eta\rangle \xi\rangle = 1 - \langle \xi, \eta\rangle^2.$$

This implies $|\langle \xi, \eta\rangle| \le 1$ and hence $|\langle x, y\rangle| \le \|x\| \, \|y\|$. In turn it follows from the Cauchy–Schwarz inequality that

$$\begin{aligned}\|x + y\|^2 &= \|x\|^2 + 2\langle x, y\rangle + \|y\|^2 \\ &\le \|x\|^2 + 2\,\|x\|\,\|y\| + \|y\|^2 \\ &= (\|x\| + \|y\|)^2.\end{aligned}$$

This proves the triangle inequality (4.19) and Lemma 4.20. □

Definition 4.21. An inner product space $(H, \langle\cdot,\cdot\rangle)$ is called a *Hilbert space* if the norm (4.17) is complete, i.e., every Cauchy sequence in H converges.

Example 4.22. Let $(X, \mathcal{A}, \mu)$ be a measure space. Then $H := L^2(\mu)$ is a Hilbert space. The inner product is induced by the bilinear map

$$\mathcal{L}^2(\mu) \times \mathcal{L}^2(\mu) \longrightarrow \mathbb{R}\colon (f, g) \longmapsto \langle f, g\rangle := \int_X fg \, d\mu. \tag{4.20}$$

It is well defined because the product of two L^2-functions $f, g\colon X \to \mathbb{R}$ is integrable by (4.3) with $p = q = 2$. That it is bilinear follows from Theorem 1.44, and that it is symmetric is obvious. In general, it is not positive definite. However, it descends to a symmetric bilinear form

$$L^2(\mu) \times L^2(\mu) \longrightarrow \mathbb{R}\colon ([f]_\mu, [g]_\mu) \longmapsto \langle f, g\rangle = \int_X fg \, d\mu. \tag{4.21}$$

by Lemma 1.48 which is positive definite by Lemma 1.49. Hence (4.21) is an inner product on $L^2(\mu)$. It is called the L^2 *inner product.* The norm associated to this inner product is

$$L^2(\mu) \longrightarrow \mathbb{R}\colon [f]_\mu \longmapsto \|f\|_2 = \left(\int_X f^2 \, d\mu\right)^{1/2} = \sqrt{\langle f, f\rangle}. \tag{4.22}$$

This is the L^2-norm in (4.5) with $p = 2$. By Theorem 4.9, $L^2(\mu)$ is complete with the norm (4.22) and hence is a Hilbert space.

Definition 4.23. Let $(V, \|\cdot\|)$ be a normed vector space. A linear functional $\Lambda\colon V \to \mathbb{R}$ is called *bounded* if there exists a constant $c \geq 0$ such that

$$|\Lambda(x)| \leq c\,\|x\| \quad \text{for all } x \in V.$$

The *norm of a bounded linear functional* $\Lambda\colon V \to \mathbb{R}$ is the smallest such constant c and will be denoted by

$$\|\Lambda\| := \sup_{0 \neq x \in V} \frac{|\Lambda(x)|}{\|x\|}. \tag{4.23}$$

The set of bounded linear functionals on V is denoted by V^* and is called the *dual space of* V.

Exercise 4.24. Prove that a linear functional on a normed vector space is bounded if and only if it is continuous.

Exercise 4.25. Let $(V, \|\cdot\|)$ be a normed vector space. Prove that the dual space V^* with the norm (4.23) is a Banach space. (See Example 1.11.)

Theorem 4.26 (Riesz). *Let H be a Hilbert space and let $\Lambda\colon H \to \mathbb{R}$ be a bounded linear functional. Then there exists a unique element $y \in H$ such that*

$$\Lambda(x) = \langle y, x\rangle \quad \textit{for all } x \in H. \tag{4.24}$$

This element $y \in H$ satisfies

$$\|y\| = \sup_{0 \neq x \in H} \frac{|\langle y, x\rangle|}{\|x\|} = \|\Lambda\|. \tag{4.25}$$

Thus the map $H \to H^*\colon y \mapsto \langle y, \cdot\rangle$ is an isometry of normed vector spaces.

Theorem 4.27. *Let H be a Hilbert space and let $E \subset H$ be a nonempty closed convex subset. Then there exists a unique element $x_0 \in E$ such that $\|x_0\| \leq \|x\|$ for all $x \in E$.*

Proof. See page 151. □

Theorem 4.27 *implies Theorem* 4.26. We prove existence. If $\Lambda = 0$, then $y = 0$ satisfies (4.24). Hence assume $\Lambda \neq 0$ and define

$$E := \{x \in H \mid \Lambda(x) = 1\}.$$

Then $E \neq \emptyset$ because there exists an element $\xi \in H$ such that $\Lambda(\xi) \neq 0$ and hence $x := \Lambda(\xi)^{-1}\xi \in E$. The set E is a closed because $\Lambda\colon H \to \mathbb{R}$ is continuous, and it is convex because Λ is linear. Hence Theorem 4.27 yields an element $x_0 \in E$ such that

$$\|x_0\| \le \|x\| \quad \text{for all } \ x \in E.$$

We prove that

$$x \in H, \quad \Lambda(x) = 0 \implies \langle x_0, x\rangle = 0. \tag{4.26}$$

To see this, fix an element $x \in H$ such that $\Lambda(x) = 0$. Then $x_0 + tx \in E$ for all $t \in \mathbb{R}$. This implies

$$\|x_0\|^2 \le \|x_0 + tx\|^2 = \|x_0\|^2 + 2t\langle x_0, x\rangle + t^2\|x\|^2 \quad \text{for all } \ t \in \mathbb{R}.$$

Thus the differentiable function $t \mapsto \|x_0 + tx\|^2$ attains its minimum at $t = 0$, and so its derivative vanishes at $t = 0$. Hence,

$$0 = \left.\frac{d}{dt}\right|_{t=0} \|x_0 + tx\|^2 = 2\langle x_0, x\rangle,$$

and this proves (4.26).

Now put

$$y := \frac{x_0}{\|x_0\|^2}.$$

Fix an element $x \in H$ and put $\lambda := \Lambda(x)$. Then $\Lambda(x - \lambda x_0) = \Lambda(x) - \lambda = 0$. Hence it follows from (4.26) that

$$0 = \langle x_0, x - \lambda x_0\rangle = \langle x_0, x\rangle - \lambda\|x_0\|^2.$$

This implies

$$\langle y, x\rangle = \frac{\langle x_0, x\rangle}{\|x_0\|^2} = \lambda = \Lambda(x).$$

Thus y satisfies (4.24).

We prove (4.25). Assume $y \in H$ satisfies (4.24). If $y = 0$, then $\Lambda = 0$, and so $\|y\| = 0 = \|\Lambda\|$. Hence assume $y \ne 0$. Then

$$\|y\| = \frac{\|y\|^2}{\|y\|} = \frac{\Lambda(y)}{\|y\|} \le \sup_{0 \ne x \in H} \frac{|\Lambda(x)|}{\|x\|} = \|\Lambda\|.$$

Conversely, it follows from the Cauchy–Schwarz inequality that

$$|\Lambda(x)| = |\langle y, x\rangle| \le \|y\|\|x\|$$

for all $x \in H$ and hence $\|\Lambda\| \le \|y\|$. This proves (4.25).

We prove uniqueness. Assume $y, z \in H$ satisfy

$$\langle y, x\rangle = \langle z, x\rangle = \Lambda(x)$$

for all $x \in H$. Then $\langle y - z, x\rangle = 0$ for all $x \in H$. Take $x := y - z$ to obtain

$$\|y - z\|^2 = \langle y - z, y - z\rangle = 0,$$

and hence $y - z = 0$. This proves Theorem 4.26, assuming Theorem 4.27. □

Proof of Theorem 4.27. Define

$$\delta := \inf\{\|x\| \mid x \in E\}.$$

We prove uniqueness. Let $x_0, x_1 \in E$ such that

$$\|x_0\| = \|x_1\| = \delta.$$

Then $\frac{1}{2}(x_0 + x_1) \in E$ because E is convex, and so $\|x_0 + x_1\| \geq 2\delta$. Thus

$$\|x_0 - x_1\|^2 = 2\|x_0\|^2 + 2\|x_1\|^2 - \|x_0 + x_1\|^2 = 4\delta^2 - \|x_0 + x_1\|^2 \leq 0$$

and therefore $x_0 = x_1$.

We prove existence. Choose a sequence $x_i \in E$ such that

$$\lim_{i\to\infty} \|x_i\| = \delta.$$

We claim that x_i is a Cauchy sequence. Indeed, fix a constant $\varepsilon > 0$. Then there exists an integer $i_0 \in \mathbb{N}$ such that

$$i \in \mathbb{N}, \quad i \geq i_0 \implies \|x_i\|^2 < \delta^2 + \frac{\varepsilon}{4}.$$

Let $i, j \in \mathbb{N}$ such that $i \geq i_0$ and $j \geq i_0$. Then $\frac{1}{2}(x_i + x_j) \in E$ because E is convex, and hence $\|x_i + x_j\| \geq 2\delta$. This implies

$$\begin{aligned}\|x_i - x_j\|^2 &= 2\|x_i\|^2 + 2\|x_j\|^2 - \|x_i + x_j\|^2 \\ &< 4\Big(\delta^2 + \frac{\varepsilon}{4}\Big) - 4\delta^2 \\ &= \varepsilon.\end{aligned}$$

Thus x_i is a Cauchy sequence. Since H is complete, the limit

$$x_0 := \lim_{i\to\infty} x_i$$

exists. Moreover, $x_0 \in E$ because E is closed and $\|x_0\| = \delta$ because the norm function (4.17) is continuous. This proves Theorem 4.27. □

Corollary 4.28. *Let $(X, \mathcal{A}, \mu)$ be a measure space and let $\Lambda\colon L^2(\mu) \to \mathbb{R}$ be a bounded linear functional. Then there exists a function $g \in \mathcal{L}^2(\mu)$, unique up to equality almost everywhere, such that*

$$\Lambda([f]_\mu) = \int_X fg\, d\mu \quad \text{for all } f \in \mathcal{L}^2(\mu).$$

Moreover $\|\Lambda\| = \|g\|_2$. Thus $L^2(\mu)^*$ is isomorphic to $L^2(\mu)$.

Proof. This follows immediately from Theorem 4.26 and Example 4.22. □

4.5 The dual space of $L^p(\mu)$

We wish to extend Corollary 4.28 to the L^p-spaces in Definition 4.3 and equation (4.8) (for $1 \le p < \infty$) and in Definition 4.7 (for $p = \infty$). When $1 < p < \infty$ it turns out that the dual space of $L^p(\mu)$ is always isomorphic to $L^q(\mu)$, where $1/p + 1/q = 1$. For $p = \infty$ the natural homomorphism $L^1(\mu) \to L^\infty(\mu)^*$ is an isometric embedding, however, in most cases the dual space of $L^\infty(\mu)$ is much larger than $L^1(\mu)$. For $p = 1$ the situation is more subtle. The natural homomorphism $L^\infty(\mu) \to L^1(\mu)^*$ need not be injective or surjective. However, it is bijective for a large class of measure spaces and one can characterize those measure spaces for which it is injective, respectively bijective. This requires the following definition.

Definition 4.29. A measure space $(X, \mathcal{A}, \mu)$ is called *σ-finite* if there exists a sequence of measurable subsets $X_i \in \mathcal{A}$ such that

$$X = \bigcup_{i=1}^{\infty} X_i, \quad X_i \subset X_{i+1}, \quad \mu(X_i) < \infty \quad \text{for all } i \in \mathbb{N}. \tag{4.27}$$

It is called *semi-finite* if every measurable set $A \in \mathcal{A}$ satisfies

$$\mu(A) > 0 \implies \begin{array}{l}\text{there exists } E \in \mathcal{A} \text{ such that } E \subset A \\ \text{and } 0 < \mu(E) < \infty.\end{array} \tag{4.28}$$

It is called *localizable* if it is semi-finite and, for every collection of measurable sets $\mathcal{E} \subset \mathcal{A}$, there is a set $H \in \mathcal{A}$ satisfying the following two conditions:

(L1) $\mu(E \setminus H) = 0$ for all $E \in \mathcal{E}$;

(L2) if $G \in \mathcal{A}$ satisfies $\mu(E \setminus G) = 0$ for all $E \in \mathcal{E}$, then $\mu(H \setminus G) = 0$.

A measurable set H satisfying (L1) and (L2) is called an *envelope* of $\mathcal{E}$.

The geometric intuition behind the definition of *localizable* is as follows. The collection $\mathcal{E} \subset \mathcal{A}$ will typically be uncountable, so one cannot expect its union to be measurable. The *envelope* H is a measurable set that replaces the union of the sets in $\mathcal{E}$. It covers each set $E \in \mathcal{E}$ up to a set of measure zero and, if any other measurable set G covers each set $E \in \mathcal{E}$ up to a set of measure zero, it also covers H up to a set of measure zero. The next lemma clarifies the notion of semi-finiteness.

Lemma 4.30. *Let $(X, \mathcal{A}, \mu)$ be a measure space:*

(i) *$(X, \mathcal{A}, \mu)$ is semi-finite if and only if*

$$\mu(A) = \sup\{\mu(E) \mid E \in \mathcal{A},\ E \subset A,\ \mu(E) < \infty\} \tag{4.29}$$

for every measurable set $A \in \mathcal{A}$;

(ii) *if $(X, \mathcal{A}, \mu)$ is σ-finite, then it is semi-finite.*

Proof. (i) Assume $(X, \mathcal{A}, \mu)$ is semi-finite, let $A \in \mathcal{A}$, and define

$$a := \sup\{\mu(E) \mid E \in \mathcal{A},\ E \subset A,\ \mu(E) < \infty\}.$$

Then $a \le \mu(A)$ and we must prove that $a = \mu(A)$. This is obvious when $a = \infty$. Hence assume $a < \infty$. Choose a sequence of measurable sets $E_i \subset A$ such that $\mu(E_i) < \infty$ and $\mu(E_i) > a - 2^{-i}$ for all i. Define

$$B_i := E_1 \cup \cdots \cup E_i, \quad B := \bigcup_{i=1}^{\infty} B_i = \bigcup_{i=1}^{\infty} E_i.$$

Then $B_i \in \mathcal{A}$, $E_i \subset B_i \subset A$, and $\mu(B_i) < \infty$. Hence $\mu(E_i) \le \mu(B_i) \le a$ for all $i \in \mathbb{N}$, and hence

$$\mu(B) = \lim_{i\to\infty} \mu(B_i) = a < \infty.$$

If $\mu(A \setminus B) > 0$, then since $(X, \mathcal{A}, \mu)$ is semi-finite, there exists a measurable set $F \in \mathcal{A}$ such that $F \subset A \setminus B$ and $0 < \mu(F) < \infty$, and hence

$$B \cup F \subset A, \quad a < \mu(B \cup F) = \mu(B) + \mu(F) < \infty,$$

contradicting the definition of a. This shows that $\mu(A \setminus B) = 0$ and hence $\mu(A) = \mu(B) + \mu(A \setminus B) = a$, as claimed. Thus we have proved that every semi-finite measure space satisfies (4.29). The converse is obvious and this proves part (i).

(ii) Assume that $(X, \mathcal{A}, \mu)$ is σ-finite and choose a sequence of measurable sets $X_i \in \mathcal{A}$ that satisfies (4.27). If $A \in \mathcal{A}$, then it follows from Theorem 1.28 that $\mu(A) = \lim_{i\to\infty} \mu(A \cap X_i)$. Since $\mu(A \cap X_i) < \infty$ for all i, this shows that every measurable set A satisfies (4.29) and so $(X, \mathcal{A}, \mu)$ is semi-finite. This proves Lemma 4.30. □

It is also true that every σ-finite measure space is localizable. This can be derived as a consequence of Theorem 4.35 (see Corollary 5.9 below). A more direct proof is outlined in Exercise 4.58.

Example 4.31. Define $(X, \mathcal{A}, \mu)$ by

$$X := \{a, b\},$$

$$\mathcal{A} := 2^X,$$

$$\mu(\{a\}) := 1, \quad \mu(\{b\}) := \infty.$$

This measure space is not semi-finite. Thus the linear map $L^\infty(\mu) \to L^1(\mu)^*$ in Theorem 4.33 below is not injective. In fact, $L^\infty(\mu)$ has dimension two and $L^1(\mu)$ has dimension one.

Example 4.32. Let X be an uncountable set, let $\mathcal{A} \subset 2^X$ be the σ-algebra of all subsets $A \subset X$ such that A or A^c is countable, and let $\mu: \mathcal{A} \to [0, \infty]$ be the counting measure. Then $(X, \mathcal{A}, \mu)$ is semi-finite, but it is not localizable. For example, let $H \subset X$ be an uncountable set with an uncountable complement and let $\mathcal{E}$ be the collection of all finite subsets of H. Then the only possible envelope of $\mathcal{E}$ would be the set H itself, which is not measurable. Thus Theorem 4.33 below shows that the map $L^\infty(\mu) \to L^1(\mu)^*$ is injective, while Theorem 4.35 below shows that it is not surjective. An example of a bounded linear functional

$$\Lambda: L^1(\mu) \longrightarrow \mathbb{R}$$

that cannot be represented by an L^∞-function is given by

$$\Lambda(f) := \sum_{x \in H} f(x)$$

for $f \in \mathcal{L}^1(\mu) = L^1(\mu)$.

The next theorem is the first step towards understanding the dual space of $L^p(\mu)$ and is a fairly easy consequence of the Hölder inequality. It asserts that for $1/p + 1/q = 1$ every element of $L^q(\mu)$ determines a bounded linear functional on $L^p(\mu)$ and that the resulting map $L^q(\mu) \to L^p(\mu)^*$ is an isometric embedding (for $p = 1$ under the *semi-finite* hypothesis). The key question is then whether every bounded linear functional on $L^p(\mu)$ is of that form. That this is indeed the case for $1 < p < \infty$ (and for $p = 1$ under the *localizable* hypothesis) is the content of Theorem 4.35 below. This is a much deeper theorem whose proof for $p \neq 2$ requires the Radon–Nikodým theorem and will be carried out in Chapter 5.

Theorem 4.33. *Let $(X, \mathcal{A}, \mu)$ be a measure space and fix constants*

$$1 \le p \le \infty, \quad 1 \le q \le \infty, \quad \frac{1}{p} + \frac{1}{q} = 1. \tag{4.30}$$

Then the following holds.

(i) *Let $g \in \mathcal{L}^q(\mu)$. Then the formula*

$$\Lambda_g([f]_\mu) := \int_X fg \, d\mu \quad \text{for } f \in \mathcal{L}^p(\mu) \tag{4.31}$$

defines a bounded linear functional

$$\Lambda_g : L^p(\mu) \longrightarrow \mathbb{R}$$

and

$$\|\Lambda_g\| = \sup_{f \in \mathcal{L}^p(\mu), \|f\|_p \neq 0} \frac{\left|\int_X fg \, d\mu\right|}{\|f\|_p} \le \|g\|_q. \tag{4.32}$$

(ii) *The map $g \mapsto \Lambda_g$ in (4.31) descends to a bounded linear operator*

$$L^q(\mu) \longrightarrow L^p(\mu)^* : [g]_\mu \longmapsto \Lambda_g. \tag{4.33}$$

(iii) *Assume $1 < p \le \infty$ Then*

$$\|\Lambda_g\| = \|g\|_q \quad \text{for all } g \in \mathcal{L}^q(\mu).$$

(iv) *Assume $p = 1$. Then the map $L^\infty(\mu) \to L^1(\mu)^*$ in (4.33) is injective if and only if it is an isometric embedding if and only if $(X, \mathcal{A}, \mu)$ is semi-finite.*

Proof. See page 158. □

The heart of the proof is the next lemma. It is slightly stronger than what is required to prove Theorem 4.33 in that the hypothesis on g to be q-integrable is dropped in part (iii) and replaced by the assumption that the measure space is semi-finite. In this form Lemma 4.34 is needed in the proof of Theorem 4.35 and will also be useful for proving the Minkowski and Calderón–Zygmund inequalities in Theorems 7.19 and 7.43 below.

Lemma 4.34. *Let $(X, \mathcal{A}, \mu)$ be a measure space and let p, q be as in (4.30). Let $g: X \to [0, \infty]$ be a measurable function and suppose that there exists a constant $c \geq 0$ such that*

$$f \in \mathcal{L}^p(\mu), \quad f \geq 0 \implies \int_X fg \, d\mu \leq c \, \|f\|_p . \tag{4.34}$$

Then the following holds:

(i) *if $q = 1$, then $\|g\|_1 \leq c$;*

(ii) *if $1 < q < \infty$ and $\|g\|_q < \infty$, then $\|g\|_q \leq c$;*

(iii) *if $1 < q < \infty$ and $(X, \mathcal{A}, \mu)$ is semi-finite, then $\|g\|_q \leq c$;*

(iv) *if $q = \infty$ and $(X, \mathcal{A}, \mu)$ is semi-finite, then $\|g\|_\infty \leq c$.*

Proof. (i) If $q = 1$ take $f \equiv 1$ in (4.34) to obtain $\|g\|_1 \leq c$.

(ii) Assume $1 < q < \infty$ and $\|g\|_q < \infty$. Then it follows from Lemma 1.47 that the set

$$A := \{x \in X \mid g(x) = \infty\}$$

has measure zero. Define the function

$$h: X \longrightarrow [0, \infty)$$

by

$$h(x) := \begin{cases} g(x) & \text{for } x \in X \setminus A, \\ 0 & \text{for } x \in A. \end{cases}$$

Then h is measurable and

$$\|h\|_q = \|g\|_q < \infty, \qquad \int_X fh \, d\mu = \int_X fg \, d\mu \leq c \, \|f\|_p$$

for all $f \in \mathcal{L}^p(\mu)$ with $f \geq 0$ by Lemma 1.48. Define

$$f: X \longrightarrow [0, \infty)$$

by

$$f(x) := h(x)^{q-1} \quad \text{for } x \in X.$$

Then

$$f^p = h^{p(q-1)} = h^q = fh$$

and hence

$$\|f\|_p = \left(\int_X h^q \, d\mu\right)^{1-1/q} = \|h\|_q^{q-1}, \quad \int_X fh \, d\mu = \|h\|_q^q.$$

Thus $f \in \mathcal{L}^p(\mu)$ and so

$$\|h\|_q^q = \int_X fh \, d\mu \leq c \, \|f\|_p = c \, \|h\|_q^{q-1}.$$

Since $\|h\|_q < \infty$, it follows that $\|g\|_q = \|h\|_q \leq c$, and this proves part (ii).

(iii) Assume $(X, \mathcal{A}, \mu)$ is semi-finite and $1 < q < \infty$. Suppose, by contradiction, that $\|g\|_q > c$. We will prove that there exists a measurable function $h \colon X \to [0, \infty)$ such that

$$0 \leq h \leq g, \quad c < \|h\|_q < \infty. \tag{4.35}$$

By (4.34), this function h satisfies

$$\int_X fh \, d\mu \leq \int_X fg \, d\mu \leq c \, \|f\|_p$$

for all $f \in \mathcal{L}^p(\mu)$ with $f \geq 0$. Since $\|h\|_q < \infty$ it follows from part (ii) that $\|h\|_q \leq c$, which contradicts the inequality $\|h\|_q > c$ in (4.35).

It remains to prove the existence of h. Since $\|g\|_q > c$, it follows from Definition 1.34 that there exists a measurable step function $s \colon X \to [0, \infty)$ such that $0 \leq s \leq g$ and $\int_X s^q \, d\mu > c^q$. If $\|s\|_q < \infty$, take $h := s$. If $\|s\|_q = \infty$, there exist a measurable set $A \subset X$ and a constant $\delta > 0$ such that $\mu(A) = \infty$ and $\delta \chi_A \leq s \leq g$. Since $(X, \mathcal{A}, \mu)$ is semi-finite, Lemma 4.30 asserts that there exists a measurable set $E \in \mathcal{A}$ such that $E \subset A$ and $c^q < \delta^q \mu(E) < \infty$. Then the function $h := \delta \chi_E \colon X \to [0, \infty)$ satisfies $0 \leq h \leq g$ and $\|h\|_q = \delta \mu(E)^{1/q} > c$, as required. This proves part (iii).

(iv) Let $q = \infty$ and assume $(X, \mathcal{A}, \mu)$ is semi-finite. Suppose, by contradiction, that $\|g\|_\infty > c$. Then there exists a constant $\delta > 0$ such that the set

$$A := \{x \in X \mid g(x) \geq c + \delta\}$$

has positive measure. Since $(X, \mathcal{A}, \mu)$ is semi-finite there exists a measurable set $E \subset A$ such that $0 < \mu(E) < \infty$. Hence, $f := \chi_E \in \mathcal{L}^1(\mu)$ and we get $\int_X fg \, d\mu \geq (c + \delta)\mu(E) > c\mu(E) = c \, \|f\|_1$, in contradiction to (4.34). This proves (iv) and Lemma 4.34. □

Proof of Theorem 4.33. The proof has four steps.

Step 1. *Let* $f \in \mathcal{L}^p(\mu)$, $g \in \mathcal{L}^q(\mu)$. *Then* $fg \in \mathcal{L}^1(\mu)$ *and* $\|fg\|_1 \le \|f\|_p \|g\|_q$.

If $1 < p < \infty$, then $\int_X |fg|\, d\mu \le \|f\|_p \|g\|_q$ by the Hölder inequality (4.3). If $p = 1$, then $|fg| \le |f| \|g\|_\infty$ almost everywhere by Lemma 4.8, so $fg \in \mathcal{L}^1(\mu)$ and $\|fg\|_1 \le \|f\|_1 \|g\|_\infty$. If $p = \infty$, interchange the pairs (f, p) and (g, q).

Step 2. *We prove* (i) *and* (ii).

By Step 1, the right hand side of (4.31) is well defined, and by Lemma 1.48 it depends only on the equivalence class of f under equality almost everywhere. Hence Λ_g is well defined. It is linear by Theorem 1.44 and $\|\Lambda_g\| \le \|g\|_q$ by Step 1. This proves (i). The bounded linear functional $\Lambda_g\colon L^p(\mu) \to \mathbb{R}$ depends only on the equivalence class of g, again by Lemma 1.48. Hence the map (4.33) is well defined. By Theorem 1.44 and (4.32), it is a bounded linear operator of norm less than or equal to one. This proves (ii).

Step 3. *If* $1 < p \le \infty$, *then* $\|\Lambda_g\| = \|g\|_q$ *for all* $g \in \mathcal{L}^q(\mu)$. *This continues to hold for* $p = 1$ *when* $(X, \mathcal{A}, \mu)$ *is semi-finite.*

Let $g \in \mathcal{L}^q(\mu)$. For $t \in \mathbb{R}$ define $\mathrm{sign}(t) \in \{-1, 0, 1\}$ by $\mathrm{sign}(t) := 1$ for $t > 0$, $\mathrm{sign}(t) := -1$ for $t < 0$, and $\mathrm{sign}(0) = 0$. If $f \in \mathcal{L}^p(\mu)$ is nonnegative, then the function $f\,\mathrm{sign}(g)\colon X \to \mathbb{R}$ is p-integrable and

$$\int_X f|g|\, d\mu = \Lambda_g(f\,\mathrm{sign}(g)) \le \|\Lambda_g\| \, \|f\,\mathrm{sign}(g)\|_p \le \|\Lambda_g\| \, \|f\|_p .$$

Hence $\|g\|_q \le \|\Lambda_g\|$ by Lemma 4.34 and so $\|\Lambda_g\| = \|g\|_q$ by Step 2.

Step 4. *If the map* $L^\infty(\mu) \to L^1(\mu)^*$ *is injective, then* $(X, \mathcal{A}, \mu)$ *is semi-finite.*

Let $A \in \mathcal{A}$ such that $\mu(A) > 0$ and define $g := \chi_A$. Then $\Lambda_g\colon L^1(\mu) \to \mathbb{R}$ is nonzero by assumption. Hence there is an $f \in \mathcal{L}^1(\mu)$ such that

$$0 < \Lambda_g(f) = \int_X fg\, d\mu = \int_A f\, d\mu. \tag{4.36}$$

For $i \in \mathbb{N}$ define $E_i := \{x \in A \mid f(x) > 2^{-i}\}$. Then $E_i \in \mathcal{A}$, $E_i \subset A$, and

$$\mu(E_i) \le 2^i \int_{E_i} f\, d\mu \le 2^i \|f\|_1 < \infty.$$

Moreover, $E := \bigcup_{i=1}^\infty E_i = \{x \in A \mid f(x) > 0\}$ is not a null set by (4.36). Hence one of the sets E_i has positive measure. Thus $(X, \mathcal{A}, \mu)$ is semi-finite. This proves Step 4 and Theorem 4.33. □

The next theorem asserts that, for $1 < p < \infty$, every bounded linear functional on $L^p(\mu)$ has the form (4.31) for some $g \in \mathcal{L}^q(\mu)$. For $p \neq 2$ this is a much deeper result than Corollary 4.28. The proof requires the Radon–Nikodým theorem and will be deferred to the next chapter.

Theorem 4.35 (the dual space of L^p). *Let $(X, \mathcal{A}, \mu)$ be a measure space and fix constants*

$$1 \le p < \infty, \quad 1 < q \le \infty, \quad \frac{1}{p} + \frac{1}{q} = 1.$$

Then the following holds.

(i) *Assume $1 < p < \infty$. Then the map $L^q(\mu) \to L^p(\mu)^*\colon [g]_\mu \mapsto \Lambda_g$, defined by (4.31) is bijective and hence is a Banach space isometry.*

(ii) *Assume $p = 1$. Then the map $L^\infty(\mu) \to L^1(\mu)^*\colon [g]_\mu \mapsto \Lambda_g$ defined by (4.31) is bijective if and only if $(X, \mathcal{A}, \mu)$ is localizable.*

Proof. See page 195. □

This next example shows that, in general, Theorem 4.35 does not extend to the case $p = \infty$ (regardless of whether or not the measure space $(X, \mathcal{A}, \mu)$ is σ-finite). By Theorem 4.33, the Banach space $L^1(\mu)$ is equipped with an isometric inclusion $L^1(\mu) \to L^\infty(\mu)^*$, however, the dual space of $L^\infty(\mu)$ is typically much larger than $L^1(\mu)$.

Example 4.36. Let $\mu\colon 2^{\mathbb{N}} \to [0, \infty]$ be the counting measure on the positive integers. Then

$$\ell^\infty := L^\infty(\mu) = \mathcal{L}^\infty(\mu)$$

is the Banach space of bounded sequences $x = (x_n)_{n\in\mathbb{N}}$ of real numbers, equipped with the supremum norm

$$\|x\|_\infty := \sup_{n\in\mathbb{N}} |x_n|.$$

An interesting closed subspace of ℓ^∞ is

$$c := \{x = (x_n)_{n\in\mathbb{N}} \in \ell^\infty \mid x \text{ is a Cauchy sequence}\}.$$

It is equipped with a bounded linear functional $\Lambda_0\colon c \to \mathbb{R}$, defined by

$$\Lambda_0(x) := \lim_{n\to\infty} x_n \quad \text{for } x = (x_n)_{n\in\mathbb{N}} \in c.$$

The *Hahn–Banach theorem,* one of the fundamental principles of *functional analysis,* asserts that every bounded linear functional on a linear subspace of a Banach space extends to a bounded linear functional on the entire Banach space (whose norm is no larger than the norm of the original bounded linear functional on the subspace). In the case at hand this means that there is a bounded linear functional $\Lambda\colon \ell^\infty \to \mathbb{R}$ such that $\Lambda|_c = \Lambda_0$. This linear functional cannot have the form (4.31) for any $g \in \mathcal{L}^1(\mu)$. To see this, note that $\ell^1 := L^1(\mu) = \mathcal{L}^1(\mu)$ is the space of summable sequences of real numbers. Let $y = (y_n)_{n\in\mathbb{N}} \in \ell^1$ be a sequence of real numbers such that $\sum_{n=1}^\infty |y_n| < \infty$, and define the linear functional

$$\Lambda_y\colon \ell^\infty \longrightarrow \mathbb{R}$$

by

$$\Lambda_y(x) := \sum_{n=1}^\infty x_n y_n \quad \text{for } x = (x_n)_{n\in\mathbb{N}} \in \ell^\infty.$$

Choose $N \in \mathbb{N}$ such that $\sum_{n=N}^\infty |y_n| =: \alpha < 1$ and define $x = (x_n)_{n\in\mathbb{N}} \in c$ by $x_n := 0$ for $n < N$ and $x_n := 1$ for $n \geq N$. Then $\Lambda_y(x) \leq \alpha < 1 = \Lambda(x)$ and hence $\Lambda_y \neq \Lambda$. This shows that Λ does not belong to the image of the isometric inclusion $\ell^1 \hookrightarrow (\ell^\infty)^*$.

Exercise 4.37. Let $\Lambda_0\colon c \to \mathbb{R}$ be the functional in Example 4.36 and denote $c_0 := \ker \Lambda_0$. Thus c_0 is the set of all sequences of real numbers that converge to zero, i.e.,

$$c_0 = \{x = (x_n)_{n\in\mathbb{N}} \in \ell^\infty \mid \lim_{n\to\infty} x_n = 0\}.$$

Prove that c_0 is a closed linear subspace of ℓ^∞ and that ℓ^1 is naturally isomorphic to the dual space of c_0. Thus

$$\ell^1 \cong (c_0)^*, \quad c_0 \subsetneq \ell^\infty \cong (\ell^1)^* \cong (c_0)^{**}, \quad \ell^1 \subsetneq (\ell^\infty)^* \cong (\ell^1)^{**}.$$

In the language of *functional analysis,* this means that the Banach spaces c_0 and ℓ^1 are not *reflexive,* and neither is ℓ^∞.

We close this section with two results that will be needed in the proof of Theorem 4.35. The first asserts that every bounded linear functional on $L^p(\mu)$ can be written as the difference of two positive bounded linear functionals (Theorem 4.39). The second asserts that every positive bounded linear functional on $L^p(\mu)$ is supported on a σ-finite subset of X (Theorem 4.40). When $\Lambda\colon L^p(\mu) \to \mathbb{R}$ is a bounded linear functional it will be convenient to abuse notation and write

$$\Lambda(f) := \Lambda([f]_\mu) \quad \text{for } f \in \mathcal{L}^p(\mu).$$

Definition 4.38. Let $(X, \mathcal{A}, \mu)$ be a measure space and let $1 \le p < \infty$. A bounded linear functional $\Lambda\colon L^p(\mu) \to \mathbb{R}$ is called *positive* if

$$f \ge 0 \implies \Lambda(f) \ge 0$$

for all $f \in \mathcal{L}^p(\mu)$.

Theorem 4.39. *Let $(X, \mathcal{A}, \mu)$ be a measure space and let $1 \le p < \infty$. Consider a bounded linear functional $\Lambda\colon L^p(\mu) \to \mathbb{R}$. Define*

$$\lambda^{\pm}\colon \mathcal{A} \longrightarrow [0, \infty]$$

by

$$\lambda^{\pm}(A) := \sup\{\Lambda(\pm\chi_E) \mid E \in \mathcal{A},\ E \subset A,\ \mu(E) < \infty\} \tag{4.37}$$

Then the maps $\lambda^{\pm}$ are measures, $\mathcal{L}^p(\mu) \subset \mathcal{L}^1(\lambda^+) \cap \mathcal{L}^1(\lambda^-)$, and the formulas

$$\Lambda^{\pm}(f) := \int_X f\, d\lambda^{\pm} \quad \textit{for } f \in \mathcal{L}^p(\mu) \tag{4.38}$$

define positive bounded linear functionals $\Lambda^{\pm}\colon L^p(\mu) \to \mathbb{R}$ such that

$$\Lambda = \Lambda^+ - \Lambda^-, \quad \|\Lambda^+\| + \|\Lambda^-\| = \|\Lambda\|. \tag{4.39}$$

Proof. The proof is divided into four steps.

Step 1. *The maps $\lambda^{\pm}\colon \mathcal{A} \to [0, \infty]$ in (4.37) are measures.*

It follows directly from the definition that $\lambda^{\pm}(\emptyset) = 0$. We must prove that λ^+ is σ-additive. That λ^- is then also σ-additive follows by reversing the sign of Λ. Thus, let $A_i \in \mathcal{A}$ be a sequence of pairwise disjoint measurable sets and set

$$A := \bigcup_{i=1}^{\infty} A_i.$$

Let $E \in \mathcal{A}$ such that $E \subset A$ and $\mu(E) < \infty$. Then it follows from the definition of λ^+ that

$$\Lambda(\chi_{E \cap A_i}) \le \lambda^+(A_i) \quad \text{for all } i \in \mathbb{N}. \tag{4.40}$$

Moreover, the sequence of measurable functions

$$f_n := \chi_E - \sum_{i=1}^{n} \chi_{E \cap A_i} \ge 0$$

converges pointwise to zero and satisfies $0 \leq f_n^p \leq \chi_E$ for all n. Since $\mu(E) < \infty$, the function χ_E is integrable, and so it follows from the Lebesgue dominated convergence theorem (Theorem 1.45) that

$$\lim_{n\to\infty} \int_X f_n^p \, d\mu = 0,$$

i.e.,

$$\lim_{n\to\infty} \Big\| \chi_E - \sum_{i=1}^n \chi_{E\cap A_i} \Big\|_p = 0.$$

Then (4.40) yields that

$$\Lambda(\chi_E) = \lim_{n\to\infty} \sum_{i=1}^n \Lambda(\chi_{E\cap A_i}) = \sum_{i=1}^\infty \Lambda(\chi_{E\cap A_i}) \leq \sum_{i=1}^\infty \lambda^+(A_i).$$

Take the supremum over all $E \in \mathcal{A}$ with $E \subset A$ and $\mu(E) < \infty$ to obtain

$$\lambda^+(A) \leq \sum_{i=1}^\infty \lambda^+(A_i).$$

To prove the converse inequality, assume first that $\lambda^+(A_i) = \infty$ for some i; since $A_i \subset A$, this implies $\lambda^+(A) = \infty = \sum_{i=1}^\infty \lambda^+(A_i)$. Hence it suffices to assume $\lambda^+(A_i) < \infty$ for all i. Fix a constant $\varepsilon > 0$ and choose a sequence of measurable sets $E_i \in \mathcal{A}$ such that $E_i \subset A_i$ and $\Lambda(\chi_{E_i}) > \lambda^+(A_i) - 2^{-i}\varepsilon$ for all i. Since $E_1 \cup \cdots \cup E_n \subset A$, it follows from the definition of λ^+ that

$$\lambda^+(A) \geq \Lambda(\chi_{E_1\cup\cdots\cup E_n}) = \sum_{i=1}^n \Lambda(\chi_{E_i}) > \sum_{i=1}^n \lambda^+(A_i) - \varepsilon.$$

Take the limit $n \to \infty$ to obtain

$$\lambda^+(A) \geq \sum_{i=1}^\infty \lambda^+(A_i) - \varepsilon \quad \text{for all } \varepsilon > 0,$$

so

$$\lambda^+(A) \geq \sum_{i=1}^\infty \lambda^+(A_i),$$

as claimed. Thus λ^+ is σ-additive and this proves Step 1.

Step 2. *Let $c := \|\Lambda\|$. Then every measurable function $f\colon X \to \mathbb{R}$ satisfies*

$$\int_X |f|\,d\lambda^+ + \int_X |f|\,d\lambda^- \le c\,\|f\|_p. \tag{4.41}$$

In particular, $\mathcal{L}^p(\mu) \subset \mathcal{L}^1(\lambda^+) \cap \mathcal{L}^1(\lambda^-)$.

Assume first that $f = s\colon X \to [0,\infty)$ is a measurable step function in $\mathcal{L}^p(\mu)$. Then there exist real numbers $\alpha_i > 0$ and measurable sets $A_i \in \mathcal{A}$ for $i = 1,\dots,\ell$ such that $A_i \cap A_j = \emptyset$ for $i \ne j$, $\mu(A_i) < \infty$ for all i, and $s = \sum_{i=1}^{\ell} \alpha_i \chi_{A_i}$. Now fix a real number $\varepsilon > 0$ and choose $\varepsilon_i > 0$ such that

$$\sum_{i=1}^{\ell} \alpha_i \varepsilon_i = \frac{\varepsilon}{2}.$$

For $i = 1,\dots,\ell$ choose $E_i^{\pm} \in \mathcal{A}$ such that

$$E_i^{\pm} \subset A_i, \quad \Lambda(\chi_{E_i^+}) \ge \lambda^+(A_i) - \varepsilon_i, \quad -\Lambda(\chi_{E_i^-}) \ge \lambda^-(A_i) - \varepsilon_i.$$

Then

$$\begin{aligned}
\int_X s\,d\lambda^+ + \int_X s\,d\lambda^- &= \sum_{i=1}^{\ell} \alpha_i(\lambda^+(A_i) + \lambda^-(A_i)) \\
&\le \sum_{i=1}^{\ell} \alpha_i(\Lambda(\chi_{E_i^+}) - \Lambda(\chi_{E_i^-}) + 2\varepsilon_i) \\
&= \Lambda\Big(\sum_{i=1}^{\ell} \alpha_i(\chi_{E_i^+} - \chi_{E_i^-})\Big) + \varepsilon \\
&\le c\,\Big\|\sum_{i=1}^{\ell} \alpha_i(\chi_{E_i^+} - \chi_{E_i^-})\Big\|_p + \varepsilon \\
&= c\Big(\sum_{i=1}^{\ell} \alpha_i^p(\mu(E_i^+ \setminus E_i^-) + \mu(E_i^- \setminus E_i^+))\Big)^{1/p} + \varepsilon \\
&\le c\Big(\sum_{i=1}^{\ell} \alpha_i^p \mu(A_i)\Big)^{1/p} + \varepsilon \\
&= c\,\|s\|_p + \varepsilon.
\end{aligned}$$

Take the limit $\varepsilon \to 0$ to obtain (4.41) for $f = s$. To prove (4.41) in general it suffices to assume that $f \in \mathcal{L}^p(\mu)$ is nonnegative. By Theorem 1.26, there is a sequence of measurable step functions $0 \le s_1 \le s_2 \le \cdots$ that converges pointwise to f. Then $(f - s_n)^p$ converges pointwise to zero and is bounded above by $f^p \in \mathcal{L}^1(\mu)$. Hence

$$\lim_{n\to\infty} \|f - s_n\|_p = 0$$

by the Lebesgue dominated convergence theorem (Theorem 1.45), and

$$\lim_{n\to\infty} \int_X s_n \, d\lambda^\pm = \int_X f \, d\lambda^\pm$$

by the Lebesgue monotone convergence theorem (Theorem 1.37). This proves the inequality (4.41). It follows from (4.41) that $\mathcal{L}^p(\mu) \subset \mathcal{L}^1(\lambda^+) \cap \mathcal{L}^1(\lambda^-)$ and this proves Step 2.

Step 3. *If $A \in \mathcal{A}$ and $\mu(A) < \infty$, then*

$$\lambda^\pm(A) < \infty, \quad \Lambda(\chi_A) = \lambda^+(A) - \lambda^-(A) \tag{4.42}$$

It follows from the inequality (4.41) in Step 2 that

$$\lambda^+(A) + \lambda^-(A) = \int_X \chi_A \, d\lambda^+ + \int_X \chi_A \, d\lambda^- \le c\|\chi_A\|_p = c\mu(A)^{1/p} < \infty.$$

Now let $\varepsilon > 0$ and choose $E \in \mathcal{A}$ such that $E \subset A$ and $\Lambda(\chi_E) > \lambda^+(A) - \varepsilon$. Since $-\Lambda(\chi_{A\setminus E}) \le \lambda^-(A)$, this implies

$$\Lambda(\chi_A) = \Lambda(\chi_E) + \Lambda(\chi_{A\setminus E}) > \lambda^+(A) - \lambda^-(A) - \varepsilon.$$

Since this holds for all $\varepsilon > 0$, we obtain $\Lambda(\chi_A) \ge \lambda^+(A) - \lambda^-(A)$. Reversing the sign of Λ we also obtain $-\Lambda(\chi_A) \ge \lambda^-(A) - \lambda^+(A)$ and this proves Step 3.

Step 4. *If $f \in \mathcal{L}^p(\mu)$, then*

$$\Lambda(f) = \int_X f \, d\lambda^+ - \int_X f \, d\lambda^-. \tag{4.43}$$

Let $s: X \to \mathbb{R}$ be a p-integrable step function. Then there are real numbers α_i and measurable sets $A_i \in \mathcal{A}$ for $i = 1, \dots, \ell$ such that $\mu(A_i) < \infty$ for all i and $s = \sum_{i=1}^{\ell} \alpha_i \chi_{A_i}$. Hence it follows from Step 3 that

$$\Lambda(s) = \sum_{i=1}^{\ell} \alpha_i \Lambda(\chi_{A_i}) = \sum_{i=1}^{\ell} \alpha_i (\lambda^+(A_i) - \lambda^-(A_i)) = \int_X s \, d\lambda^+ - \int_X s \, d\lambda^-.$$

This proves (4.43) for p-integrable step functions. Now let $f \in \mathcal{L}^p(\mu)$ and assume $f \geq 0$. By Theorem 1.26, there is a sequence of measurable step functions $0 \leq s_1 \leq s_2 \leq \cdots$ that converges pointwise to f. Then $(f - s_n)^p$ converges pointwise to zero and is bounded above by $f^p \in \mathcal{L}^1(\mu)$. Hence

$$\lim_{n\to\infty} \|f - s_n\|_p = 0$$

by the Lebesgue dominated convergence theorem and so

$$\lim_{n\to\infty} \Lambda(s_n) = \Lambda(f).$$

Moreover,

$$\int_X f\, d\lambda^{\pm} \leq c\,\|f\|_p < \infty$$

by Step 2, and

$$\lim_{n\to\infty} \int_X s_n\, d\lambda^{\pm} = \int_X f\, d\lambda^{\pm}$$

by the Lebesgue monotone convergence theorem. Thus every nonnegative L^p-function $f\colon X \to [0,\infty)$ satisfies (4.43). If $f \in \mathcal{L}^p(\mu)$, then $f^{\pm} \in \mathcal{L}^p(\mu)$ satisfy (4.43) by what we have just proved, and hence so does $f = f^+ - f^-$. This proves Step 4.

It follows from Steps 2 and 4 that the linear functionals $\Lambda^{\pm}\colon L^p(\mu) \to \mathbb{R}$ in (4.38) are bounded and satisfy (4.39). This proves Theorem 4.39. □

Theorem 4.40. *Let $(X, \mathcal{A}, \mu)$ be a measure space and let $1 < p < \infty$. Consider a positive bounded linear functional $\Lambda\colon L^p(\mu) \to \mathbb{R}$. Define*

$$\lambda(A) := \sup\{\Lambda(\chi_E) \mid E \in \mathcal{A},\ E \subset A,\ \mu(E) < \infty\} \tag{4.44}$$

for $A \in \mathcal{A}$. Then the map $\lambda\colon \mathcal{A} \to [0,\infty]$ is a measure, $\mathcal{L}^p(\mu) \subset \mathcal{L}^1(\lambda)$, and

$$\Lambda(f) = \int_X f\, d\lambda \quad \textit{for all } f \in \mathcal{L}^p(\mu). \tag{4.45}$$

Moreover, there are measurable sets $N \in \mathcal{A}$ and $X_n \in \mathcal{A}$ for $n \in \mathbb{N}$ such that

$$X \setminus N = \bigcup_{n=1}^{\infty} X_n, \quad \lambda(N) = 0, \quad \mu(X_n) < \infty, \quad X_n \subset X_{n+1} \tag{4.46}$$

for all $n \in \mathbb{N}$.

Proof. That λ is a measure satisfying $\mathcal{L}^p(\mu) \subset \mathcal{L}^1(\lambda)$ and (4.45) follows from Theorem 4.39 and the fact that $\lambda^+ = \lambda$ and $\lambda^- = 0$ because Λ is positive. Now define $c := \|\Lambda\|$. We prove in three steps that there exist measurable sets $N \in \mathcal{A}$ and $X_n \in \mathcal{A}$ for $n \in \mathbb{N}$ satisfying (4.46).

Step 1. *For every $\varepsilon > 0$ there exist a measurable set $A \in \mathcal{A}$ and a measurable function*

$$f: X \longrightarrow [0, \infty)$$

such that

$$f|_{X \setminus A} = 0, \quad \inf_A f > 0, \quad \|f\|_p = 1, \quad \Lambda(f) > c - \varepsilon. \tag{4.47}$$

In particular,

$$\mu(A) \leq (\inf_A f)^{-p} < \infty.$$

Choose $h \in \mathcal{L}^p(\mu)$ such that $\|h\|_p = 1$ and $\Lambda(h) > c - \varepsilon$. Assume without loss of generality that $h \geq 0$. (Otherwise replace h by $|h|$.) Define

$$A_i := \{x \in X \mid h(x) > 2^{-i}\}.$$

Then $(h - h\chi_{A_i})^p$ converges pointwise to zero as $i \to \infty$ and is bounded by the integrable function h^p. Hence it follows from the Lebesgue dominated convergence theorem (Theorem 1.45) that

$$\lim_{i \to \infty} \|h - h\chi_{A_i}\|_p = 0,$$

and therefore

$$\lim_{i \to \infty} \Lambda(h\chi_{A_i}) = \Lambda(h) > c - \varepsilon.$$

Choose $i \in \mathbb{N}$ such that

$$\Lambda(h\chi_{A_i}) > c - \varepsilon$$

and define

$$A := A_i, \quad f := \frac{h\chi_{A_i}}{\|h\chi_{A_i}\|_p}.$$

Then A and f satisfy (4.47) and so

$$\mu(A) \leq (\inf_A f)^{-p} \int_X f^p \, d\mu = (\inf_A f)^{-p}.$$

This proves Step 1.

Step 2. *Let ε, A, f be as in Step 1 and let $E \in \mathcal{A}$. Then*

$$(E \cap A = \emptyset, \mu(E) < \infty) \implies \frac{\Lambda(\chi_E)}{\mu(E)^{1/p}} < \varepsilon^{1/q}\Big(\frac{c}{p} + 1\Big), \tag{4.48}$$

where $1 < q < \infty$ is chosen such that $1/p + 1/q = 1$.

Define

$$g := f + \left(\frac{\varepsilon}{\mu(E)}\right)^{1/p} \chi_E.$$

Then

$$\|g\|_p = \left(\int_X f^p \, d\mu + \varepsilon\right)^{1/p} = (1+\varepsilon)^{1/p}$$

and, by (4.47),

$$\Lambda(g) = \Lambda(f) + \left(\frac{\varepsilon}{\mu(E)}\right)^{1/p} \Lambda(\chi_E) > c - \varepsilon + \varepsilon^{1/p} \frac{\Lambda(\chi_E)}{\mu(E)^{1/p}}.$$

Since $\Lambda(g) \le c \, \|g\|_p$, it follows that

$$c - \varepsilon + \varepsilon^{1/p} \frac{\Lambda(\chi_E)}{\mu(E)^{1/p}} < c(1+\varepsilon)^{1/p}.$$

Further since $(1+\varepsilon)^{1/p} - 1 \le \varepsilon/p$ for all $\varepsilon \ge 0$, this implies

$$\varepsilon^{1/p} \frac{\Lambda(\chi_E)}{\mu(E)^{1/p}} < c((1+\varepsilon)^{1/p} - 1) + \varepsilon \le \varepsilon\left(\frac{c}{p} + 1\right).$$

Finally since $\varepsilon^{1-1/p} = \varepsilon^{1/q}$, Step 2 is proved.

Step 3. *There exist measurable sets $N, X_1, X_2, X_3, \dots$ satisfying* (4.46).

Choose $A_n \in \mathcal{A}$ and $f_n \in \mathcal{L}^p(\mu)$ as in Step 1 with $\varepsilon = 1/n$. For $n \in \mathbb{N}$, set

$$X_n := A_1 \cup \cdots \cup A_n, \quad N := X \setminus \bigcup_{n=1}^{\infty} A_n = X \setminus \bigcup_{n=1}^{\infty} X_n.$$

By Step 2, every measurable set $E \subset N$ with $\mu(E) < \infty$ satisfies

$$\frac{\Lambda(\chi_E)}{\mu(E)^{1/p}} < \frac{1}{n^{1/q}}\left(\frac{c}{p} + 1\right)$$

for all $n \in \mathbb{N}$, and hence $\Lambda(\chi_E) = 0$. This implies $\lambda(N) = 0$ by (4.44). Moreover, $\mu(X_n) \le \sum_{i=1}^n \mu(A_i) < \infty$ for every n by Step 1. This proves Step 3 and Theorem 4.40. □

4.6 Exercises

Many of the exercises in this section are taken from Rudin [17, pages 71–75].

Exercise 4.41. Let $(X, \mathcal{A}, \mu)$ be a measure space and let

$$f = (f_1, \dots, f_n) \colon X \longrightarrow \mathbb{R}^n$$

be a measurable function such that

$$\int_X |f_i| \, d\mu < \infty$$

for $i = 1, \dots, n$. Define

$$\int_X f \, d\mu := \left(\int_X f_1 \, d\mu, \dots, \int_X f_n \, d\mu \right) \in \mathbb{R}^n.$$

Let $\mathbb{R}^n \to [0, \infty) \colon v \mapsto \|v\|$, be any norm on $\mathbb{R}^n$. Prove that the function

$$X \longrightarrow [0, \infty) \colon x \longmapsto \|f(x)\|$$

is integrable and

$$\left\| \int_X f \, d\mu \right\| \le \int_X \|f\| \, d\mu. \tag{4.49}$$

Hint. First prove the inequality first for vector-valued integrable step functions $s \colon X \to \mathbb{R}^n$. Then show that for all $\varepsilon > 0$ there is a vector-valued integrable step function $s \colon X \to \mathbb{R}^n$ such that $\|\int_X (f - s) \, d\mu\| < \varepsilon$ and $\int_X \|f - s\| \, d\mu < \varepsilon$.

Exercise 4.42. Let $(X, \mathcal{A}, \mu)$ be a measure space such that $\mu(X) = 1$. Let $f \in \mathcal{L}^1(\mu)$, and let $\phi \colon \mathbb{R} \to \mathbb{R}$ be convex. Prove *Jensen's inequality*

$$\phi\left(\int_X f \, d\mu \right) \le \int_X (\phi \circ f) \, d\mu. \tag{4.50}$$

(In particular, show that $\phi^- \circ f$ is necessarily integrable, so the right-hand side is well defined, even if $\phi \circ f$ is not integrable.) Deduce that

$$\exp\left(\int_X f \, d\mu \right) \le \int_X \exp(f) \, d\mu. \tag{4.51}$$

Deduce also the inequality

$$\sum_{i=1}^n \lambda_i = 1 \implies \prod_{i=1}^n a_i^{\lambda_i} \le \sum_{i=1}^n \lambda_i a_i \tag{4.52}$$

for all positive real numbers λ_i and a_i. In particular, $ab \le a^p/p + b^q/q$ for all positive real numbers a, b, p, q such that $1/p + 1/q = 1$.

Exercise 4.43. Let $(X, \mathcal{A}, \mu)$ be a measure space, choose $p, q, r \in [1, \infty]$ such that

$$\frac{1}{p} + \frac{1}{q} = \frac{1}{r},$$

and let $f \in \mathcal{L}^p(\mu)$ and $g \in \mathcal{L}^q(\mu)$. Prove that $fg \in \mathcal{L}^r(\mu)$ and

$$\|fg\|_r \le \|f\|_p \, \|g\|_q \,. \tag{4.53}$$

Exercise 4.44. Let $(X, \mathcal{A}, \mu)$ be a measure space, choose real numbers

$$1 \le r < p < s < \infty,$$

and let $0 < \lambda < 1$ such that

$$\frac{\lambda}{r} + \frac{1-\lambda}{s} = \frac{1}{p}.$$

Prove that every measurable function $f \colon X \to \mathbb{R}$ satisfies the inequality

$$\|f\|_p \le \|f\|_r^{\lambda} \, \|f\|_s^{1-\lambda} \,. \tag{4.54}$$

Deduce that $\mathcal{L}^r(\mu) \cap \mathcal{L}^s(\mu) \subset \mathcal{L}^p(\mu)$.

Exercise 4.45. Let $(X, \mathcal{A}, \mu)$ be a measure space and let $f \colon X \to \mathbb{R}$ be a measurable function. Define

$$I_f := \{p \in \mathbb{R} \mid 1 < p < \infty,\ f \in \mathcal{L}^p(\mu)\}.$$

Prove that I_f is an interval. Assume f does not vanish almost everywhere and define the function $\phi_f \colon (1, \infty) \to \mathbb{R}$ by

$$\phi_f(p) := p \log \|f\|_p \quad \text{for } p > 1.$$

Prove that ϕ_f is continuous and that the restriction of ϕ_f to the interior of I_f is convex. Find examples where I_f is closed, where I_f is open, and where I_f is a single point. If $I_f \neq \emptyset$ prove that

$$\lim_{p \to \infty} \|f\|_p = \|f\|_\infty \,.$$

Exercise 4.46. For each of the following three conditions find an example of measure space $(X, \mathcal{A}, \mu)$ that satisfies it for all $p, q \in [1, \infty]$.

(a) If $p < q$, then $L^p(\mu) \subsetneq L^q(\mu)$.

(b) If $p < q$, then $L^q(\mu) \subsetneq L^p(\mu)$.

(c) If $p \neq q$, then $L^p(\mu) \not\subset L^q(\mu)$ and $L^q(\mu) \not\subset L^q(\mu)$.

Exercise 4.47. Let $(X, \mathcal{U})$ be a locally compact Hausdorff space and define

$$C_0(X) := \left\{ f\colon X \to \mathbb{R} \;\middle|\; \begin{array}{l} f \text{ is continuous and} \\ \text{for all } \varepsilon > 0, \text{ there exists } K \subset X \text{ such that} \\ K \text{ is compact and } \sup_{X \setminus K} |f| < \varepsilon \end{array} \right\}.$$

Prove that X is a Banach space with respect to the sup-norm. Prove that $C_c(X)$ is dense in $C_0(X)$.

Exercise 4.48. Let $(X, \mathcal{A}, \mu)$ be a measure space such that $\mu(X) = 1$ and let $f, g\colon X \to [0, \infty]$ be measurable functions such that $fg \geq 1$. Prove that

$$\|f\|_1 \, \|g\|_1 \geq 1.$$

Exercise 4.49. Let $(X, \mathcal{A}, \mu)$ be a measure space such that $\mu(X) = 1$ and let $f\colon X \to [0, \infty]$ be a measurable function. Prove that

$$\sqrt{1 + \|f\|_1^2} \leq \int_X \sqrt{1 + f^2} \, d\mu \leq 1 + \|f\|_1 . \tag{4.55}$$

Find a geometric interpretation of this inequality when μ is the restriction of the Lebesgue measure to the unit interval $X = [0, 1]$ and $f = F'$ is the derivative of a continuously differentiable function $F\colon [0, 1] \to \mathbb{R}$. Under which conditions does equality hold in either of the two inequalities in (4.55)?

Exercise 4.50. Let $(X, \mathcal{A}, \mu)$ be a measure space and let $f\colon X \to \mathbb{R}$ be a measurable function such that $f > 0$ and $\int_X f \, d\mu = 1$. Let $E \subset X$ be a measurable set such that $0 < \mu(E) < \infty$. Prove that

$$\int_E \log(f) \, d\mu \leq \mu(E) \log\left(\frac{1}{\mu(E)}\right) \tag{4.56}$$

and

$$\int_E f^p \, d\mu \leq \mu(E)^{1-p} \quad \text{for } 0 < p < 1. \tag{4.57}$$

Exercise 4.51. Let $f\colon [0, 1] \to (0, \infty)$ be Lebesgue measurable. Prove that

$$\int_0^1 f(s) \, ds \int_0^1 \log(f(t)) \, dt \leq \int_0^1 f(x) \log(f(x)) \, dx. \tag{4.58}$$

Exercise 4.52. Fix two constants $1 < p < \infty$ and $a > 0$.

(i) Let $f\colon (0,\infty) \to \mathbb{R}$ be Lebesgue measurable and suppose that the function $(0,\infty) \to \mathbb{R}\colon x \mapsto x^{p-1-a}|f(x)|^p$, is integrable. Show that the restriction of f to each interval $(0,x]$ is integrable and prove *Hardy's inequality*

$$\left(\int_0^\infty x^{-1-a}\left|\int_0^x f(t)\,dt\right|^p dx\right)^{1/p} \le \frac{p}{a}\left(\int_0^\infty x^{p-1-a}|f(x)|^p\,dx\right)^{1/p}. \tag{4.59}$$

Show that equality holds in (4.59) if and only if $f = 0$ almost everywhere. *Hint.* Assume first that f is nonnegative with compact support and define

$$F(x) := \frac{1}{x}\int_0^x f(t)\,dt \quad \text{for } x > 0.$$

Use integration by parts to obtain

$$\int_0^\infty x^{p-1-a}F(x)^p\,dx = \frac{p}{a}\int_0^\infty x^{p-1-a}F(x)^{p-1}f(x)\,dx.$$

Use Hölder's inequality.

(ii) Show that the constant p/a in Hardy's inequality is sharp. *Hint.* Choose $\lambda < 1 - a/p$ and take $f(x) := x^{-\lambda}$ for $x \le 1$ and $f(x) := 0$ for $x > 1$.

(iii) Prove that every sequence $(a_n)_{n\in\mathbb{N}}$ of positive real numbers satisfies

$$\sum_{N=1}^\infty \left(\frac{1}{N}\sum_{n=1}^N a_n\right)^p \le \left(\frac{p}{p-1}\right)^p \sum_{n=1}^\infty a_n^p. \tag{4.60}$$

Hint. If a_n is nonincreasing, then (4.60) follows from (4.59) with $a = p - 1$ for a suitable function f. Deduce the general case from the special case.

(iv) Let $f\colon (0,\infty) \to \mathbb{R}$ be Lebesgue measurable and suppose that the function $(0,\infty) \to \mathbb{R}\colon x \mapsto x^{p-1+a}|f(x)|^p$, is integrable. Show that the restriction of f to each interval $[x,\infty)$ is integrable and prove the inequality

$$\left(\int_0^\infty x^{a-1}\left|\int_x^\infty f(t)\,dt\right|^p dx\right)^{1/p} \le \frac{p}{a}\left(\int_0^\infty x^{p-1+a}|f(x)|^p\,dx\right)^{1/p}. \tag{4.61}$$

Hint. Apply the inequality (4.59) to the function $g(x) := x^{-2}f(x^{-1})$.

Exercise 4.53. Let $(X, \mathcal{U})$ be a locally compact Hausdorff space and let $\mu: \mathcal{B} \to [0, \infty]$ be an outer regular Borel measure on X that is inner regular on open sets. Let $g \in \mathcal{L}^1(\mu)$. Prove that the following are equivalent:

(i) the function g vanishes μ-almost everywhere;

(ii) $\int_X fg \, d\mu = 0$ for all $f \in C_c(X)$.

Hint. Assume (ii). Let $K \subset X$ be compact. Use Urysohn's lemma (Theorem A.1) to show that there is a sequence $f_n \in C_c(X)$ such that $0 \le f_n \le 1$ and f_n converges almost everywhere to χ_K. Deduce that $\int_K g \, d\mu = 0$. Then prove that $\int_U g \, d\mu = 0$ for every open set $U \subset X$ and $\int_B g \, d\mu = 0$ for all $B \in \mathcal{B}$.

Warning! The regularity hypotheses on μ cannot be removed. Find an example of a Borel measure where (ii) does not imply (i). (See Example 4.16.)

Exercise 4.54. Egoroff's theorem. *Let $(X, \mathcal{A}, \mu)$ be a measure space such that $\mu(X) < \infty$ and let $f_n: X \to \mathbb{R}$ be a sequence of measurable functions that converges pointwise to $f: X \to \mathbb{R}$. Fix a constant $\varepsilon > 0$. Then there exists a measurable set $E \in \mathcal{A}$ such that*

$$\mu(X \setminus E) < \varepsilon$$

and $f_n|_E$ converges uniformly to $f|_E$.

Hint. Define

$$S(k, n) := \{x \in X \mid |f_i(x) - f_j(x)| < 1/k, \text{ for all } i, j > n\} \quad \text{for } k, n \in \mathbb{N}.$$

Prove that

$$\lim_{n \to \infty} \mu(S(k, n)) = \mu(X) \quad \text{for all } k \in \mathbb{N}.$$

Deduce that there is a sequence $n_k \in \mathbb{N}$ such that $E := \bigcap_{k \in \mathbb{N}} S(k, n_k)$ satisfies the required conditions. Show that Egoroff's theorem does not extend to σ-finite measure spaces.

Exercise 4.55. Let $(X, \mathcal{A}, \mu)$ be a measure space and let $1 < p < \infty$. Let $f \in \mathcal{L}^p(\mu)$ and let $f_n \in \mathcal{L}^p(\mu)$ be a sequence such that

$$\lim_{n \to \infty} \|f_n\|_p = \|f\|_p$$

and f_n converges to f almost everywhere. Prove that

$$\lim_{n \to \infty} \|f - f_n\|_p = 0.$$

Prove that the hypothesis

$$\lim_{n\to\infty} \|f_n\|_p = \|f\|_p$$

cannot be removed.

Hint 1. Fix a constant $\varepsilon > 0$. Use Egoroff's theorem to construct disjoint measurable sets $A, B \in \mathcal{A}$ such that $X = A \cup B$, $\int_A |f|^p \, d\mu < \varepsilon$, $\mu(B) < \infty$, and f_n converges to f uniformly on B. Use Fatou's lemma (Theorem 1.41) to prove that $\limsup_{n\to\infty} \int_A |f_n|^p \, d\mu < \varepsilon$.

Hint 2. Let $g_n := 2^{p-1}(|f_n|^p + |f|^p) - |f - f_n|^p$ and use Fatou's lemma (Theorem 1.41) as in the proof of the Lebesgue dominated convergence theorem (Theorem 1.45).

Exercise 4.56. Let $(X, \mathcal{A}, \mu)$ be a measure space, let $f_n\colon X \to \mathbb{R}$ be a sequence of measurable functions, and let $f\colon X \to \mathbb{R}$ be a measurable function. The sequence f_n is said to *converge in measure* to f if

$$\lim_{n\to\infty} \mu(\{x \in X \mid |f_n(x) - f(x)| > \varepsilon\}) = 0$$

for all $\varepsilon > 0$. (On page 53 this is called *convergence in probability.*) Assume $\mu(X) < \infty$ and prove the following.

(i) If f_n converges to f almost everywhere, then f_n converges to f in measure. *Hint.* See page 53.

(ii) If f_n converges to f in measure, then a subsequence of f_n converges to f almost everywhere.

(iii) If $1 \le p \le \infty$ and $f_n, f \in \mathcal{L}^p(\mu)$ satisfy

$$\lim_{n\to\infty} \|f_n - f\|_p = 0,$$

then f_n converges to f in measure.

Exercise 4.57. Let $(X, \mathcal{U})$ be a compact Hausdorff space and $\mu\colon \mathcal{B} \to [0, \infty]$ be a Borel measure. Let $C(X) = C_c(X)$ be the space of continuous real-valued functions on X. Consider the following conditions:

(a) every nonempty open subset of X has positive measure;

(b) there exists a Borel set $E \subset X$ and an element $x_0 \in X$ such that every open neighborhood U of x_0 satisfies $\mu(U \cap E) > 0$ and $\mu(U \setminus E) > 0$;

(c) μ is outer regular and is inner regular on open sets.

Prove the following.

(i) Assume (a). Then the map $C(X) \to L^\infty(\mu)$ in (b) is an isometric embedding and hence its image is a closed linear subspace of $L^\infty(\mu)$.

(ii) Assume (a) and (b). Then there is a nonzero bounded linear functional

$$\Lambda\colon L^\infty(\mu) \longrightarrow \mathbb{R}$$

that vanishes on the image of the inclusion $C(\mu) \to L^\infty(\mu)$. *Hint.* If $f = \chi_E$ almost everywhere, then f is discontinuous at x_0.

(iii) Assume (a), (b), and (c). Then the isometric embedding $L^1(\mu) \to L^\infty(\mu)^*$ of Theorem 4.33 is not surjective. *Hint.* Use part (ii) and Exercise 4.53.

(iv) The Lebesgue measure on [0, 1] satisfies (a), (b), and (c).

Exercise 4.58. Prove that every σ-finite measure space $(X, \mathcal{A}, \mu)$ is localizable. *Hint.* Assume first that $\mu(X) < \infty$. Let $\mathcal{E} \subset \mathcal{A}$ and define

$$c := \sup\{\mu(E_1 \cup \cdots \cup E_n) \mid n \in \mathbb{N},\ E_1, \ldots, E_n \in \mathcal{E}\}.$$

Show that there is a sequence $E_i \in \mathcal{E}$ such that $\mu(\bigcup_{i=1}^\infty E_i) = c$. Prove that $H := \bigcup_{i=1}^\infty E_i$ is an envelope of $\mathcal{E}$.

Exercise 4.59. Let $(X, \mathcal{A}, \mu)$ be a localizable measure space. Prove that it satisfies the following.

(F) *Let $\mathcal{F}$ be a collection of measurable functions $f\colon A_f \to \mathbb{R}$, each defined on a measurable set $A_f \in \mathcal{A}$. Suppose that any two functions $f_1, f_2 \in \mathcal{F}$ agree almost everywhere on $A_{f_1} \cap A_{f_2}$. Then there exists a measurable function $g\colon X \to \mathbb{R}$ such that $g|_{A_f} = f$ almost everywhere for all $f \in \mathcal{F}$.*

We will see in the next chapter that condition (F) is equivalent to localizability for semi-finite measure spaces. *Hint.* Let $\mathcal{F}$ be a collection of measurable functions as in (F). For $a \in \mathbb{R}$ and $f \in \mathcal{F}$ define

$$A_f^a := \{x \in A_f \mid f(x) < a\}.$$

For $q \in \mathbb{Q}$, let $H^q \in \mathcal{A}$ be an envelope of the collection

$$\mathcal{E}^q := \{A_f^q \mid f \in \mathcal{F}\}.$$

Define the measurable sets

$$X^a := \bigcup_{\substack{q \in \mathbb{Q} \\ q < a}} H^q, \quad a \in \mathbb{R}.$$

Prove the following.

(i) If $a < b$, then $X^a \subset X^b$.

(ii) For every $a \in \mathbb{R}$ the measurable set X^a is an envelope of the collection

$$\mathcal{E}^a := \{A_f^a \mid f \in \mathcal{F}\}.$$

Thus $\mu(A_f^a \setminus X^a) = 0$ for all $f \in \mathcal{F}$ and, if $G \in \mathcal{A}$, then

$$\mu(A_f^a \setminus G) = 0 \text{ for all } f \in \mathcal{F} \implies \mu(X^a \setminus G) = 0.$$

(iii) If $a \in \mathbb{R}$ and $E \in \mathcal{A}$, then

$$\mu(A_f^a \cap E) = 0 \text{ for all } f \in \mathcal{F} \implies \mu(X^a \cap E) = 0.$$

(iv) $\mu(X^a \cap A_f \setminus A_f^a) = 0$ for all $f \in \mathcal{F}$ and all $a \in \mathbb{R}$.

(v) Define

$$E_0 := \Big(\bigcap_{r \in \mathbb{R}} X^r\Big) \cup \Big(X \setminus \bigcup_{s \in \mathbb{R}} X^s\Big).$$

Then E_0 is measurable and $\mu(A_f \cap E_0) = 0$ for all $f \in \mathcal{F}$.

(vi) For $f \in \mathcal{F}$ define the measurable set $E_f \subset A_f$ by

$$E_f := (A_f \cap E_0) \cup \bigcup_{q \in \mathbb{Q}} (A_f^q \setminus X^q) \cup \bigcup_{q \in \mathbb{Q}} (X^q \cap A_f \setminus A_f^q).$$

Then $\mu(E_f) = 0$.

(vii) Define $g \colon X \to \mathbb{R}$ by

$$g(x) := \begin{cases} 0 & \text{if } x \in E_0, \\ a & \text{if } x \in X^s \text{ for all } s > a \text{ and } x \notin X^r \text{ for all } r < a. \end{cases} \tag{4.62}$$

Then g is well defined and measurable and $g = f$ on $A_f \setminus E_f$ for all $f \in \mathcal{F}$.

Example 4.60. This example is closely related to Exercise 3.24, however, it requires a considerable knowledge of *functional analysis* and the details go much beyond the scope of the present manuscript. It introduces the *Stone–Čech compactification* X of the natural numbers. This is a compact Hausdorff space containing $\mathbb{N}$ and satisfying the *universality property* that every continuous map from $\mathbb{N}$ to another compact Hausdorff space Y extends uniquely to a continuous map from X to Y.

The space $C(X)$ of continuous functions on X can be naturally identified with the space ℓ^∞. Hence the space of positive bounded linear functionals on ℓ^∞ is isomorphic to the space of Radon measures on X by Theorem 3.15. Thus the Stone–Čech compactification of $\mathbb{N}$ can be used to understand the dual space of ℓ^∞. Moreover, it gives rise to an interesting example of a Radon measure which is not outer regular (explained to me by Theo Buehler).

Consider the inclusion

$$\mathbb{N} \longrightarrow (\ell^\infty)^*: n \longmapsto \Lambda_n$$

which assigns to each natural number $n \in \mathbb{N}$ the bounded linear functional

$$\Lambda_n: \ell^\infty \longrightarrow \mathbb{R}$$

defined by

$$\Lambda_n(\xi) := \xi_n \quad \text{for } \xi = (\xi_i)_{i\in\mathbb{N}} \in \ell^\infty.$$

This functional has norm one. Now the space of all bounded linear functionals on ℓ^∞ of norm at most one, i.e., the unit ball in $(\ell^\infty)^*$, is compact with respect to the weak-$*$ topology by the Banach–Alaoglu theorem. Define X to be the closure of the set $\{\Lambda_n \mid n \in \mathbb{N}\}$ in $(\ell^\infty)^*$ with respect to the weak-$*$ topology. Thus

$$X := \left\{ \Lambda \in (\ell^\infty)^* \;\middle|\; \begin{array}{l} \text{For all finite sequences } c^1, \dots, c^\ell \in \mathbb{R} \\ \text{and } \xi^1 = (\xi_i^1)_{i\in\mathbb{N}}, \dots, \xi^\ell = (\xi_i^\ell)_{i\in\mathbb{N}} \in \ell^\infty \\ \text{satisfying } \Lambda(\xi^j) < c^j \text{ for } j = 1, \dots, \ell, \\ \text{there exists an } n \in \mathbb{N} \text{ such that} \\ \xi_n^j < c^j \text{ for } j = 1, \dots, \ell \end{array} \right\}.$$

The weak-$*$ topology $\mathcal{U} \subset 2^X$ is the smallest topology such that the map

$$f_\xi: X \longrightarrow \mathbb{R}, \quad f_\xi(\Lambda) := \Lambda(\xi),$$

is continuous for each $\xi \in \ell^\infty$. The topological space $(X, \mathcal{U})$ is a separable compact Hausdorff space, called the *Stone–Čech compactification* of $\mathbb{N}$. It is not second countable and one can show that the complement of a point in X that is not equal to one of the Λ_n is not σ-compact. The only continuous functions on X are those of the form f_ξ, so the map

$$\ell^\infty \longrightarrow C(X): \xi \longmapsto f_\xi$$

is a Banach space isometry. (Verify that $\|f_\xi\| := \sup_{\Lambda\in X} |f_\xi(\Lambda)| = \|\xi\|_\infty$ for all $\xi \in \ell^\infty$.) Thus the dual space of ℓ^∞ can be understood in terms of the Borel measures on X.

By Theorem 3.18, every Radon measure on X is regular. However, the Borel σ-algebra $\mathcal{B} \subset 2^X$ does carry σ-finite measures $\mu\colon \mathcal{B} \to [0,\infty]$ that are inner regular, but not outer regular (and must necessarily satisfy $\mu(X) = \infty$). Here is an example pointed out to me by Theo Buehler. Define

$$\mu(B) := \sum_{\substack{n\in\mathbb{N} \\ \Lambda_n \in B}} \frac{1}{n}$$

for every Borel set $B \subset X$. This measure is σ-finite and inner regular, but not outer regular. (The set $U := \{\Lambda_n \mid n \in \mathbb{N}\}$ is open, its complement $K := X \setminus U$ is compact and has measure zero, and every open set containing K misses only a finite subset of U and hence has infinite measure.) Now let $X_0 \subset X$ be the union of all open sets in X with finite measure. Then X_0 is not σ-compact and the restriction of μ to the Borel σ-algebra of X_0 is a Radon measure, but is not outer regular.

Chapter 5

The Radon–Nikodým theorem

Recall from Theorem 1.40 that every measurable function $f: X \to [0, \infty)$ on a measure space $(X, \mathcal{A}, \mu)$ determines a measure $\mu_f: \mathcal{A} \to [0, \infty]$ defined by

$$\mu_f(A) := \int_A f \, d\mu \quad \text{for } A \in \mathcal{A}.$$

By Theorem 1.35 it satisfies $\mu_f(A) = 0$ whenever $\mu(A) = 0$. A measure with this property is called *absolutely continuous* with respect to μ. The Radon–Nikodým theorem asserts that, when μ is σ-finite, every σ-finite measure that is absolutely continuous with respect to μ has the form μ_f for some measurable function $f: X \to [0, \infty)$. It was proved by Johann Radon in 1913 for the Lebesgue measure space and extended by Otton Nikodým in 1930 to general σ-finite measure spaces. A proof is given in Section 5.1. Consequences of the Radon–Nikodým theorem include the proof of Theorem 4.35 about the dual space of $L^p(\mu)$ (Section 5.2) and the decomposition theorems of Lebesgue, Hahn, and Jordan for signed measures (Section 5.3). An extension of the Radon–Nikodým theorem to general measure spaces is proved in Section 5.4.

5.1 Absolutely continuous measures

Definition 5.1. Let $(X, \mathcal{A}, \mu)$ be a measure space. A measure $\lambda: \mathcal{A} \to [0, \infty)$ is *absolutely continuous with respect to* μ if

$$\mu(A) = 0 \implies \lambda(A) = 0 \quad \text{for all } A \in \mathcal{A}.$$

λ is *singular with respect to* μ if there exists a measurable set A such that

$$\lambda(A) = 0 \quad \text{and} \quad \mu(A^c) = 0.$$

In this case we also say that λ and μ are *mutually singular*. We write

$$\lambda \ll \mu$$

if λ is absolutely continuous with respect to μ and $\lambda \perp \mu$ if λ and μ are mutually singular.

Lemma 5.2. *Let $(X, \mathcal{A})$ be a measurable space and let $\mu, \lambda, \lambda_1, \lambda_2$ be measures on $\mathcal{A}$. Then the following holds*:

(i) *if $\lambda_1 \perp \mu$ and $\lambda_2 \perp \mu$, then $\lambda_1 + \lambda_2 \perp \mu$;*

(ii) *if $\lambda_1 \ll \mu$ and $\lambda_2 \ll \mu$, then $\lambda_1 + \lambda_2 \ll \mu$;*

(iii) *if $\lambda_1 \ll \mu$ and $\lambda_2 \perp \mu$, then $\lambda_1 \perp \lambda_2$;*

(iv) *if $\lambda \ll \mu$ and $\lambda \perp \mu$, then $\lambda = 0$.*

Proof. (i) Suppose that $\lambda_1 \perp \mu$ and $\lambda_2 \perp \mu$. Then there exist measurable sets $A_i \in \mathcal{A}$ such that $\lambda_i(A_i) = 0$ and $\mu(A_i^c) = 0$ for $i = 1, 2$. Define $A := A_1 \cap A_2$. Then $A^c = A_1^c \cup A_2^c$ is a null set for μ and A is a null set for both λ_1 and λ_2, and hence also for $\lambda_1 + \lambda_2$. Thus $\lambda_1 + \lambda_2 \perp \mu$ and this proves (i).

(ii) Suppose that $\lambda_1 \ll \mu$ and $\lambda_2 \ll \mu$. If $A \in \mathcal{A}$ satisfies $\mu(A) = 0$, then $\lambda_1(A) = \lambda_2(A) = 0$ and so $(\lambda_1 + \lambda_2)(A) = \lambda_1(A) + \lambda_2(A) = 0$. Thus $\lambda_1 + \lambda_2 \ll \mu$ and this proves (ii).

(iii) Suppose that $\lambda_1 \ll \mu$ and $\lambda_2 \perp \mu$. Since $\lambda_2 \perp \mu$, there exists a measurable set $A \in \mathcal{A}$ such that $\lambda_2(A) = 0$ and $\mu(A^c) = 0$. Since $\lambda_1 \ll \mu$ it follows that $\lambda_1(A^c) = 0$ and hence $\lambda_1 \perp \lambda_2$.This proves (iii).

(iv) Suppose that $\lambda \ll \mu$ and $\lambda \perp \mu$. Since $\lambda \perp \mu$, there exists a measurable set $A \in \mathcal{A}$ such that $\lambda(A) = 0$ and $\mu(A^c) = 0$. Since $\lambda \ll \mu$, it follows that $\lambda(A^c) = 0$ and hence $\lambda(X) = \lambda(A) + \lambda(A^c) = 0$. This proves (iv) and Lemma 5.2. □

Theorem 5.3 (Lebesgue decomposition theorem). *Let $(X, \mathcal{A}, \mu)$ be a σ-finite measure space and let λ be a σ-finite measure on $\mathcal{A}$. Then there exist unique measures $\lambda_a, \lambda_s \colon \mathcal{A} \to [0, \infty]$ such that*

$$\lambda = \lambda_a + \lambda_s, \quad \lambda_a \ll \mu, \quad \lambda_s \perp \mu. \tag{5.1}$$

Proof. See page 186. □

Theorem 5.4 (Radon–Nikodým). *Let $(X, \mathcal{A}, \mu)$ be a σ-finite measure space and let $\lambda \colon \mathcal{A} \to [0, \infty]$ be a measure. The following are equivalent*:

(i) *λ is σ-finite and absolutely continuous with respect to μ;*

(ii) *there exists a measurable function $f \colon X \to [0, \infty)$ such that*

$$\lambda(A) = \int_A f \, d\mu \quad \text{for all } A \in \mathcal{A}. \tag{5.2}$$

If (i) *holds, then equation* (5.2) *determines f uniquely up to equality μ-almost everywhere. Moreover, $f \in \mathcal{L}^1(\mu)$ if and only if $\lambda(X) < \infty$.*

Proof. The last assertion follows by taking $A = X$ in (5.2).

We prove that (ii) $\Longrightarrow$ (i). Thus assume that there exists a measurable function $f: X \to [0, \infty)$ such that λ is given by (5.2). Then λ is absolutely continuous with respect to μ by Theorem 1.35. Since μ is σ-finite, there exists a sequence of measurable sets $X_1 \subset X_2 \subset X_3 \subset \cdots$ such that $\mu(X_n) < \infty$ and $X = \bigcup_{n=1}^{\infty} X_n$. Define

$$A_n := \{x \in X_n \mid f(x) \le n\}.$$

Then $A_n \subset A_{n+1}$ and $\lambda(A_n) \le n\mu(X_n) < \infty$ for all n and $X = \bigcup_{n=1}^{\infty} A_n$. Thus λ is σ-finite and this shows that (ii) $\Longrightarrow$ (i).

It remains to prove that (i) $\Longrightarrow$ (ii) and that f is uniquely determined by (5.2) up to equality μ-almost everywhere. This is proved in three steps. The first step is uniqueness, the second step is existence under the assumption $\lambda(X) < \infty$ and $\mu(X) < \infty$, and the last step establishes existence in general.

Step 1. *Let $(X, \mathcal{A}, \mu)$ be a measure space, let $\lambda: \mathcal{A} \to [0, \infty]$ be a σ-finite measure, and let $f, g: X \to [0, \infty)$ be two measurable functions such that*

$$\lambda(A) = \int_A f \, d\mu = \int_A g \, d\mu \quad \text{for all } A \in \mathcal{A}. \tag{5.3}$$

Then f and g agree μ-almost everywhere.

Since $(X, \mathcal{A}, \lambda)$ is a σ-finite measure space, there exists a sequence of measurable sets $A_1 \subset A_2 \subset A_3 \subset \cdots$ such that $\lambda(A_n) < \infty$ for all $n \in \mathbb{N}$ and $X = \bigcup_{n=1}^{\infty} A_n$. For $n \in \mathbb{N}$ put

$$\mathcal{A}_n := \{E \in \mathcal{A} \mid E \subset A_n\}, \qquad \mu_n := \mu|_{\mathcal{A}_n}.$$

Take $A = A_n$ in (5.3) to obtain $f, g \in \mathcal{L}^1(\mu_n)$ for all n. Thus

$$f - g \in \mathcal{L}^1(\mu_n), \qquad \int_E (f - g) \, d\mu_n = 0 \quad \text{for all } E \in \mathcal{A}_n.$$

Hence $f - g$ vanishes μ_n-almost everywhere by Lemma 1.49. Thus the set

$$E_n := \{x \in A_n \mid f(x) \ne g(x)\}$$

satisfies $\mu(E_n) = \mu_n(E_n) = 0$, and hence the set

$$E := \{x \in X \mid f(x) \ne g(x)\} = \bigcup_{n=1}^{\infty} E_n$$

satisfies $\mu(E) = 0$. This proves Step 1.

Step 2. *Let $(X, \mathcal{A})$ be a measurable space and let $\lambda, \mu \colon \mathcal{A} \to [0, \infty]$ be measures such that $\lambda(X) < \infty$, $\mu(X) < \infty$, and $\lambda \ll \mu$. Then there exists a measurable function*

$$h \colon X \longrightarrow [0, \infty)$$

such that

$$\lambda(A) = \int_A h \, d\mu \quad \textit{for all } A \in \mathcal{A}.$$

By assumption, $\lambda + \mu \colon \mathcal{A} \to [0, \infty]$ is a finite measure defined by

$$(\lambda + \mu)(A) := \lambda(A) + \mu(A) \quad \text{for } A \in \mathcal{A}.$$

Since $(\lambda + \mu)(X) < \infty$, it follows from the Cauchy–Schwarz inequality that

$$H := L^2(\lambda + \mu) \subset L^1(\lambda + \mu).$$

Namely, if $f \in \mathcal{L}^2(\lambda + \mu)$, then

$$\int_X |f| \, d(\lambda + \mu) \le c \sqrt{\int_X |f|^2 \, d(\lambda + \mu)} < \infty, \quad c := \sqrt{\lambda(X) + \mu(X)}.$$

Define

$$\Lambda \colon L^2(\lambda + \mu) \longrightarrow \mathbb{R}$$

by

$$\Lambda(f) := \int_X f \, d\lambda.$$

for $f \in L^2(\lambda + \mu)$. (Here we abuse notation and use the same letter f for a function in $\mathcal{L}^2(\lambda + \mu)$ and its equivalence class in $L^2(\lambda + \mu)$.) Then

$$|\Lambda(f)| \le \int_X |f| \, d\lambda \le \int_X |f| \, d(\lambda + \mu) \le c \, \|f\|_{L^2(\lambda + \mu)}$$

for all $f \in L^2(\lambda + \mu)$. Thus Λ is a bounded linear functional on $L^2(\lambda + \mu)$ and it follows from Corollary 4.28 that there exists an L^2-function $g \in \mathcal{L}^2(\lambda + \mu)$ such that

$$\int_X f \, d\lambda = \int_X fg \, d(\lambda + \mu) \tag{5.4}$$

for all $f \in \mathcal{L}^2(\lambda + \mu)$. This implies

$$\begin{aligned} \int_X f(1 - g) \, d(\lambda + \mu) &= \int_X f \, d(\lambda + \mu) - \int_X fg \, d(\lambda + \mu) \\ &= \int_X f \, d(\lambda + \mu) - \int_X f \, d\lambda \qquad (5.5) \\ &= \int_X f \, d\mu \end{aligned}$$

for all $f \in \mathcal{L}^2(\lambda + \mu)$.

We claim that the inequalities $0 \le g < 1$ hold $(\lambda + \mu)$-almost everywhere. To see this, consider the measurable sets

$$E_0 := \{x \in X \mid g(x) < 0\}, \quad E_1 := \{x \in X \mid g(x) \ge 1\}.$$

Then it follows from (5.4) with $f := \chi_{E_0}$ that

$$0 \le \lambda(E_0) = \int_X \chi_{E_0}\, d\lambda = \int_X \chi_{E_0} g\, d(\lambda + \mu) \le 0.$$

Hence

$$\int_X \chi_{E_0} g\, d(\lambda + \mu) = 0$$

and Lemma 1.49 shows that the function $f := -\chi_{E_0} g$ vanishes $(\lambda + \mu)$-almost everywhere. Hence $(\lambda + \mu)(E_0) = 0$. Likewise, (5.5) with $f := \chi_{E_1}$ yields that

$$\mu(E_1) = \int_X \chi_{E_1}\, d\mu = \int_{E_1} (1 - g)\, d(\lambda + \mu) \le 0.$$

Hence $\mu(E_1) = 0$. Since λ is absolutely continuous with respect to μ, it follows that $\lambda(E_1) = 0$ and hence $(\lambda + \mu)(E_1) = 0$ as claimed. Assume from now on that $0 \le g(x) < 1$ for all $x \in X$. (Namely, redefine $g(x) := 0$ for $x \in E_0 \cup E_1$ without changing the identities (5.4) and (5.5).)

Apply identity (5.5) to the characteristic function $f := \chi_A \in \mathcal{L}^2(\lambda + \mu)$ of a measurable set A to obtain the identity

$$\mu(A) = \int_A (1 - g)\, d(\lambda + \mu) \quad \text{for all } A \in \mathcal{A}.$$

By Theorem 1.40 this implies that (5.5) continues to hold for every measurable function $f\colon X \to [0, \infty)$, whether or not it belongs to $\mathcal{L}^2(\lambda + \mu)$. Now define the measurable function

$$h\colon X \longrightarrow [0, \infty)$$

by

$$h(x) := \frac{g(x)}{1 - g(x)} \quad \text{for } x \in X.$$

By (5.4) with $f = \chi_A$ and (5.5) with $f = \chi_A h$ it satisfies

$$\begin{aligned}\lambda(A) &= \int_X \chi_A \, d\lambda \\ &= \int_X \chi_A g \, d(\lambda + \mu) \\ &= \int_X \chi_A h(1-g) \, d(\lambda + \mu) \\ &= \int_X \chi_A h \, d\mu \\ &= \int_A h \, d\mu\end{aligned}$$

for all $A \in \mathcal{A}$. This proves Step 2.

Step 3. *We prove that* (i) $\Longrightarrow$ (ii).

Since λ and μ are σ-finite measures, there exist sequences of measurable sets $A_n, B_n \in \mathcal{A}$ such that $A_n \subset A_{n+1}$, $\lambda(A_n) < \infty$, $B_n \subset B_{n+1}$, $\mu(B_n) < \infty$ for all n, and $X = \bigcup_{n=1}^\infty A_n = \bigcup_{n=1}^\infty B_n$. Define $X_n := A_n \cap B_n$. Then

$$X_n \subset X_{n+1}, \quad \lambda(X_n) < \infty, \quad \mu(X_n) < \infty$$

for all n, and $X = \bigcup_{n=1}^\infty X_n$. Thus, by Step 2, there exists a sequence of measurable functions $f_n \colon X_n \to [0,\infty)$ such that

$$\lambda(A) = \int_A f_n \, d\mu \quad \text{for all } n \in \mathbb{N} \text{ and all } A \in \mathcal{A} \text{ such that } A \subset X_n. \tag{5.6}$$

It follows from Step 1 that the restriction of f_{n+1} to X_n agrees with f_n μ-almost everywhere. Thus, modifying f_{n+1} on a set of measure zero if necessary, we may assume without loss of generality that $f_{n+1}|_{X_n} = f_n$ for all $n \in \mathbb{N}$. With this understood, define

$$f \colon X \longrightarrow [0,\infty)$$

by

$$f|_{X_n} := f_n \quad \text{for } n \in \mathbb{N}.$$

This function is measurable because

$$f^{-1}([0,c]) = \bigcup_{n=1}^{\infty}(X_n \cap f^{-1}([0,c])) = \bigcup_{n=1}^{\infty} f_n^{-1}([0,c]) \in \mathcal{A}$$

for all $c \geq 0$. Now let $E \in \mathcal{A}$ and define $E_n := E \cap X_n \in \mathcal{A}$ for $n \in \mathbb{N}$. Then

$$E_1 \subset E_2 \subset E_3 \subset \cdots, \quad E = \bigcup_{n=1}^{\infty} E_n.$$

Hence it follows from part (iv) of Theorem 1.28 that

$$\begin{aligned}\lambda(E) &= \lim_{n\to\infty} \lambda(E_n) \\ &= \lim_{n\to\infty} \int_{E_n} f \, d\mu \\ &= \lim_{n\to\infty} \int_X \chi_{E_n} f \, d\mu \\ &= \int_X \chi_E f \, d\mu \\ &= \int_E f \, d\mu.\end{aligned}$$

Here the last but one equality follows from the Lebesgue monotone convergence theorem (Theorem 1.37). This proves Step 3 and Theorem 5.4. □

Example 5.5. Let X be a one-element set and let $\mathcal{A} := 2^X$. Define the measure $\mu: 2^X \to [0,\infty]$ by $\mu(\emptyset) := 0$ and $\mu(X) := \infty$.

(i) Choose $\lambda(\emptyset) := 0$ and $\lambda(X) := 1$. Then $\lambda \ll \mu$, but there does not exist a (measurable) function $f: X \to [0,\infty]$ such that $\int_X f \, d\mu = \lambda(X)$. Thus the hypothesis that $(X, \mathcal{A}, \mu)$ is σ-finite cannot be removed in Theorem 5.4.

(ii) Choose $\lambda := \mu$. Then we get $\lambda(A) = \int_A f \, d\mu$ for every nonzero function $f: X \to [0,\infty)$. Thus the hypothesis that $(X, \mathcal{A}, \lambda)$ is σ-finite cannot be removed in Step 1 in the proof of Theorem 5.4.

Example 5.6. Let X be an uncountable set and denote by $\mathcal{A} \subset 2^X$ the set of all subsets $A \subset X$ such that either A or A^c is countable. Choose an uncountable subset $H \subset X$ with an uncountable complement and define $\lambda, \mu, \nu: \mathcal{A} \to [0,\infty]$ by

$$\lambda(A) := \begin{cases} 0 & \text{if } A \text{ is countable}, \\ 1 & \text{if } A^c \text{ is countable}, \end{cases} \qquad \mu(A) := \#(A \cap H), \quad \nu(A) := \#A.$$

Then $\lambda \ll \mu \ll \nu$ and μ and ν are not σ-finite. There does not exist any measurable function $f: X \to [0,\infty]$ such that $\lambda(X) = \int_X f\,d\mu$. Nor is there any measurable function $h: X \to \mathbb{R}$ such that $\mu(A) = \int_A h\,d\nu$ for all $A \in \mathcal{A}$. (The only possible such function would be $h := \chi_H$ which is not measurable.)

Proof of Theorem 5.3. We prove uniqueness. Let $\lambda_{\mathrm{a}}, \lambda_{\mathrm{s}}, \lambda'_{\mathrm{a}}, \lambda'_{\mathrm{s}}: \mathcal{A} \to [0,\infty]$ be measures such that

$$\lambda = \lambda_{\mathrm{a}} + \lambda_{\mathrm{s}} = \lambda'_{\mathrm{a}} + \lambda'_{\mathrm{s}}, \quad \lambda_{\mathrm{a}} \ll \mu, \quad \lambda'_{\mathrm{a}} \ll \mu, \quad \lambda_{\mathrm{s}} \perp \mu, \quad \lambda'_{\mathrm{s}} \perp \mu.$$

Then there exist measurable sets $A, A' \in \mathcal{A}$ such that

$$\lambda_{\mathrm{s}}(A) = 0, \quad \mu(X \setminus A) = 0, \quad \lambda'_{\mathrm{s}}(A') = 0, \quad \mu(X \setminus A') = 0.$$

Since $X \setminus (A \cap A') = (X \setminus A) \cup (X \setminus A')$, this implies

$$\mu(X \setminus (A \cap A')) = 0.$$

Let $E \in \mathcal{A}$. Then

$$\lambda_{\mathrm{s}}(E \cap A \cap A') = 0 = \lambda'_{\mathrm{s}}(E \cap A \cap A')$$

and hence

$$\lambda_{\mathrm{a}}(E \cap A \cap A') = \lambda(E \cap A \cap A') = \lambda'_{\mathrm{a}}(E \cap A \cap A').$$

Moreover $\mu(E \setminus (A \cap A')) = 0$, hence

$$\lambda_{\mathrm{a}}(E \setminus (A \cap A')) = 0 = \lambda'_{\mathrm{a}}(E \setminus (A \cap A'))$$

and so

$$\lambda_{\mathrm{s}}(E \setminus (A \cap A')) = \lambda(E \setminus (A \cap A')) = \lambda'_{\mathrm{s}}(E \setminus (A \cap A')).$$

This implies

$$\begin{aligned}\lambda_{\mathrm{a}}(E) &= \lambda_{\mathrm{a}}(E \cap A \cap A') = \lambda'_{\mathrm{a}}(E \cap A \cap A') = \lambda'_{\mathrm{a}}(E)\\ \lambda_{\mathrm{s}}(E) &= \lambda_{\mathrm{s}}(E \setminus (A \cap A')) = \lambda'_{\mathrm{s}}(E \setminus (A \cap A')) = \lambda'_{\mathrm{s}}(E).\end{aligned}$$

This proves uniqueness.

We prove existence. The measure

$$\nu := \lambda + \mu: \mathcal{A} \longrightarrow [0,\infty]$$

is σ-finite. Hence it follows from the Radon–Nikodým theorem (Theorem 5.4) that there exist measurable functions $f, g: X \to [0,\infty)$ such that

$$\lambda(E) = \int_E f\,d\nu, \quad \mu(E) = \int_E g\,d\nu \quad \text{for all } E \in \mathcal{A}. \tag{5.7}$$

Define

$$A := \{x \in X \mid g(x) > 0\} \tag{5.8}$$

and

$$\lambda_a(E) := \lambda(E \cap A), \quad \lambda_s(E) := \lambda(E \cap A^c) \quad \text{for } E \in \mathcal{A}. \tag{5.9}$$

Then it follows directly from (5.9) that the maps $\lambda_a, \lambda_s: \mathcal{A} \to [0, \infty]$ are measures and satisfy $\lambda_a + \lambda_s = \lambda$. Moreover, by (5.9),

$$\lambda_s(A) = \lambda(A \cap A^c) = \lambda(\emptyset) = 0$$

and from (5.8) that $g|_{A^c} = 0$, so by (5.7)

$$\mu(A^c) = \int_{A^c} g \, d\nu = 0.$$

This shows that $\lambda_s \perp \mu$. It remains to prove that λ_a is absolutely continuous with respect to μ. To see this, let $E \in \mathcal{A}$ such that $\mu(E) = 0$. Then by (5.7)

$$\int_X \chi_E g \, d\nu = \int_E g \, d\nu = \mu(E) = 0.$$

Hence it follows from Lemma 1.49 that $\chi_E g$ vanishes ν-almost everywhere. Thus $\chi_{E \cap A} g = \chi_A \chi_E g$ vanishes ν-almost everywhere. Since $g(x) > 0$ for all $x \in E \cap A$, this implies

$$\nu(E \cap A) = 0.$$

Hence

$$\lambda_a(E) = \lambda(E \cap A) = \int_{E \cap A} f \, d\nu = 0.$$

This shows that $\lambda_a \ll \mu$ and completes the proof of Theorem 5.3. □

5.2 The dual space of $L^p(\mu)$ revisited

This section is devoted to the proof of Theorem 4.35. Assume throughout that $(X, \mathcal{A}, \mu)$ is a measure space and fix two constants

$$1 \le p < \infty, \quad 1 < q \le \infty, \quad \frac{1}{p} + \frac{1}{q} = 1. \tag{5.10}$$

As in Section 4.5 we abuse notation and write

$$\Lambda(f) := \Lambda([f]_\mu)$$

for the value of a bounded linear functional $\Lambda: L^p(\mu) \to \mathbb{R}$ on the equivalence class of a function $f \in \mathcal{L}^p(\mu)$. Recall from Theorem 4.33 that every $g \in \mathcal{L}^q(\mu)$ determines a bounded linear functional

$$\Lambda_g: L^p(\mu) \longrightarrow \mathbb{R}$$

via

$$\Lambda_g(f) := \int_X fg\, d\mu \quad \text{for } f \in \mathcal{L}^p(\mu).$$

The next result proves Theorem 4.35 in σ-finite case.

Theorem 5.7. *Assume $(X, \mathcal{A}, \mu)$ is σ-finite and let $\Lambda: L^p(\mu) \to \mathbb{R}$ be a bounded linear functional. Then there exists a function $g \in \mathcal{L}^q(\mu)$ such that*

$$\Lambda_g = \Lambda.$$

Proof. Assume first that Λ is positive. We prove in six steps that there exists a function $g \in \mathcal{L}^q(\mu)$ such that $g \geq 0$ and $\Lambda_g = \Lambda$.

Step 1. *Define*

$$\lambda(A) := \sup\{\Lambda(\chi_E) \mid E \in \mathcal{A},\ E \subset A,\ \mu(E) < \infty\} \tag{5.11}$$

for $A \in \mathcal{A}$. Then the map $\lambda: \mathcal{A} \to [0, \infty]$ is a measure, $\mathcal{L}^p(\mu) \subset \mathcal{L}^1(\lambda)$, and

$$\Lambda(f) = \int_X f\, d\lambda \quad \textit{for all } f \in \mathcal{L}^p(\mu).$$

This follows directly from Theorem 4.40.

Step 2. *Let λ be as in Step 1 and define $c := \|\Lambda\|$. Then $\lambda(A) \leq c\mu(A)^{1/p}$ for all $A \in \mathcal{A}$.*

By assumption, $\Lambda(f) \leq c\,\|f\|_p$ for all $f \in \mathcal{L}^p(\mu)$. Take $f := \chi_E$ to obtain $\Lambda(\chi_E) \leq c\mu(E)^{1/p} \leq c\mu(A)^{1/p}$ for all $E \in \mathcal{A}$ with $E \subset A$ and $\mu(E) < \infty$. Take the supremum over all such E to obtain $\lambda(A) \leq c\mu(A)^{1/p}$ by (5.11).

Step 3. *Let λ be as in Step 1. Then there exists a measurable function*

$$g: X \longrightarrow [0, \infty)$$

such that

$$\lambda(A) = \int_A g\, d\mu \quad \textit{for all } A \in \mathcal{A}.$$

By Step 2, λ is σ-finite and $\lambda \ll \mu$. Hence Step 3 follows from the Radon–Nikodým theorem (Theorem 5.4) for σ-finite measure spaces.

Step 4. *Let λ be as in Step 1 and g be as in Step 3. Then*

$$\int_X fg\,d\mu = \int_X f\,d\lambda$$

for every measurable function $f: X \to [0, \infty)$.

This follows immediately from Step 3 and Theorem 1.40.

Step 5. *Let c be as in Step 2 and g be as in Step 3. Then $\|g\|_q \leq c$.*

Let λ be as in Step 1 and let $f \in \mathcal{L}^p(\mu)$ such that $f \geq 0$. Then

$$\int_X fg\,d\mu \overset{\text{Step 4}}{=} \int_X f\,d\lambda \overset{\text{Step 1}}{=} \Lambda(f) \leq c\,\|f\|_p. \tag{5.12}$$

Moreover, the measure space $(X, \mathcal{A}, \mu)$ is semi-finite by Lemma 4.30. Hence it follows from parts (iii) and (iv) of Lemma 4.34 that $\|g\|_q \leq c$.

Step 6. *Let g be as in Step 3. Then*

$$\Lambda = \Lambda_g.$$

Since $g \in \mathcal{L}^q(\mu)$ by Step 5, the function g determines a bounded linear functional $\Lambda_g: L^p(\mu) \to \mathbb{R}$ via $\Lambda_g(f) := \int_X fg\,d\mu$ for $f \in \mathcal{L}^p(\mu)$. By (5.12), it satisfies $\Lambda_g(f) = \Lambda(f)$ for all $f \in \mathcal{L}^p(\mu)$ with $f \geq 0$. Apply this identity to the functions $f^\pm: X \to [0, \infty)$ for all $f \in \mathcal{L}^p(\mu)$ to obtain $\Lambda = \Lambda_g$. This proves the assertion of Theorem 5.7 for every positive bounded linear functional $\Lambda: L^p(\mu) \to \mathbb{R}$.

Let $\Lambda: L^p(\mu) \to \mathbb{R}$ be any bounded linear functional. By Theorem 4.39 there exist positive bounded linear functionals $\Lambda^\pm: L^p(\mu) \to \mathbb{R}$ such that $\Lambda = \Lambda^+ - \Lambda^-$. Hence, by what we have just proved, there exist functions $g^\pm \in \mathcal{L}^q(\mu)$ such that $g^\pm \geq 0$ and $\Lambda^\pm = \Lambda_{g^\pm}$. Define

$$g := g^+ - g^-.$$

Then $g \in \mathcal{L}^q(\mu)$ and

$$\Lambda_g = \Lambda_{g^+} - \Lambda_{g^-} = \Lambda^+ - \Lambda^- = \Lambda.$$

This proves Theorem 5.7. □

The next result proves Theorem 4.35 in the case $p = 1$.

Theorem 5.8. *Assume $p = 1$. Then the following are equivalent:*

(i) *the measure space $(X, \mathcal{A}, \mu)$ is localizable;*

(ii) *the measure space $(X, \mathcal{A}, \mu)$ is semi-finite and satisfies condition* (F) *in Exercise 4.59, i.e., if $\mathcal{F}$ is a collection of measurable functions $f: A_f \to \mathbb{R}$, each defined on a measurable set $A_f \in \mathcal{A}$, such that any two functions $f_1, f_2 \in \mathcal{F}$ agree almost everywhere on $A_{f_1} \cap A_{f_2}$, then there exists a measurable function $g: X \to \mathbb{R}$ such that $g|_{A_f} = f$ almost everywhere for all $f \in \mathcal{F}$;*

(iii) *the linear map*

$$L^\infty(\mu) \longrightarrow L^1(\mu)^*: g \longmapsto \Lambda_g \tag{5.13}$$

is bijective.

Proof. The proof that (i) $\Longrightarrow$ (ii) is outlined in Exercise 4.59.

We prove that (ii) $\Longrightarrow$ (iii). Since $(X, \mathcal{A}, \mu)$ is semi-finite, the linear map (5.13) is injective by Theorem 4.33. We must prove that it is surjective. Assume first that $\Lambda: L^1(\mu) \to \mathbb{R}$ is a positive bounded linear functional. Define

$$\mathcal{E} := \{E \in \mathcal{A} \mid \mu(E) < \infty\}$$

and, for $E \in \mathcal{E}$, define

$$\mathcal{A}_E := \{A \in \mathcal{A} \mid A \subset E\}, \quad \mu_E := \mu|_{\mathcal{A}_E}. \tag{5.14}$$

Then there is an extension operator

$$\iota_E: \mathcal{L}^1(\mu_E) \longrightarrow \mathcal{L}^1(\mu)$$

defined by

$$\iota_E(f)(x) := \begin{cases} f(x) & \text{for } x \in E, \\ 0 & \text{for } x \in X \setminus E. \end{cases} \tag{5.15}$$

It descends to a bounded linear operator from $L^1(\mu_E)$ to $L^1(\mu)$ which will still be denoted by ι_E. Define

$$\Lambda_E = \Lambda \circ \iota_E: L^1(\mu_E) \longrightarrow \mathbb{R}.$$

This is a positive bounded linear functional for every $E \in \mathcal{E}$. Hence it follows from Theorem 5.7 (and the Axiom of Choice) that there is a collection of bounded measurable functions

$$g_E: E \longrightarrow [0, \infty), \quad E \in \mathcal{E},$$

such that

$$\Lambda_E(f) = \int_E f g_E \, d\mu_E \quad \text{for all } E \in \mathcal{E} \text{ and all } f \in \mathcal{L}^1(\mu_E).$$

If $E, F \in \mathcal{E}$, then $E \cap F \in \mathcal{E}$ and the functions $g_E|_{E \cap F}$, $g_F|_{E \cap F}$, and $g_{E \cap F}$ all represent the same bounded linear functional $\Lambda_{E \cap F}: L^1(\mu_{E \cap F}) \to \mathbb{R}$. Hence they agree almost everywhere by Theorem 4.33. This shows that the collection

$$\mathcal{F} := \{g_E \mid E \in \mathcal{E}\}$$

satisfies the hypotheses of condition (F) on page 174. Thus it follows from (ii) that there exists a measurable function $g: X \to \mathbb{R}$ such that, for all $E \in \mathcal{E}$, the restriction $g|_E$ agrees with g_E almost everywhere on E.

We claim that $g \geq 0$ almost everywhere. Suppose on the contrary that the set

$$A^- := \{x \in X \mid g(x) < 0\}$$

has positive measure. Since $(X, \mathcal{A}, \mu)$ is semi-finite, there exists a set $E \in \mathcal{E}$ such that $E \subset A^-$ and $\mu(E) > 0$. Since $g(x) < 0 \leq g_E(x)$ for all $x \in E$ it follows that $g|_E$ does not agree with g_E almost everywhere, a contradiction. This shows that $g \geq 0$ almost everywhere.

We prove that $g \leq \|\Lambda\|$ almost everywhere. Suppose otherwise that the set

$$A^+ := \{x \in X \mid g(x) > \|\Lambda\|\}$$

has positive measure. Since $(X, \mathcal{A}, \mu)$ is semi-finite, there exists a set $E \in \mathcal{E}$ such that $E \subset A^+$ and $\mu(E) > 0$. Since $\|g_E\|_\infty = \|\Lambda_E\| \leq \|\Lambda\|$, it follows from Lemma 4.8 that $g_E(x) \leq \|\Lambda\| < g(x)$ for almost every $x \in E$. Hence $g|_E$ does not agree with g_E almost everywhere, a contradiction. This shows that $g \leq \|\Lambda\|$ almost everywhere and we may assume without loss of generality that

$$0 \leq g(x) \leq \|\Lambda\|$$

for all $x \in X$.

We prove that $\Lambda_g = \Lambda$. Fix a function $f \in \mathcal{L}^1(\mu)$ such that $f \geq 0$. Then there exists a sequence $E_i \in \mathcal{E}$ such that

$$E_1 \subset E_2 \subset E_3 \subset \cdots$$

and $\chi_{E_i} f$ converges pointwise to f. Namely, by Theorem 1.26, there exists a sequence of measurable step functions

$$s_i: X \longrightarrow [0, \infty)$$

such that

$$0 \le s_1 \le s_2 \le \cdots$$

and s_i converges pointwise to f. Since

$$\int_X s_i \, d\mu \le \int_X f \, d\mu < \infty \quad \text{for all } i,$$

the sets

$$E_i := \{x \in X \mid s_i(x) > 0\}$$

have finite measure and

$$0 \le s_i \le \chi_{E_i} f \le f \quad \text{for all } i.$$

Thus the E_i are as required. Since the sequence $|f - \chi_{E_i} f|$ converges pointwise to zero and is bounded above by the integrable function f, it follows from the Lebesgue dominated convergence theorem (Theorem 1.45) that

$$\lim_{i\to\infty} \|f - \chi_{E_i} f\|_1 = 0.$$

Hence

$$\begin{aligned}
\Lambda(f) &= \lim_{i\to\infty} \Lambda(\chi_{E_i} f) \\
&= \lim_{i\to\infty} \Lambda_{E_i}(f|_{E_i}) \\
&= \lim_{i\to\infty} \int_{E_i} f g_{E_i} \, d\mu \\
&= \lim_{i\to\infty} \int_{E_i} f g \, d\mu \\
&= \int_X f g \, d\mu.
\end{aligned}$$

Here the last step follows from the Lebesgue monotone convergence theorem (Theorem 1.37). This shows that $\Lambda(f) = \Lambda_g(f)$ for every nonnegative integrable function $f\colon X \to [0, \infty)$. Consequently,

$$\Lambda(f) = \Lambda(f^+) - \Lambda(f^-) = \Lambda_g(f^+) - \Lambda_g(f^-) = \Lambda_g(f)$$

for all $f \in \mathcal{L}^1(\mu)$. Thus $\Lambda = \Lambda_g$ as claimed.

This shows that every positive bounded linear functional on $L^1(\mu)$ belongs to the image of the map (5.13). Since every bounded linear functional on $L^1(\mu)$ is the difference of two positive bounded linear functionals by Theorem 4.39, it follows that the map (5.13) is surjective. Thus we have proved that (ii) $\Longrightarrow$ (iii).

We prove that (iii) $\Longrightarrow$ (i). Assume that the map (5.13) is bijective. Then $(X, \mathcal{A}, \mu)$ is semi-finite by part (iv) of Theorem 4.33. Now let $\mathcal{E} \subset \mathcal{A}$ be any collection of measurable sets. Assume without loss of generality that

$$E_1, \ldots, E_\ell \in \mathcal{E} \implies E_1 \cup \cdots \cup E_\ell \in \mathcal{E}.$$

(Otherwise, replace $\mathcal{E}$ by the collection $\mathcal{E}'$ of all finite unions of elements of $\mathcal{E}$; then every measurable envelope of $\mathcal{E}'$ is also an envelope of $\mathcal{E}$.) For $E \in \mathcal{E}$ define $\mathcal{A}_E$ and μ_E by (5.14) and define the bounded linear functional

$$\Lambda_E \colon L^1(\mu_E) \longrightarrow \mathbb{R}$$

by

$$\Lambda_E(f) := \int_E f \, d\mu_E \quad \text{for } f \in \mathcal{L}^1(\mu_E). \tag{5.16}$$

Then for all $E, F \in \mathcal{A}$ and $f \in \mathcal{L}^1(\mu)$

$$E \subset F, \quad f \geq 0 \implies \Lambda_E(f) \leq \Lambda_F(f). \tag{5.17}$$

Define

$$\Lambda \colon L^1(\mu) \longrightarrow \mathbb{R}$$

by

$$\Lambda(f) := \sup_{E \in \mathcal{E}} \Lambda_E(f^+|_E) - \sup_{E \in \mathcal{E}} \Lambda_E(f^-|_E). \tag{5.18}$$

We prove that this is a well defined bounded linear functional with $\|\Lambda\| \leq 1$. To see this, note that

$$\Lambda_E(f|_E) \leq \int_X f \, d\mu$$

for every nonnegative function $f \in \mathcal{L}^1(\mu)$, and so

$$|\Lambda(f)| \leq \int_X f^+ \, d\mu + \int_X f^- \, d\mu = \|f\|_1 \quad \text{for all } f \in \mathcal{L}^1(\mu).$$

Moreover, it follows directly from the definition that $\Lambda(cf) = c\Lambda(f)$ for all $c \geq 0$ and $\Lambda(-f) = -\Lambda(f)$. Now let $f, g \in \mathcal{L}^1(\mu)$ be nonnegative integrable functions.

Then

$$\begin{aligned}\Lambda(f+g) &= \sup_{E\in\mathcal{E}} \Lambda_E(f|_E + g|_E)\\ &\le \sup_{E\in\mathcal{E}} \Lambda_E(f|_E) + \sup_{E\in\mathcal{E}} \Lambda_E(g|_E)\\ &= \Lambda(f) + \Lambda(g).\end{aligned}$$

To prove the converse inequality, let $\varepsilon > 0$ and choose $E, F \in \mathcal{E}$ such that

$$\Lambda_E(f|_E) > \Lambda(f) - \varepsilon, \quad \Lambda_F(g|_F) > \Lambda(g) - \varepsilon.$$

Then $E \cup F \in \mathcal{E}$ and it follows from (5.17) that

$$\begin{aligned}\Lambda_{E\cup F}((f+g)|_{E\cup F}) &= \Lambda_{E\cup F}(f|_{E\cup F}) + \Lambda_{E\cup F}(g|_{E\cup F})\\ &\ge \Lambda_E(f|_E) + \Lambda_F(g|_F)\\ &> \Lambda(f) + \Lambda(g) - 2\varepsilon.\end{aligned}$$

Hence

$$\Lambda(f+g) > \Lambda(f) + \Lambda(g) - 2\varepsilon \quad \text{for all } \varepsilon > 0$$

and so

$$\Lambda(f) + \Lambda(g) \le \Lambda(f+g) \le \Lambda(f) + \Lambda(g).$$

This shows that

$$\Lambda(f+g) = \Lambda(f) + \Lambda(g)$$

for all $f, g \in \mathcal{L}^1(\mu)$ such that $f, g \ge 0$. If $f, g \in \mathcal{L}^1(\mu)$, then

$$(f+g)^+ + f^- + g^- = (f+g)^- + f^+ + g^+,$$

and hence

$$\Lambda((f+g)^+) + \Lambda(f^-) + \Lambda(g^-) = \Lambda((f+g)^-) + \Lambda(f^+) + \Lambda(g^+)$$

by what we have just proved. Since $\Lambda(f) = \Lambda(f^+) - \Lambda(f^-)$ by definition, it follows that

$$\Lambda(f+g) = \Lambda(f) + \Lambda(g) \quad \text{for all } f, g \in \mathcal{L}^1(\mu).$$

This shows that $\Lambda\colon L^1 \to \mathbb{R}$ is a positive bounded linear functional of norm $\|\Lambda\| \le 1$.

With this understood, it follows from (iii) that there exists a function $g \in \mathcal{L}^\infty(\mu)$ such that $\Lambda = \Lambda_g$. Define

$$H := \{x \in X \mid g(x) > 0\}.$$

We prove that H is an envelope of $\mathcal{E}$. Fix a set $E \in \mathcal{E}$ and suppose, by contradiction, that $\mu(E \setminus H) > 0$. Then, since $(X, \mathcal{A}, \mu)$ is semi-finite, there exists a measurable set $A \in \mathcal{A}$ such that $0 < \mu(A) < \infty$ and $A \subset E \setminus H$. Since $g(x) \leq 0$ for all $x \in A$, it follows that

$$0 < \mu(A) = \Lambda_E(\chi_A|_E) = \int_A g\, d\mu \leq 0,$$

a contradiction. This shows that our assumption $\mu(E \setminus H) > 0$ was wrong. Hence $\mu(E \setminus H) = 0$ for all $E \in \mathcal{E}$, as claimed.

Now let $G \in \mathcal{A}$ be any measurable set such that $\mu(E \setminus G) = 0$ for all $E \in \mathcal{E}$. We must prove that $\mu(H \setminus G) = 0$. Suppose, by contradiction, that $\mu(H \setminus G) > 0$. Since $(X, \mathcal{A}, \mu)$ is semi-finite there exists a measurable set $A \in \mathcal{A}$ such that $0 < \mu(A) < \infty$ and $A \subset H \setminus G$. Then

$$\int_A g\, d\mu = \Lambda(\chi_A) = \sup_{E \in \mathcal{E}} \Lambda_E(\chi_A|_E) = \sup_{E \in \mathcal{E}} \mu(E \cap A) = 0.$$

Here the second equality follows from (5.18), the third follows from (5.16), and the last follows from the fact that $E \cap A \subset E \setminus G$ for all $E \in \mathcal{E}$. Since $g > 0$ on A, it follows from Lemma 1.49 that $\mu(A) = 0$, a contradiction. This shows that our assumption that $\mu(H \setminus G) > 0$ was wrong and so $\mu(H \setminus G) = 0$ as claimed. Thus we have proved that every collection of measurable sets $\mathcal{E} \subset \mathcal{A}$ has a measurable envelope, and this completes the proof of Theorem 5.8. □

Now we are in a position to prove Theorem 4.35 in general.

Proof of Theorem 4.35. For $p = 1$ the assertion of Theorem 4.35 follows from the equivalence of (i) and (iii) in Theorem 5.8. Hence assume $p > 1$. We must prove that the linear map $L^q(\mu) \to L^p(\mu)^*: g \mapsto \Lambda_g$ is surjective. Let $\Lambda: L^p(\mu) \to \mathbb{R}$ be a positive bounded linear functional and define

$$\lambda(A) := \sup\{\Lambda(\chi_E) \mid E \in \mathcal{A},\ E \subset A,\ \mu(E) < \infty\} \quad \text{for } A \in \mathcal{A}.$$

Then $\lambda: \mathcal{A} \to [0, \infty]$ is a measure by Theorem 4.40 and

$$\mathcal{L}^p(\mu) \subset \mathcal{L}^1(\lambda), \quad \Lambda(f) = \int_X f\, d\lambda \quad \text{for all } f \in \mathcal{L}^p(\mu).$$

Theorem 4.40 also asserts that there exists a measurable set $N \in \mathcal{A}$ such that $\lambda(N) = 0$ and the restriction of μ to $X \setminus N$ is σ-finite. Define

$$X_0 := X \setminus N, \quad \mathcal{A}_0 := \{A \in \mathcal{A} \mid A \subset X_0\}, \quad \mu_0 := \mu|_{\mathcal{A}_0}$$

as in (5.14), let $\iota_0 : L^p(\mu_0) \to L^p(\mu)$ be the extension operator as in (5.15), and define

$$\Lambda_0 := \Lambda \circ \iota_0 \colon L^p(\mu_0) \longrightarrow \mathbb{R}.$$

Then Λ_0 is a positive bounded linear functional on $L^p(\mu_0)$ and

$$\Lambda(f) = \int_X f\, d\lambda = \int_{X \setminus N} f\, d\lambda = \Lambda_0(f|_{X_0}) \quad \text{for all } f \in \mathcal{L}^p(\mu).$$

Since $(X_0, \mathcal{A}_0, \mu_0)$ is σ-finite, Theorem 5.7 shows that there exists a function $g_0 \in \mathcal{L}^q(\mu_0)$ such that $g_0 \geq 0$ and

$$\Lambda_0(f_0) = \int_{X_0} f_0 g_0\, d\mu_0 \quad \text{for all } f_0 \in \mathcal{L}^p(\mu_0).$$

Define

$$g \colon X \longrightarrow [0, \infty)$$

by

$$g(x) := \begin{cases} g_0(x) & \text{for } x \in X_0 = X \setminus N, \\ 0 & \text{for } x \in N. \end{cases}$$

Then, by Theorem 4.33,

$$\|g\|_{L^q(\mu)} = \|g_0\|_{L^q(\mu_0)} = \|\Lambda_0\| = \|\Lambda\|$$

and, for all $f \in \mathcal{L}^p(\mu)$,

$$\Lambda(f) = \Lambda_0(f|_{X_0}) = \int_{X_0} f g_0\, d\mu_0 = \int_X f g\, d\mu.$$

This proves the assertion for positive bounded linear functionals. Since every bounded linear functional $\Lambda \colon L^p(\mu) \to \mathbb{R}$ is the difference of two positive bounded linear functionals by Theorem 4.39, this proves Theorem 4.35. □

Corollary 5.9. *Every σ-finite measure space is localizable.*

Proof. Let $(X, \mathcal{A}, \mu)$ be a σ-finite measure space. Then $(X, \mathcal{A}, \mu)$ is semi-finite by Lemma 4.30. Hence the map $L^\infty(\mu) \to L^1(\mu)^* \colon g \mapsto \Lambda_g$ in (4.31) is injective by Theorem 4.33 and is surjective by Theorem 5.7. Hence it follows from Theorem 5.8 that $(X, \mathcal{A}, \mu)$ is localizable. □

5.3 Signed measures

Throughout this section $(X, \mathcal{A})$ is a measurable space, i.e., X is a set and $\mathcal{A} \subset 2^X$ is a σ-algebra. The following definition extends the notion of a measure on $(X, \mathcal{A})$ to a *signed measure* which can have positive and negative values. As an example from physics one can think of electrical charge.

Definition 5.10. A function $\lambda\colon \mathcal{A} \to \mathbb{R}$ is called a *signed measure* if it is *σ-additive,* i.e., for every sequence $E_i \in \mathcal{A}$ of pairwise disjoint measurable sets, it holds that

$$\sum_{i=1}^{\infty} |\lambda(E_i)| < \infty, \quad \lambda\Big(\bigcup_{i=1}^{\infty} E_i\Big) = \sum_{i=1}^{\infty} \lambda(E_i). \tag{5.19}$$

Lemma 5.11. *Every signed measure $\lambda\colon \mathcal{A} \to \mathbb{R}$ satisfies the following:*

(1) $\mu(\emptyset) = 0$;

(2) *if $E_1, \dots, E_\ell \in \mathcal{A}$ are pairwise disjoint, then*

$$\lambda\Big(\bigcup_{i=1}^{\ell} E_i\Big) = \sum_{i=1}^{\ell} \lambda(E_i).$$

Proof. To prove (i) take $E_i := \emptyset$ in (5.19). To prove (ii) take $E_i := \emptyset$ for all $i > \ell$. □

Given a signed measure $\lambda\colon \mathcal{A} \to \mathbb{R}$, it is a natural question to ask whether it can be written as the difference of two measures $\lambda^{\pm}\colon \mathcal{A} \to [0, \infty)$. Closely related to this is the question whether there exists a measure $\mu\colon \mathcal{A} \to [0, \infty)$ that satisfies

$$|\lambda(A)| \le \mu(A) \quad \text{for all } A \in \mathcal{A}. \tag{5.20}$$

If such a measure exists, it must satisfy

$$(E, F \in \mathcal{A}, E \cap F = \emptyset) \implies \lambda(E) - \lambda(F) \le \mu(E \cup F)$$

Thus a lower bound for $\mu(A)$ is the supremum of the numbers $\lambda(E) - \lambda(F)$ over all decompositions of A into pairwise disjoint measurable sets E and F. The next theorem shows that this supremum defines the smallest measure that satisfies (5.20).

Theorem 5.12. *Let $\lambda\colon \mathcal{A} \to \mathbb{R}$ be a signed measure and define*

$$|\lambda|(A) := \sup\left\{ \lambda(E) - \lambda(F) \;\middle|\; \begin{array}{l} E, F \in \mathcal{A}, \\ E \cap F = \emptyset, \\ E \cup F = A \end{array} \right\} \quad \text{for } A \in \mathcal{A}. \tag{5.21}$$

Then $|\lambda(A)| \le |\lambda|(A) < \infty$ for all $A \in \mathcal{A}$ and $|\lambda|\colon \mathcal{A} \to [0, \infty)$ is a measure, called the ***total variation of*** *λ.*

Proof. We prove that $|\lambda|$ is a measure. If follows directly from the definition that $|\lambda|(\emptyset) = 0$ and $|\lambda|(A) \geq |\lambda(A)| \geq 0$ for all $A \in \mathcal{A}$. We must prove that the function $|\lambda|: \mathcal{A} \to [0, \infty]$ is σ-additive. Let $A_i \in \mathcal{A}$ be a sequence of pairwise disjoint measurable sets and put

$$A := \bigcup_{i=1}^{\infty} A_i.$$

Let $E, F \in \mathcal{A}$ are measurable sets such that

$$E \cap F = \emptyset, \quad E \cup F = A. \tag{5.22}$$

Then

$$E = \bigcup_{i=1}^{\infty} (E \cap A_i), \quad F = \bigcup_{i=1}^{\infty} (F \cap A_i).$$

Hence

$$\begin{aligned} \lambda(E) - \lambda(F) &= \sum_{i=1}^{\infty} \lambda(E \cap A_i) - \sum_{i=1}^{\infty} \lambda(F \cap A_i) \\ &= \sum_{i=1}^{\infty} (\lambda(E \cap A_i) - \lambda(F \cap A_i)) \\ &\leq \sum_{i=1}^{\infty} |\lambda|(A_i). \end{aligned}$$

Take the supremum over all pairs of measurable sets E, F satisfying (5.22) to obtain

$$|\lambda|(A) \leq \sum_{i=1}^{\infty} |\lambda|(A_i) \tag{5.23}$$

To prove the converse inequality, fix a constant $\varepsilon > 0$. Then there are sequences of measurable sets $E_i, F_i \in \mathcal{A}$ such that

$$E_i \cap F_i = \emptyset, \quad E_i \cup F_i = A_i, \quad \lambda(E_i) - \lambda(F_i) > |\lambda|(A_i) - \frac{\varepsilon}{2^i}$$

for all $i \in \mathbb{N}$. The sets

$$E := \bigcup_{i=1}^{\infty} E_i \quad \text{and} \quad F := \bigcup_{i=1}^{\infty} F_i$$

satisfy (5.22) and so

$$|\lambda|(A) \geq \lambda(E) - \lambda(F) = \sum_{i=1}^{\infty}(\lambda(E_i) - \lambda(F_i)) > \sum_{i=1}^{\infty}|\lambda|(A_i) - \varepsilon.$$

Hence

$$|\lambda|(A) > \sum_{i=1}^{\infty}|\lambda|(A_i) - \varepsilon \quad \text{for all } \varepsilon > 0.$$

Thus

$$|\lambda|(A) \geq \sum_{i=1}^{\infty}|\lambda|(A_i)$$

and so

$$|\lambda|(A) = \sum_{i=1}^{\infty}|\lambda|(A_i)$$

by (5.23). This shows that $|\lambda|$ is a measure.

It remains to prove that $|\lambda|(X) < \infty$. Suppose, by contradiction, that

$$|\lambda|(X) = \infty.$$

Claim. *Let $A \in \mathcal{A}$ such that $|\lambda|(X \setminus A) = \infty$. Then there exists a measurable set $B \in \mathcal{A}$ such that $A \subset B$, $|\lambda(B \setminus A)| \geq 1$, and $|\lambda|(X \setminus B) = \infty$.*

There exist measurable sets E, F such that $E \cap F = \emptyset$, $E \cup F = X \setminus A$, and

$$\lambda(E) - \lambda(F) \geq 2 + |\lambda(X \setminus A)|,$$

$$\lambda(E) + \lambda(F) = \lambda(X \setminus A).$$

Take the sum, respectively the difference, of these (in)equalities to obtain

$$2\lambda(E) \geq 2 + |\lambda(X \setminus A)| + \lambda(X \setminus A) \geq 2,$$

$$2\lambda(F) \leq \lambda(X \setminus A) - 2 - |\lambda(X \setminus A)| \leq -2,$$

and hence $|\lambda(E)| \geq 1$ and $|\lambda(F)| \geq 1$. Since $|\lambda|(E) + |\lambda|(F) = |\lambda|(X \setminus A) = \infty$, it follows that $|\lambda|(E) = \infty$ or $|\lambda|(F) = \infty$. If $|\lambda|(E) = \infty$ choose $B := A \cup F$, and if $|\lambda|(F) = \infty$ choose $B := A \cup E$. This proves the claim.

It follows from the claim by induction that there exists a sequence of measurable sets $\emptyset := A_0 \subset A_1 \subset A_2 \subset \cdots$ such that $|\lambda(A_n \setminus A_{n-1})| \geq 1$ for all $n \in \mathbb{N}$. Hence $E_n := A_n \setminus A_{n-1}$ is a sequence of pairwise disjoint measurable sets such that $\sum_{n=1}^{\infty}|\lambda(E_n)| = \infty$, in contradiction to Definition 5.10. This contradiction shows that the assumption that $|\lambda|(X) = \infty$ was wrong. Hence $|\lambda|(X) < \infty$ and thus $|\lambda|(A) < \infty$ for all $A \in \mathcal{A}$. This proves Theorem 5.12. □

Definition 5.13. Let $\lambda: \mathcal{A} \to \mathbb{R}$ be a signed measure and let $|\lambda|: \mathcal{A} \to [0, \infty)$ the measure in Theorem 5.12. The *Jordan decomposition* of λ is the representation of λ as the difference of two measures $\lambda^{\pm}$ whose sum is equal to $|\lambda|$. The measures

$$\lambda^{\pm}: \mathcal{A} \longrightarrow [0, \infty)$$

are defined by

$$\lambda^{\pm}(A) := \frac{|\lambda|(A) \pm \lambda(A)}{2} = \sup\{\pm\lambda(E) \mid E \in \mathcal{A},\ E \subset A\} \tag{5.24}$$

for $A \in \mathcal{A}$ and they satisfy

$$\lambda^+ - \lambda^- = \lambda, \quad \lambda^+ + \lambda^- = |\lambda|. \tag{5.25}$$

Exercise 5.14. Let $(X, \mathcal{A}, \mu)$ be a measure space, let $f \in \mathcal{L}^1(\mu)$, and define

$$\lambda(A) := \int_A f \, d\mu \quad \text{for } A \in \mathcal{A}.$$

Prove that λ is a signed measure and

$$|\lambda|(A) = \int_A |f| \, d\mu, \quad \lambda^{\pm}(A) = \int_A f^{\pm} \, d\mu \quad \text{for all } A \in \mathcal{A}. \tag{5.26}$$

Definition 5.15. Let $\mu: \mathcal{A} \to [0, \infty]$ be a measure and let $\lambda, \lambda_1, \lambda_2: \mathcal{A} \to \mathbb{R}$ be signed measures.

(i) λ is called *absolutely continuous* with respect to μ (notation "$\lambda \ll \mu$") if $\mu(E) = 0$ implies $\lambda(E) = 0$ for all $E \in \mathcal{A}$.

(ii) λ is called *concentrated* on $A \in \mathcal{A}$ if $\lambda(E) = \lambda(E \cap A)$ for all $E \in \mathcal{A}$.

(iii) λ is called *singular* with respect to μ (notation "$\lambda \perp \mu$") if there exists a measurable set A such that $\mu(A) = 0$ and λ is concentrated on A.

(iv) λ_1 and λ_2 are called *mutually singular* (notation "$\lambda_1 \perp \lambda_2$") if there are measurable sets A_1, A_2 such that $A_1 \cap A_2 = \emptyset$, $A_1 \cup A_2 = X$, and λ_i is concentrated on A_i for $i = 1, 2$.

Lemma 5.16. *Let μ be a measure on $\mathcal{A}$ and let $\lambda, \lambda_1, \lambda_2$ be signed measures on $\mathcal{A}$. Then the following holds:*

(i) $\lambda \ll \mu$ *if and only if* $|\lambda| \ll \mu$;

(ii) $\lambda \perp \mu$ *if and only if* $|\lambda| \perp \mu$;

(iii) $\lambda_1 \perp \lambda_2$ *if and only if* $|\lambda_1| \perp |\lambda_2|$.

Proof. The proof has four steps.

Step 1. *Let $A \in \mathcal{A}$. Then $|\lambda|(A) = 0$ if and only if $\lambda(E) = 0$ for all measurable sets $E \subset A$.*

If $|\lambda|(A) = 0$, then $|\lambda(E)| \leq |\lambda|(E) \leq |\lambda|(A) = 0$ for all measurable sets $E \subset A$. The converse implication follows directly from the definition.

Step 2. *λ is concentrated on $A \in \mathcal{A}$ if and only if $|\lambda|(X \setminus A) = 0$.*

The signed measure λ is concentrated on A if and only if $\lambda(E) = \lambda(E \cap A)$ for all $E \in \mathcal{A}$, or equivalently $\lambda(E \setminus A) = 0$ for all $E \in \mathcal{A}$. By Step 1, this holds if and only if $|\lambda|(X \setminus A) = 0$.

Step 3. *We prove* (i).

Assume $\lambda \ll \mu$. If $E \in \mathcal{A}$ satisfies $\mu(E) = 0$, then every measurable set $F \in \mathcal{A}$ with $F \subset E$ satisfies $\mu(F) = 0$ and so $\lambda(F) = 0$; hence $|\lambda|(E) = 0$ by Step 1. Thus $|\lambda| \ll \mu$. The converse follows from the fact that $\lambda \ll |\lambda|$.

Step 4. *We prove* (ii) *and* (iii).

$\lambda \perp \mu$ if and only if there is a measurable sets $A \in \mathcal{A}$ such that $\mu(A) = 0$ and λ is concentrated on A. By Step 2, the latter holds if and only if $|\lambda|(X \setminus A) = 0$ or, equivalently, $|\lambda| \perp \mu$. This proves (ii). Assertion (iii) follows from Step 2 by the same argument and this proves Lemma 5.16. □

Theorem 5.17 (Lebesgue decomposition). *Let $(X, \mathcal{A}, \mu)$ be a σ-finite measure space and let $\lambda : \mathcal{A} \to \mathbb{R}$ be a signed measure. Then there exists a unique pair of signed measures*

$$\lambda_a, \lambda_s : \mathcal{A} \longrightarrow \mathbb{R}$$

such that

$$\lambda = \lambda_a + \lambda_s, \quad \lambda_a \ll \mu, \quad \lambda_s \perp \mu. \tag{5.27}$$

Proof. We prove existence. Let $\lambda^{\pm} : \mathcal{A} \to [0, \infty)$ be the measures defined by (5.24). By Theorem 5.3, there exist measures $\lambda_a^{\pm} : \mathcal{A} \to [0, \infty)$ and $\lambda_s^{\pm} : \mathcal{A} \to [0, \infty)$ such that $\lambda_a^{\pm} \ll \mu$, $\lambda_s^{\pm} \perp \mu$, and $\lambda^{\pm} = \lambda_a^{\pm} + \lambda_s^{\pm}$. Hence the signed measures

$$\lambda_a := \lambda_a^+ - \lambda_a^- \quad \text{and} \quad \lambda_s := \lambda_s^+ - \lambda_s^-$$

satisfy (5.27).

We prove uniqueness. Assume $\lambda = \lambda_{\rm a} + \lambda_{\rm s} = \lambda'_{\rm a} + \lambda'_{\rm s}$ where $\lambda_{\rm a}, \lambda_{\rm s}, \lambda'_{\rm a}, \lambda'_{\rm s}$ are signed measures on $\mathcal{A}$ such that $\lambda_{\rm a}, \lambda'_{\rm a} \ll \mu$ and $\lambda_{\rm s}, \lambda'_{\rm s} \perp \mu$. Then $|\lambda_{\rm a}|, |\lambda'_{\rm a}| \ll \mu$ and $|\lambda_{\rm s}|, |\lambda'_{\rm s}| \perp \mu$ by Lemma 5.16. This implies $|\lambda_{\rm a}|+|\lambda'_{\rm a}| \ll \mu$ and $|\lambda_{\rm s}|+|\lambda'_{\rm s}| \perp \mu$ by parts (i) and (ii) of Lemma 5.2. Moreover,

$$|\lambda_{\rm a} - \lambda'_{\rm a}| \ll |\lambda_{\rm a}| + |\lambda'_{\rm a}|, \quad |\lambda'_{\rm a} - \lambda_{\rm a}| = |\lambda_{\rm s} - \lambda'_{\rm s}| \ll |\lambda_{\rm s}| + |\lambda'_{\rm s}|.$$

Hence $|\lambda_{\rm a} - \lambda'_{\rm a}| \ll \mu$ and $|\lambda_{\rm a} - \lambda'_{\rm a}| \perp \mu$ by part (iii) of Lemma 5.2. Thus, by part (iv) of Lemma 5.2, $|\lambda_{\rm a} - \lambda'_{\rm a}| = 0$, and therefore $\lambda_{\rm a} = \lambda'_{\rm a}$ and $\lambda_{\rm s} = \lambda'_{\rm s}$. This proves Theorem 5.17. □

Theorem 5.18 (Radon–Nikodým). *Let $(X, \mathcal{A}, \mu)$ be a σ-finite measure space and let $\lambda: \mathcal{A} \to \mathbb{R}$ be a signed measure. Then $\lambda \ll \mu$ if and only if there exists a μ-integrable function $f: X \to \mathbb{R}$ such that*

$$\lambda(A) = \int_A f \, d\mu \quad \text{for all } A \in \mathcal{A}. \tag{5.28}$$

f is determined uniquely by (5.28) *up to equality μ-almost everywhere.*

Proof. If λ is given by (5.28) for some $f \in \mathcal{L}^1(\mu)$, then $\lambda \ll \mu$ by part (vi) of Theorem 1.44. Conversely, assume $\lambda \ll \mu$ and let $|\lambda|, \lambda^+, \lambda^-: \mathcal{A} \to [0, \infty)$ be the measures defined by (5.21) and (5.24). Then $|\lambda| \ll \mu$ by part (i) of Lemma 5.16, and so $\lambda^\pm \ll \mu$. Hence it follows from Theorem 5.4 that there exist μ-integrable functions $f^\pm: \mathcal{A} \to [0, \infty)$ such that $\lambda^\pm(A) = \int_A f^\pm \, d\mu$ for all $A \in \mathcal{A}$. Hence the function $f := f^+ - f^- \in \mathcal{L}^1(\mu)$ satisfies (5.28). The uniqueness of f, up to equality μ-almost everywhere, follows from Lemma 1.49. This proves Theorem 5.18. □

Theorem 5.19 (Hahn decomposition). *Let $\lambda: \mathcal{A} \to \mathbb{R}$ be a signed measure. Then there exists a measurable set $P \in \mathcal{A}$ such that*

$$\lambda(A \cap P) \geq 0, \quad \lambda(A \setminus P) \leq 0 \quad \text{for all } A \in \mathcal{A}. \tag{5.29}$$

Moreover, there exists a measurable function $h: X \to \{1, -1\}$ such that

$$\lambda(A) = \int_A h \, d|\lambda| \quad \text{for all } A \in \mathcal{A}. \tag{5.30}$$

Proof. By Theorem 5.12, the function $\mu := |\lambda|: \mathcal{A} \to [0, \infty)$ in (5.21) is a finite measure and satisfies $|\lambda(A)| \leq \mu(A)$ for all $A \in \mathcal{A}$. Hence $\lambda \ll \mu$ and it follows from Theorem 5.18 that there exists a function $h \in \mathcal{L}^1(\mu)$ such that (5.30) holds.

We prove that $h(x) \in \{1, -1\}$ for μ-almost every $x \in X$. To see this, fix a real number $0 < r < 1$ and set

$$A_r := \{x \in X \mid |h(x)| \leq r\}.$$

If $E, F \in \mathcal{A}$ such that $E \cap F = \emptyset$ and $E \cup F = A_r$, then

$$\begin{aligned} \lambda(E) - \lambda(F) &= \int_E h\, d\mu - \int_F h\, d\mu \\ &\leq \int_E |h|\, d\mu + \int_F |h|\, d\mu \\ &\leq r\mu(A_r). \end{aligned}$$

Take the supremum over all pairs $E, F \in \mathcal{A}$ such that $E \cap F = \emptyset$ and $E \cup F = A_r$ to obtain $\mu(A_r) \leq r\mu(A_r)$, and hence $\mu(A_r) = 0$. Since this holds for all $r < 1$, it follows that $|h| \geq 1$ μ-almost everywhere. Modifying h on a set of measure zero, if necessary, we may assume without loss of generality that $|h(x)| \geq 1$ for all $x \in X$. Define

$$\begin{aligned} P &:= \{x \in X \mid h(x) \geq 1\}, \\ N &:= \{x \in X \mid h(x) \leq -1\}. \end{aligned}$$

Then $P \cap N = \emptyset$, $P \cup N = X$, and

$$\mu(P) \leq \int_P h\, d\mu = \lambda(P) \leq \mu(P),$$

$$-\mu(N) \leq \lambda(N) = \int_N h\, d\mu \leq -\mu(N).$$

Hence

$$\int_P (h-1)\, d\mu = \lambda(P) - \mu(P) = 0,$$

$$\int_N (h+1)\, d\mu = \lambda(N) + \mu(N) = 0.$$

By Lemma 1.49 this implies $h = 1$ μ-almost everywhere on P and $h = -1$ μ-almost everywhere on N. Modify h again on a set of measure zero, if necessary, to obtain $h(x) = 1$ for all $x \in P$ and $h(x) = -1$ for all $x \in N$. This proves Theorem 5.19. □

Theorem 5.20 (Jordan decomposition). *Let $(X, \mathcal{A})$ be a measurable space, let $\lambda: \mathcal{A} \to \mathbb{R}$ be a signed measure, and let $\lambda^{\pm}: \mathcal{A} \to [0, \infty)$ be finite measures such that $\lambda = \lambda^+ - \lambda^-$. Then the following are equivalent:*

(i) $\lambda^+ + \lambda^- = |\lambda|$;

(ii) $\lambda^+ \perp \lambda^-$;

(iii) *there exists a measurable set $P \in \mathcal{A}$ such that $\lambda^+(A) = \lambda(A \cap P)$ and $\lambda^-(A) = -\lambda(A \setminus P)$ for all $A \in \mathcal{A}$.*

Moreover, for every signed measure λ, there is a unique pair of measures $\lambda^{\pm}$ satisfying $\lambda = \lambda^+ - \lambda^-$ and these equivalent conditions.

Proof. (i) $\Longrightarrow$ (ii). By Theorem 5.19, there exists a measurable function

$$h: X \longrightarrow \{\pm 1\}$$

such that

$$\lambda(A) = \int_A h \, d|\lambda|$$

for all $A \in \mathcal{A}$. Set

$$P := \{x \in X \mid h(x) = 1\}.$$

Then it follows from (i) that

$$\lambda^+(P^c) = \frac{|\lambda|(P^c) + \lambda(P^c)}{2} = \int_{P^c} \frac{1+h}{2} \, d|\lambda| = 0,$$

$$\lambda^-(P) = \frac{|\lambda|(P) - \lambda(P)}{2} = \int_P \frac{1-h}{2} \, d|\lambda| = 0.$$

Hence $\lambda^+ \perp \lambda^-$.

(ii) $\Longrightarrow$ (iii). By (ii), there exists a measurable set $P \in \mathcal{A}$ such that

$$\lambda^+(P^c) = 0 \quad \text{and} \quad \lambda^-(P) = 0.$$

Therefore,

$$\lambda^+(A) = \lambda^+(A \cap P) = \lambda^+(A \cap P) - \lambda^-(A \cap P) = \lambda(A \cap P),$$

$$\lambda^-(A) = \lambda^-(A \setminus P) = \lambda^-(A \setminus P) - \lambda^+(A \setminus P) = -\lambda(A \setminus P)$$

for all $A \in \mathcal{A}$.

(iii) $\Longrightarrow$ (i). Assume (iii) and fix a set $A \in \mathcal{A}$. Then

$$\lambda^+(A) + \lambda^-(A) = \lambda(A \cap P) - \lambda(A \setminus P) \leq |\lambda|(A).$$

Now choose $E, F \in \mathcal{A}$ such that $E \cap F = \emptyset$ and $E \cup F = A$. Then

$$\begin{aligned}
\lambda(E) - \lambda(F) &= \lambda(E \cap P) + \lambda(E \setminus P) - \lambda(F \cap P) - \lambda(F \setminus P) \\
&\leq \lambda(E \cap P) - \lambda(E \setminus P) + \lambda(F \cap P) - \lambda(F \setminus P) \\
&= \lambda(A \cap P) - \lambda(A \setminus P) \\
&= \lambda^+(A) + \lambda^-(A).
\end{aligned}$$

Take the supremum over all such pairs $E, F \in \mathcal{A}$ to obtain the inequality

$$|\lambda|(A) \leq \lambda^+(A) + \lambda^-(A)$$

for all $A \in \mathcal{A}$ and hence $|\lambda| = \lambda^+ + \lambda^-$.

Thus we have proved that assertions (i), (ii), and (iii) are equivalent. Existence and uniqueness of $\lambda^\pm$ now follows from (i) with $\lambda^\pm = \frac{1}{2}(|\lambda| \pm \lambda)$. This proves Theorem 5.20. □

5.4 Radon–Nikodým generalized

This section discusses an extension of the Radon–Nikodým theorem (Theorem 5.18) for signed measures to all measure spaces. Thus we drop the hypothesis that μ is σ-finite. In this case Examples 5.5 and 5.6 show that absolute continuity of λ with respect to μ is not sufficient for obtaining the conclusion of the Radon–Nikodým theorem and a stronger condition is needed. In [4, Theorem 232B] Fremlin introduces the notion "truly continuous", which is equivalent to "absolutely continuous" whenever μ is σ-finite. In [8] König reformulates Fremlin's criterion in terms of "inner regularity of λ with respect to μ". We shall discuss both conditions below, show that they are equivalent, and prove the generalized Radon–Nikodým theorem. As a warmup we rephrase absolute continuity in the familiar ε-δ language of analysis.

Standing assumption. Throughout this section $(X, \mathcal{A}, \mu)$ is a measure space and $\lambda\colon \mathcal{A} \to \mathbb{R}$ is a signed measure.

Lemma 5.21 (absolute continuity). *The following are equivalent.*

(i) λ *is absolutely continuous with respect to* μ.

(ii) *For every* $\varepsilon > 0$ *there exists a constant* $\delta > 0$ *such that*

$$A \in \mathcal{A}, \quad \mu(A) < \delta \implies |\lambda(A)| < \varepsilon.$$

Proof. That (ii) implies (i) is obvious. Conversely, assume (i). Then $|\lambda| \ll \mu$ by Lemma 5.16. Assume by contradiction that (ii) does not hold. Then there exist a constant $\varepsilon > 0$ and a sequence of measurable sets $A_i \in \mathcal{A}$ such that

$$\mu(A_i) \le 2^{-i}, \quad |\lambda(A_i)| \ge \varepsilon \quad \text{for all } i \in \mathbb{N}.$$

For $n \in \mathbb{N}$ define

$$B_n := \bigcup_{i=n}^{\infty} A_i, \quad B := \bigcap_{n=1}^{\infty} B_n.$$

Then

$$B_n \supset B_{n+1}, \quad \mu(B_n) \le \frac{1}{2^{n-1}}, \quad |\lambda|(B_n) \ge |\lambda|(A_n) \ge |\lambda(A_n)| \ge \varepsilon$$

for all $n \in \mathbb{N}$. Hence $\mu(B) = 0$ and $|\lambda|(B) = \lim_{n\to\infty}|\lambda|(B_n) \ge \varepsilon$ by part (v) of Theorem 1.28. This contradicts the fact that $|\lambda| \ll \mu$ and shows that our assumption that (ii) does not hold was wrong. Thus (i) implies (ii) and this proves Lemma 5.21. □

Definition 5.22. The signed measure λ is called *truly continuous with respect to* μ if, for every $\varepsilon > 0$, there exist a constant $\delta > 0$ and a measurable set $E \in \mathcal{A}$ such that $\mu(E) < \infty$ and

$$A \in \mathcal{A}, \quad \mu(A \cap E) < \delta \implies |\lambda(A)| < \varepsilon. \tag{5.31}$$

Lemma 5.23. *The following are equivalent*:

(i) λ *is truly continuous with respect to* μ;

(ii) $|\lambda|$ *is truly continuous with respect to* μ;

(iii) λ^+ *and* λ^- *are truly continuous with respect to* μ.

Proof. Assume (i), fix a constant $\varepsilon > 0$, and choose $\delta > 0$ and $E \in \mathcal{A}$ such that $\mu(E) < \infty$ and (5.31) holds. Let $A \in \mathcal{A}$ such that $\mu(A \cap E) < \delta$. Then $\lambda(B) - \lambda(A \setminus B) < 2\varepsilon$ for every measurable set $B \subset A$, and hence $|\lambda|(A) \le 2\varepsilon$ by Theorem 5.12. This shows that (i) $\implies$ (ii). That (ii) $\implies$ (iii) and (iii) $\implies$ (i) follows directly from the definitions. □

Definition 5.22 is due to Fremlin [4, Chapter 23]. If the measure space $(X, \mathcal{A}, \mu)$ is σ-finite, then λ is truly continuous with respect to μ if and only if it is absolutely continuous with respect to μ by Theorem 5.26 below. However, for general measure spaces the condition of true continuity is stronger than absolute continuity. The reader may verify that, when $(X, \mathcal{A}, \mu)$ and λ are as in part (i) of Example 5.5 or Example 5.6, the finite measure λ is absolutely continuous with respect to μ, but is not truly continuous with respect to μ. Fremlin's condition was reformulated by König [8] in terms of *inner regularity of λ with respect to μ*. This notion can be defined in several equivalent ways. To formulate the conditions it is convenient to introduce the notation

$$\mathcal{E} := \{E \in \mathcal{A} \mid \mu(E) < \infty\}.$$

Lemma 5.24. *The following are equivalent*:

(i) *for all $A \in \mathcal{A}$,*

$$\lambda(A \cap E) = 0 \textit{ for all } E \in \mathcal{E} \implies \lambda(A) = 0;$$

(ii) *for all $A \in \mathcal{A}$,*

$$|\lambda|(A \cap E) = 0 \textit{ for all } E \in \mathcal{E} \implies |\lambda|(A) = 0;$$

(iii) *for all $A \in \mathcal{A}$,*

$$|\lambda|(A) = \sup_{E \in \mathcal{E}} |\lambda|(A \cap E) = \sup_{\substack{E \in \mathcal{E} \\ E \subset A}} |\lambda|(E).$$

Proof. By Theorem 5.19, there exists a set $P \in \mathcal{A}$ such that

$$\lambda(A \cap P) \ge 0, \quad \lambda(A \setminus P) \le 0, \quad |\lambda|(P) = \lambda(A \cap P) - \lambda(A \setminus P) \tag{5.32}$$

for all $A \in \mathcal{A}$. Such a measurable set P will be fixed throughout the proof.

We prove that (i) implies (ii). Fix a set $A \in \mathcal{A}$ such that $|\lambda|(A \cap E) = 0$ for all $E \in \mathcal{E}$. Then it follows from (5.32) that $\lambda(A \cap E \cap P) = \lambda(A \cap E \setminus P) = 0$ for all $E \in \mathcal{E}$. By (i), this implies $\lambda(A \cap P) = \lambda(A \setminus P) = 0$ and hence $|\lambda|(A) = 0$ by (5.32). This shows that (i) implies (ii).

We prove that (ii) implies (i). Fix a set $A \in \mathcal{A}$ such that $\lambda(A \cap E) = 0$ for all $E \in \mathcal{E}$. Since $E \cap P \in \mathcal{E}$ and $E \setminus P \in \mathcal{E}$ for all $E \in \mathcal{E}$, this implies $\lambda(A \cap E \cap P) = \lambda(A \cap E \setminus P) = 0$ for all $E \in \mathcal{E}$. Hence it follows from (5.32) that $|\lambda|(A \cap E) = 0$ for all $E \in \mathcal{E}$. By (ii), this implies $|\lambda|(A) = 0$ and hence $\lambda(A) = 0$ because $|\lambda(A)| \le |\lambda|(A)$. This shows that (ii) implies (i).

We prove that (ii) implies (iii). Fix a set $A \in \mathcal{A}$ and define

$$c := \sup_{\substack{E \in \mathcal{E} \\ E \subset A}} |\lambda|(E) \leq |\lambda|(A). \tag{5.33}$$

Choose a sequence $E_i \in \mathcal{E}$ such that $E_i \subset A$ for all i and

$$\lim_{i \to \infty} |\lambda|(E_i) = c.$$

For $i \in \mathbb{N}$ define

$$F_i := E_1 \cup E_2 \cup \cdots \cup E_i.$$

Then

$$F_i \in \mathcal{E}, \quad F_i \subset F_{i+1} \subset A, \quad |\lambda|(E_i) \leq |\lambda|(F_i) \leq c \tag{5.34}$$

for all i, and hence

$$\lim_{i \to \infty} |\lambda|(F_i) = c. \tag{5.35}$$

Put

$$B := A \setminus F, \quad F := \bigcup_{i=1}^{\infty} F_i. \tag{5.36}$$

Then

$$|\lambda|(F) = \lim_{i \to \infty} |\lambda|(F_i) = c$$

by part (iv) of Theorem 1.28 and hence

$$|\lambda|(B) = |\lambda|(A) - |\lambda|(F) = |\lambda|(A) - c. \tag{5.37}$$

Let $E \in \mathcal{E}$ such that $E \subset B$. Then $E \cap F_i = \emptyset$, $E \cup F_i \in \mathcal{E}$, and $E \cup F_i \subset A$ for all i by (5.36). Hence

$$|\lambda|(E) + |\lambda|(F_i) = |\lambda|(E \cup F_i) \leq c$$

for all i by (5.33). This implies

$$|\lambda|(E) \leq \lim_{i \to \infty} (c - |\lambda|(F_i)) = 0$$

by (5.35). Hence $|\lambda|(E) = 0$ for all $E \in \mathcal{E}$ with $E \subset B$ and it follows from (ii) that $|\lambda|(B) = 0$. Therefore, (5.37) yields $|\lambda|(A) = c$. This shows that (ii) $\Longrightarrow$ (iii). That (iii) $\Longrightarrow$ (ii) is obvious and this proves Lemma 5.24. □

Definition 5.25. The signed measure λ is called *inner regular with respect to* μ if it satisfies the equivalent conditions of Lemma 5.24.

Theorem 5.26 (generalized Radon–Nikodým theorem). *Let $(X, \mathcal{A}, \mu)$ be a measure space and let $\lambda: \mathcal{A} \to \mathbb{R}$ be a signed measure. Then the following are equivalent*:

(i) *λ is truly continuous with respect to μ*;

(ii) *λ is absolutely continuous and inner regular with respect to μ*;

(iii) *there exists a function $f \in \mathcal{L}^1(\mu)$ such that* (5.28) *holds.*

If these equivalent conditions are satisfied, then the function f in (iii) *is uniquely determined by* (5.28) *up to equality μ-almost everywhere.*

First proof of Theorem 5.26. This proof is due to König [8]. It has the advantage that it reduces the proof of the generalized Radon–Nikodým theorem to the standard Radon–Nikodým theorem (Theorem 5.18) for σ-finite measure spaces.

(i) $\Longrightarrow$ (ii). To see that λ is absolutely continuous with respect to μ, fix a measurable set $A \in \mathcal{A}$ such that $\mu(A) = 0$ and fix a constant $\varepsilon > 0$. Choose $\delta > 0$ and $E \in \mathcal{A}$ such that $\mu(E) < \infty$ and (5.31) holds. Then we obtain $\mu(A \cap E) \leq \mu(A) = 0 < \delta$, and hence $|\lambda(A)| < \varepsilon$ by (5.31). Thus $|\lambda(A)| < \varepsilon$ for all $\varepsilon > 0$, and hence $\lambda(A) = 0$. This shows that $\lambda \ll \mu$.

We prove that λ is inner regular with respect to μ by verifying that λ satisfies condition (i) in Lemma 5.24. Fix a set $A \in \mathcal{A}$ such that

$$E \in \mathcal{A}, \quad \mu(E) < \infty \implies \lambda(A \cap E) = 0.$$

We must prove that $\lambda(A) = 0$. Let $\varepsilon > 0$ and choose $\delta > 0$ and $E \in \mathcal{A}$ such that $\mu(E) < \infty$ and (5.31) holds. Then $\mu((A \setminus E) \cap E) = 0 < \delta$, hence $|\lambda(A \setminus E)| < \varepsilon$ by (5.31), so

$$|\lambda(A)| = |\lambda(A \setminus E) + \lambda(A \cap E)| = |\lambda(A \setminus E)| < \varepsilon.$$

This shows that $|\lambda(A)| < \varepsilon$ for all $\varepsilon > 0$ and so $\lambda(A) = 0$, as claimed. Thus we have proved that (i) $\Longrightarrow$ (ii).

(ii) $\Longrightarrow$ (iii). Since λ is inner regular with respect to μ, there exists a sequence of measurable sets $E_i \in \mathcal{A}$ such that $E_i \subset E_{i+1}$ and $\mu(E_i) < \infty$ for all $i \in \mathbb{N}$ and $|\lambda|(X) = \lim_{i \to \infty} |\lambda|(E_i)$. Define

$$X_0 := \bigcup_{i=1}^{\infty} E_i, \quad \mathcal{A}_0 := \{A \in \mathcal{A} \mid A \subset X_0\}, \quad \mu_0 := \mu|_{\mathcal{A}_0}, \quad \lambda_0 := \lambda|_{\mathcal{A}_0}.$$

Then $(X_0, \mathcal{A}_0, \mu_0)$ is a σ-finite measure space and $\lambda_0: \mathcal{A}_0 \to \mathbb{R}$ is a signed measure that is absolutely continuous with respect to μ_0. Hence the Radon–Nikodým theorem (Theorem 5.18) for σ-finite measure spaces asserts that there exists a function

$f_0 \in \mathcal{L}^1(\mu_0)$ such that

$$\lambda_0(A) = \int_A f_0 \, d\mu_0 \quad \text{for all } A \in \mathcal{A}_0.$$

Define

$$f : X \longrightarrow \mathbb{R}$$

by

$$f|_{X_0} := f_0 \quad \text{and} \quad f|_{X \setminus X_0} := 0.$$

Then $f \in \mathcal{L}^1(\mu)$. Choose a measurable set $A \in \mathcal{A}$. Then it follows from part (v) of Theorem 1.28 that

$$|\lambda(A \setminus X_0)| \le |\lambda|(A \setminus X_0) \le |\lambda|(X \setminus X_0) = \lim_{i\to\infty} |\lambda|(X \setminus E_i) = 0.$$

Hence

$$\lambda(A) = \lambda_0(A \cap X_0) = \int_{A\cap X_0} f_0 \, d\mu_0 = \int_A f \, d\mu$$

for all $A \in \mathcal{A}$. This shows that (i) $\Longrightarrow$ (iii). The uniqueness of f up to equality μ-almost everywhere follows immediately from Lemma 1.49.

(iii) $\Longrightarrow$ (i). Choose $f \in \mathcal{L}^1(\mu)$ such that (5.28) holds. Set

$$c := |\lambda|(X) = \int_X |f| \, d\mu$$

and

$$E_n := \{x \in X \mid 2^{-n} \le |f(x)| \le 2^n\},$$

$$E_\infty := \{x \in X \mid f(x) \ne 0\} = \bigcup_{n\in\mathbb{N}} E_n.$$

Then $2^{-n}\mu(E_n) \le |\lambda|(E_n) \le c$ and hence $\mu(E_n) \le 2^n c < \infty$ for all $n \in \mathbb{N}$. Moreover,

$$c = |\lambda|(X) = |\lambda|(E_\infty) = \lim_{n\to\infty} |\lambda|(E_n).$$

Now fix a constant $\varepsilon > 0$. Choose $n \in \mathbb{N}$ such that $|\lambda|(E_n) > c - \varepsilon/2$ and put

$$\delta := 2^{-n-1}\varepsilon.$$

If $A \in \mathcal{A}$ such that $\mu(A \cap E_n) < \delta$, then

$$\begin{aligned} |\lambda|(A) &= |\lambda|(A \setminus E_n) + |\lambda|(A \cap E_n) \\ &\le |\lambda|(X \setminus E_n) + 2^n \mu(A \cap E_n) \\ &< \frac{\varepsilon}{2} + 2^n \delta = \varepsilon. \end{aligned}$$

This shows that λ is truly continuous with respect to μ. This completes the first proof of Theorem 5.26. □

Second proof of Theorem 5.26. This proof is due to Fremlin [4, Chapter 23]. It shows directly that (i) implies (iii) and has the advantage that it only uses the Hahn decomposition theorem (Theorem 5.19). It thus also provides an alternative proof of Theorem 5.18 (assuming the Hahn decomposition theorem) which is of interest on its own. By Lemma 5.23, it suffices to consider the case where $\lambda\colon \mathcal{A} \to [0,\infty)$ is a finite measure that is truly continuous with respect to μ.

Consider the set

$$\mathcal{F} := \left\{ f\colon X \to [0,\infty) \;\middle|\; \begin{array}{l} f \text{ is measurable and} \\ \int_A f\,d\mu \le \lambda(A) \text{ for all } A \in \mathcal{A} \end{array} \right\}.$$

This set is nonempty because $0 \in \mathcal{F}$. Moreover,

$$\int_X f\,d\mu \le \lambda(X) < \infty \quad \text{for all } f \in \mathcal{F},$$

and[1]

$$f, g \in \mathcal{F} \implies \max\{f,g\} \in \mathcal{F}. \tag{5.38}$$

Now put

$$c := \sup_{f\in\mathcal{F}} \int_X f\,d\mu \le \lambda(X)$$

and choose a sequence $g_i \in \mathcal{F}$ such that

$$\lim_{i\to\infty} \int_X g_i\,d\mu = c.$$

Then it follows from (5.38) that

$$f_i := \max\{g_1, g_2, \dots, g_i\} \in \mathcal{F}$$

and

$$\int_X g_i\,d\mu \le \int_X f_i\,d\mu \le c \quad \text{for all } i \in \mathbb{N}.$$

[1]Let $f, g \in \mathcal{F}$ and $A \in \mathcal{A}$ and define the sets

$$A_f := \{x \in A \mid f(x) > g(x)\}$$

and

$$A_g := \{x \in A \mid g(x) \ge f(x)\};$$

then $A_f, A_g \in \mathcal{A}$, $A_f \cap A_g = \emptyset$, and $A_f \cup A_g = A$; hence

$$\int_A \max\{f,g\}\,d\mu = \int_{A_f} f\,d\mu + \int_{A_g} g\,d\mu \le \lambda(A_f) + \lambda(A_g) = \lambda(A).$$

Hence $f_i \leq f_{i+1}$ for all i and

$$\lim_{i\to\infty} \int_X f_i \, d\mu = c.$$

Define the function

$$f : X \longrightarrow [0, \infty]$$

by

$$f(x) := \lim_{i\to\infty} f_i(x) \quad \text{for } x \in X.$$

Then it follows from the Lebesgue monotone convergence Theorem 1.37 that

$$\int_X f \, d\mu = \lim_{i\to\infty} \int_X f_i \, d\mu = c,$$

and

$$\int_A f \, d\mu = \lim_{i\to\infty} \int_A f_i \, d\mu \leq \lambda(A) \quad \text{for all } A \in \mathcal{A}.$$

Hence $f < \infty$ μ-almost everywhere by Lemma 1.47 and we may assume without loss of generality that $f(x) < \infty$ for all $x \in X$. Thus $f \in \mathcal{F}$.

We claim that

$$\int_A f \, d\mu = \lambda(A) \quad \text{for all } A \in \mathcal{A}.$$

Suppose on the contrary that there exists a set $A_0 \in \mathcal{A}$ such that

$$\int_{A_0} f \, d\mu < \lambda(A_0).$$

Then the formula

$$\lambda'(A) := \lambda(A) - \int_A f \, d\mu \quad \text{for } A \in \mathcal{A} \tag{5.39}$$

defines a finite measure by Theorem 1.40.

We prove that there is a measurable function $h \colon X \to [0, \infty)$ such that

$$\int_X h \, d\mu > 0, \qquad \int_A h \, d\mu \leq \lambda'(A) \quad \text{for all } A \in \mathcal{A}. \tag{5.40}$$

Define

$$\varepsilon := \frac{\lambda'(A_0)}{3} > 0. \tag{5.41}$$

Since λ is truly continuous with respect to μ, so is λ'. Hence there are a $\delta > 0$ and a set $E \in \mathcal{A}$ such that $\mu(E) < \infty$ and

$$A \in \mathcal{A}, \quad \mu(A \cap E) < \delta \implies \lambda'(A) < \varepsilon. \tag{5.42}$$

Take $A := X \setminus E$ to obtain $\lambda'(X \setminus E) < \varepsilon$ and hence

$$\lambda'(E) \geq \lambda'(A_0 \cap E) = \lambda'(A_0) - \lambda'(A_0 \setminus E) = 3\varepsilon - \lambda'(A_0 \setminus E) > 2\varepsilon.$$

Then take $A := A_0$. Since $\lambda'(A_0) = 3\varepsilon \geq \varepsilon$ by (5.41), it follows from (5.42) that

$$\mu(E) \geq \mu(A_0 \cap E) \geq \delta > 0.$$

Define the signed measure

$$\lambda'' : \mathcal{A} \longrightarrow \mathbb{R}$$

by

$$\lambda''(A) := \lambda'(A) - \varepsilon \frac{\mu(A \cap E)}{\mu(E)} \quad \text{for } A \in \mathcal{A}. \tag{5.43}$$

Then

$$\lambda''(E) = \lambda'(E) - \varepsilon \geq \varepsilon.$$

By the Hahn decomposition theorem (Theorem 5.19), there exists a measurable set $P \in \mathcal{A}$ such that

$$\lambda''(A \cap P) \geq 0, \quad \lambda''(A \setminus P) \leq 0 \quad \text{for all } A \in \mathcal{A}.$$

Since $\lambda''(E \setminus P) \leq 0$, it follows that

$$\varepsilon \leq \lambda''(E) \leq \lambda''(E \cap P) \leq \lambda'(E \cap P).$$

Hence $\mu(E \cap P) \geq \delta$ by (5.42). Now put

$$h := \frac{\varepsilon}{\mu(E)} \chi_{E \cap P}. \tag{5.44}$$

Then

$$\int_X h \, d\mu > 0.$$

Moreover, if $A \in \mathcal{A}$, then $\lambda''(A \cap P) \geq 0$ and so, by (5.43),

$$\lambda'(A \cap P) \geq \varepsilon \frac{\mu(A \cap P \cap E)}{\mu(E)} = \int_A h \, d\mu.$$

Thus

$$\int_A h \, d\mu \leq \lambda'(A) \quad \text{for all } A \in \mathcal{A},$$

and so h satisfies (5.40), as claimed.

It follows from (5.40) that

$$\int_A (f+h)\,d\mu \le \int_A f\,d\mu + \lambda'(A) = \lambda(A)$$

for all $A \in \mathcal{A}$ and hence $f+h \in \mathcal{F}$. Since

$$\int_X (f+h)\,d\mu = c + \int_X h\,d\mu > c,$$

this contradicts the definition of c. Thus we have proved that $\int_A f\,d\mu = \lambda(A)$ for all $A \in \mathcal{A}$, and hence f satisfies (5.28). This completes the second proof of Theorem 5.26. □

5.5 Exercises

Exercise 5.27. Let $(X, \mathcal{A}, \mu)$ be a measure space such that $\mu(X) < \infty$. Define

$$\rho(A,B) := \mu(A \setminus B) + \mu(B \setminus A) \quad \text{for } A, B \in \mathcal{A}. \tag{5.45}$$

Define an equivalence relation on $\mathcal{A}$ by $A \sim B$ if $\rho(A,B) = 0$. Prove that ρ descends to a function

$$\rho\colon \mathcal{A}/{\sim} \times \mathcal{A}/{\sim} \longrightarrow [0,\infty)$$

(denoted by the same letter) and that the pair $(\mathcal{A}/{\sim}, \rho)$ is a complete metric space. Prove that the function

$$\mathcal{A} \longrightarrow \mathbb{R}\colon A \longmapsto \int_A f\,d\mu$$

descends to a continuous function on $\mathcal{A}/{\sim}$ for every $f \in \mathcal{L}^1(\mu)$.

Exercise 5.28 (Rudin [17, page 133]). Let $(X, \mathcal{A}, \mu)$ be a measure space. A subset $\mathcal{F} \subset \mathcal{L}^1(\mu)$ is called *uniformly integrable* if, for every $\varepsilon > 0$, there is a constant $\delta > 0$ such that, for all $E \in \mathcal{A}$ and all $f \in \mathcal{F}$,

$$\mu(E) < \delta \implies \left|\int_E f\,d\mu\right| < \varepsilon.$$

Prove the following.

(i) Every finite subset of $\mathcal{L}^1(\mu)$ is uniformly integrable. *Hint.* Lemma 5.21.

(ii) VITALI'S THEOREM. *Assume $\mu(X) < \infty$, let $f: X \to \mathbb{R}$ be measurable, and let $f_n \in \mathcal{L}^1(\mu)$ be a uniformly integrable sequence that converges almost everywhere to f. Then $f \in \mathcal{L}^1(\mu)$ and*

$$\lim_{n\to\infty} \int_X |f - f_n| \, d\mu = 0.$$

Hint. Use Egoroff's theorem in Exercise 4.54.

(iii) The hypothesis $\mu(X) < \infty$ cannot be omitted in Vitali's theorem.

Hint. Consider the Lebesgue measure on $\mathbb{R}$. Find a uniformly integrable sequence $f_n \in \mathcal{L}^1(\mathbb{R})$ that converges pointwise to the constant function $f \equiv 1$.

(iv) Vitali's theorem implies the Lebesgue dominated convergence theorem (Theorem 1.45) under the assumption $\mu(X) < \infty$.

(v) Find an example where Vitali's theorem applies although the hypotheses of the Lebesgue dominated convergence theorem are not satisfied.

(vi) Find an example of a measure space $(X, \mathcal{A}, \mu)$ with $\mu(X) < \infty$ and a sequence $f_n \in \mathcal{L}^1(\mu)$ that is not uniformly integrable, converges pointwise to zero, and satisfies

$$\lim_{n\to\infty} \int_X f_n \, d\mu = 0.$$

Hint. Consider the Lebesgue measure on $X = [0, 1]$.

(vii) CONVERSE OF VITALI'S THEOREM. *Assume $\mu(X) < \infty$ and let f_n be a sequence in $\mathcal{L}^1(\mu)$ such that the limit $\lim_{n\to\infty} \int_A f_n \, d\mu$ exists for all $A \in \mathcal{A}$. Then the sequence f_n is uniformly integrable.*

Hint. Let $\varepsilon > 0$. Prove that there exist a constant $\delta > 0$, an integer $n_0 \in \mathbb{N}$, and a measurable set $E_0 \in \mathcal{E}$ such that, for all $E \in \mathcal{A}$ and all $n \in \mathbb{N}$,

$$\rho(E, E_0) < \delta, \quad n \geq n_0 \implies \left| \int_E (f_n - f_{n_0}) \, d\mu \right| < \varepsilon. \tag{5.46}$$

(Here $\rho(E, E_0)$ is defined by (5.45) as in Exercise 5.27.)

If $A \in \mathcal{A}$ satisfies $\mu(A) < \delta$, then the sets

$$E := E_0 \setminus A \quad \text{and} \quad E := E_0 \cup A$$

both satisfy $\rho(E, E_0) < \delta$. Deduce that, for all $A \in \mathcal{A}$ and all $n \in \mathbb{N}$,

$$\mu(A) < \delta, \quad n \geq n_0 \implies \left| \int_A (f_n - f_{n_0})\, d\mu \right| < 2\varepsilon. \tag{5.47}$$

Now use part (i) to find a constant $\delta' > 0$ such that, for all $A \in \mathcal{A}$,

$$\mu(A) < \delta' \implies \sup_{n \in \mathbb{N}} \left| \int_A f_n \, d\mu \right| < 3\varepsilon. \tag{5.48}$$

Exercise 5.29 (Rudin [17, page 134]). Let $(X, \mathcal{A}, \mu)$ be a measure space such that $\mu(X) < \infty$ and fix a real number $p > 1$. Let $f\colon X \to \mathbb{R}$ be a measurable function and let $f_n \in \mathcal{L}^1(\mu)$ be a sequence that converges pointwise to f and satisfies

$$\sup_{n \in \mathbb{N}} \int_X |f_n|^p \, d\mu < \infty.$$

Prove that

$$f \in \mathcal{L}^1(\mu), \quad \lim_{n \to \infty} \int_X |f - f_n| \, d\mu = 0.$$

Hint. Use Vitali's theorem in Exercise 5.28.

Exercise 5.30. Let $X := \mathbb{R}$, denote by $\mathcal{B} \subset 2^X$ the Borel σ-algebra, and let $\mu\colon \mathcal{B} \to [0, \infty]$ be the restriction of the Lebesgue measure to $\mathcal{B}$. Let $\lambda\colon \mathcal{B} \to [0, \infty]$ be a measure. Prove the following.

(1) If $B \in \mathcal{B}$ and $0 < c < \mu(B)$, then there exists a Borel set $A \subset B$ such that $\mu(A) = c$. *Hint.* Show that the function $f(t) := \mu(B \cap [-t, t])$ is continuous.

(2) If there exists a constant $0 < c < \infty$ such that

$$\mu(B) = c \implies \lambda(B) = c.$$

for all $B \in \mathcal{B}$, then $\lambda \ll \mu$.

Exercise 5.31. Let $X := \mathbb{R}$, denote by $\mathcal{B} \subset 2^X$ the Borel σ-algebra, let $\mu\colon \mathcal{B} \to [0, \infty]$ be the restriction of the Lebesgue measure to $\mathcal{B}$, and let $\nu\colon \mathcal{B} \to [0, \infty]$ be the counting measure.

Prove the following.

(i) $\mu \ll \nu$.

(ii) μ is not inner regular with respect to ν.

(iii) There does not exist any measurable function $f\colon X \to [0,\infty]$ such that

$$\mu(B) = \int_B f\,d\nu \quad \text{for all } B \in \mathcal{B}.$$

Exercise 5.32. Let $X := [1,\infty)$, denote by $\mathcal{B} \subset 2^X$ the Borel σ-algebra, and let $\mu\colon \mathcal{B} \to [0,\infty]$ be the restriction of the Lebesgue measure to $\mathcal{B}$. Let $\lambda\colon \mathcal{B} \to [0,\infty]$ be a Borel measure such that

$$\lambda(B) = \alpha\lambda(\alpha B) \quad \text{for all } \alpha \geq 1 \text{ and all } B \in \mathcal{B}. \tag{5.49}$$

Prove that there exists a real number $c \geq 0$ such that

$$\lambda(B) := \int_B f\,d\mu \quad \text{for all } B \in \mathcal{B}, \tag{5.50}$$

where $f\colon [1,\infty) \to [0,\infty)$ is the function given by

$$f(x) := \frac{c}{x^2} \quad \text{for } x \geq 1. \tag{5.51}$$

Hint. Show that $\lambda([1,\infty)) < \infty$, and then that $\lambda \ll \mu$.

Exercise 5.33. Let $X := [0,\infty)$, denote by $\mathcal{B} \subset 2^X$ the Borel σ-algebra, and let $\mu\colon \mathcal{B} \to [0,\infty]$ be the restriction of the Lebesgue measure to $\mathcal{B}$. Define the measures

$$\lambda_1, \lambda_2\colon \mathcal{B} \longrightarrow [0,\infty]$$

by[2]

$$\lambda_1(B) := \sum_{n=1}^{\infty} \frac{1}{n^3} \int_{B\cap[n,n+1]} x\,dx,$$

$$\lambda_2(B) := \int_{B\cap[1,\infty)} \frac{1}{x^2}\,dx$$

for $B \in \mathcal{B}$.

[2]Here we denote by

$$\int_B f(x)\,dx := \int_B f\,d\mu$$

the Lebesgue integral of a Borel measurable function $f\colon [0,\infty) \to [0,\infty)$ over a Borel set $B \in \mathcal{B}$.

Prove that λ_1 and λ_2 are finite measures that satisfy

$$\lambda_1 \ll \mu, \quad \lambda_2 \ll \mu, \quad \lambda_1 \ll \lambda_2, \quad \lambda_2 \ll \lambda_1,$$

and

$$\mu \not\ll \lambda_1, \quad \mu \not\ll \lambda_2.$$

Exercise 5.34. Let $(X, \mathcal{A}, \mu)$ be a measure space. Show that the signed measures $\lambda : \mathcal{A} \to \mathbb{R}$ form a Banach space $\mathcal{M} = \mathcal{M}(X, \mathcal{A})$ with the norm

$$\|\lambda\| := |\lambda|(X).$$

Show that the map

$$L^1(\mu) \longrightarrow \mathcal{M} : [f]_\mu \longmapsto \mu_f$$

defined by (5.53) is an isometric linear embedding and hence $L^1(\mu)$ is a closed subspace of $\mathcal{M}$.

Exercise 5.35. Let $(X, \mathcal{U})$ be a compact Hausdorff space such that every open subset of X is σ-compact, and denote by $\mathcal{B} \subset 2^X$ its Borel σ-algebra. Denote by

$$C(X) := C_c(X)$$

the space of continuous real-valued functions on X. This is a Banach space equipped with the supremum norm

$$\|f\| := \sup_{x \in X} |f(x)|.$$

Let $\mathcal{M}(X)$ denote the space of signed Borel measures as in Exercise 5.34. For $\lambda \in \mathcal{M}(X)$ define the linear functional

$$\Lambda_\lambda : C(X) \longrightarrow \mathbb{R}$$

by

$$\Lambda_\lambda(f) := \int_X f \, d\lambda.$$

Prove the following.

(i) $\|\Lambda_\lambda\| = \|\lambda\|$. *Hint.* Use the Hahn decomposition theorem (Theorem 5.19) and the fact that every Borel measure on X is regular by Theorem 3.18.

(ii) Every bounded linear functional on $C(X)$ is the difference of two positive linear functionals. *Hint.* For $f \in C(X)$ with $f \geq 0$ prove that

$$\begin{aligned} \Lambda^+(f) &:= \sup\{\Lambda(hf) \mid h \in C(X),\ 0 \leq h \leq 1\} \\ &= \sup\{\Lambda(g) \mid g \in C(X),\ 0 \leq g \leq f\}. \end{aligned} \tag{5.52}$$

Here the second supremum is obviously greater than or equal to the first. To prove the converse inequality, show that, for all $g \in C(X)$ with $0 \le g \le f$ and all $\varepsilon > 0$ there is an $h \in C(X)$ such that $0 \le h \le 1$ and $|\Lambda(g - hf)| < \varepsilon$. Namely, find $\phi \in C(X)$ such that $0 \le \phi \le 1$, $\phi(x) = 0$ when $f(x) \le \varepsilon/2\,\|\Lambda\|$ and $\phi(x) = 1$ when $f(x) \ge \varepsilon/\,\|\Lambda\|$; then define $h := \phi g/f$. Once (5.52) is established, show that Λ^+ extends to a positive linear functional on $C(X)$.

(iii) The map $\mathcal{M}(X) \to C(X)^*\colon \lambda \mapsto \Lambda_\lambda$ is bijective. *Hint.* Use the Riesz representation theorem (Theorem 3.15).

(iv) The hypothesis that every open subset of X is σ-compact cannot be removed in part (i). *Hint.* Consider Example 3.6.

Exercise 5.36. Let $(X, \mathcal{A}, \mu)$ be a measure space and let $f\colon X \to [0,\infty)$ be a measurable function. Define the measure $\mu_f\colon \mathcal{A} \to [0,\infty]$ by

$$\mu_f(A) := \int_A f\,d\mu \quad \text{for } A \in \mathcal{A}. \tag{5.53}$$

(See Theorem 1.40.)

Prove the following.

(i) If μ is σ-finite, then so is μ_f.

(ii) If μ is semi-finite, then so is μ_f.

(iii) If μ is localizable, then so is μ_f.

Note. See Theorem 5.4 for (i) and [4, Proposition 234N] for (ii) and (iii). It is essential that f does not take on the value ∞. Find an example of a measure space $(X, \mathcal{A}, \mu)$ and a measurable function $f\colon X \to [0,\infty]$ that violates assertions (i), (ii), and (iii).

Hint 1. To prove (ii), fix a set $A \in \mathcal{A}$, define $A_f := \{x \in A \mid f(x) > 0\}$, and choose a measurable set $E \in \mathcal{A}$ such that $E \subset A_f$ and $0 < \mu(E) < \infty$. Consider the sets $E_n := \{x \in E \mid f(x) \le n\}$.

Hint 2. To prove (iii), let $\mathcal{E} \subset \mathcal{A}$ be any collection of measurable sets and choose a measurable μ-envelope $H \in \mathcal{A}$ of $\mathcal{E}$. Prove that the set

$$H_f := \{x \in H \mid f(x) > 0\}$$

is a measurable μ_f-envelope of $\mathcal{E}$. In particular, if $N \in \mathcal{A}$ is a measurable set such that $\mu_f(E \cap N) = 0$ for all $E \in \mathcal{E}$, define $N_f := \{x \in N \mid f(x) > 0\}$, show that $\mu(H \cap N_f) = 0$, and deduce that $\mu_f(H_f \cap N) = \mu_f(H \cap N_f) = 0$.

Chapter 6

Differentiation

This chapter returns to the Lebesgue measure on Euclidean space $\mathbb{R}^n$ introduced in Chapter 2. The main result is the *Lebesgue differentiation theorem* (Section 6.3). It implies that if $f: \mathbb{R}^n \to \mathbb{R}$ is a Lebesgue integrable function, then for almost every element $x \in \mathbb{R}^n$, the mean value of f over a ball centered at x converges to $f(x)$ as the radius tends to zero. Essential ingredients in the proof are the *Vitali covering lemma* and the *Hardy–Littlewood maximal inequality* (Section 6.2). One of the consequences of the Lebesgue differentiation theorem is the *fundamental theorem of calculus* for absolutely continuous functions of one real variable (Section 6.4). The Lebesgue differentiation theorem also plays a central role in the proof of the Calderón–Zygmund inequality (Section 7.7). The chapter begins with a discussion of weakly integrable functions on general measure spaces.

6.1 Weakly integrable functions

Assume throughout that $(X, \mathcal{A}, \mu)$ is a measure space. Let $f: X \to \mathbb{R}$ be a measurable function. Define

$$\kappa_f: [0, \infty) \longrightarrow [0, \infty]$$

by

$$\kappa_f(t) := \kappa(t, f) := \mu(A(t, f)), \quad A(t, f) := \{x \in X \mid |f(x)| > t\}, \tag{6.1}$$

for $t \geq 0$. The function κ_f is nonincreasing and hence Borel measurable. Define the function $f^*: [0, \infty) \to [0, \infty]$ by

$$f^*(\alpha) := \inf\{t \geq 0 \mid \kappa(t, f) \leq \alpha\} \quad \text{for } 0 \leq \alpha < \infty. \tag{6.2}$$

Thus $f^*(0) = \|f\|_\infty$ and f^* is nonincreasing and hence Borel measurable. By definition, the infimum of the empty set is infinity. Thus $f^*(\alpha) = \infty$ if and only if $\mu(A(t, f)) > \alpha$ for all $t > 0$. When $f^*(\alpha) < \infty$, it is the smallest number t such that the domain $A(t, f)$ (on which $|f| > t$) has measure at most α. This is spelled out in the next lemma.

Lemma 6.1. *Let $0 \le \alpha < \infty$ and $0 \le t < \infty$. Then the following holds:*

(i) $f^*(\alpha) = \infty$ *if and only if* $\kappa_f(s) > \alpha$ *for all* $s \ge 0$;

(ii) $f^*(\alpha) = t$ *if and only if* $\kappa_f(t) \le \alpha$ *and* $\kappa_f(s) > \alpha$ *for* $0 \le s < t$;

(iii) $f^*(\alpha) \le t$ *if and only if* $\kappa_f(t) \le \alpha$.

Proof. It follows directly from the definition of f^* in (6.2) that $f^*(\alpha) = \infty$ if and only if $\kappa(s, f) > \alpha$ for all $s \in [0, \infty)$, and this proves (i).

To prove (ii), fix a constant $0 \le t < \infty$. Assume first that

$$\kappa(t, f) \le \alpha \quad \text{and} \quad \kappa(s, f) > \alpha \quad \text{for } 0 \le s < t.$$

Since κ_f is nonincreasing, this implies

$$\kappa(s, f) \le \kappa(t, f) \le \alpha \quad \text{for all } s \ge t,$$

and hence $f^*(\alpha) = t$ by definition. Conversely, suppose that $f^*(\alpha) = t$. Then it follows from the definition of f^* that

$$\kappa(s, f) \le \alpha \quad \text{for } s > t$$

and

$$\kappa(s, t) > \alpha \quad \text{for } 0 \le s < t.$$

We must prove that $\kappa(t, f) \le \alpha$. To see this, observe that

$$A(t, f) = \bigcup_{n=1}^{\infty} A(t + 1/n, f).$$

Hence it follows from part (iv) of Theorem 1.28 that

$$\begin{aligned} \kappa_f(t) &= \mu(A(t, f)) \\ &= \lim_{n\to\infty} \mu(A(t + 1/n, f)) \\ &= \lim_{n\to\infty} \kappa(t + 1/n, f) \\ &\le \alpha. \end{aligned}$$

This proves (ii).

If $f^*(\alpha) \le t$, then $\kappa_f(t) \le \kappa_f(f^*(\alpha)) \le \alpha$ by (ii). If $\kappa_f(t) \le \alpha$, then $f^*(\alpha) \le t$ by the definition of f^*. This proves (iii) and Lemma 6.1. □

Lemma 6.2. *Let $f, g\colon X \to \mathbb{R}$ be measurable functions and let $c \in \mathbb{R}$. Then*

$$\|f\|_{1,\infty} := \sup_{\alpha>0} \alpha f^*(\alpha) = \sup_{t>0} t\kappa_f(t) \le \|f\|_1, \tag{6.3}$$

$$\|cf\|_{1,\infty} = |c|\,\|f\|_{1,\infty}, \tag{6.4}$$

$$\|f+g\|_{1,\infty} \le \frac{\|f\|_{1,\infty}}{\lambda} + \frac{\|g\|_{1,\infty}}{1-\lambda} \quad \textit{for } 0<\lambda<1, \tag{6.5}$$

$$\sqrt{\|f+g\|_{1,\infty}} \le \sqrt{\|f\|_{1,\infty}} + \sqrt{\|g\|_{1,\infty}}. \tag{6.6}$$

Moreover, $\|f\|_{1,\infty} = 0$ if and only if f vanishes almost everywhere. The inequality (6.6) *is called the* ***weak triangle inequality****.*

Proof. For $0 < t, c < \infty$ it follows from part (iii) of Lemma 6.1 that

$$\begin{aligned} t\kappa(t,f) \le c \iff& \kappa(t,f) \le ct^{-1} \\ \iff& f^*(ct^{-1}) \le t \\ \iff& ct^{-1} f^*(ct^{-1}) \le c. \end{aligned}$$

This shows that

$$\sup_{t>0} t\kappa(t,f) = \sup_{\alpha>0} \alpha f^*(\alpha).$$

Moreover,

$$t\kappa(t,f) = t\mu(A(t,f)) \le \int_{A(t,f)} |f|\,d\mu \le \int_X |f|\,d\mu \quad \text{for all } t>0.$$

This proves (6.3).

For $c > 0$ equation (6.4) follows from the fact that $A(t, cf) = A(t/c, f)$ and hence $\kappa(t, cf) = \kappa(t/c, f)$ for all $t > 0$. Since $\|-f\|_{1,\infty} = \|f\|_{1,\infty}$ by definition, this proves (6.4).

To prove (6.5), observe that $A(t, f+g) \subset A(\lambda t, f) \cup A((1-\lambda)t, g)$, hence

$$\kappa(t, f+g) \le \kappa(\lambda t, f) + \kappa((1-\lambda)t, g), \tag{6.7}$$

and so

$$t\kappa(t, f+g) \le t\kappa(\lambda t, f) + t\kappa((1-\lambda)t, g) \le \frac{\|f\|_{1,\infty}}{\lambda} + \frac{\|g\|_{1,\infty}}{1-\lambda}$$

for all $t > 0$. Take the supremum over all $t > 0$ to obtain (6.5).

The inequality (6.6) follows from (6.5) and the identity

$$\inf_{0<\lambda<1}\sqrt{\frac{a}{\lambda}+\frac{b}{1-\lambda}}=\sqrt{a}+\sqrt{b}\quad\text{for } a,b\geq 0. \tag{6.8}$$

This is obvious when $a = 0$ or $b = 0$. Hence assume a and b are positive and define the function

$$h\colon(0,1)\longrightarrow(0,\infty)$$

by

$$h(\lambda):=\frac{a}{\lambda}+\frac{b}{1-\lambda}.$$

It satisfies

$$h'(\lambda)=\frac{b}{(1-\lambda)^2}-\frac{a}{\lambda^2}$$

and hence has a unique critical point at

$$\lambda_0:=\frac{\sqrt{a}}{\sqrt{a}+\sqrt{b}}.$$

Since $h(\lambda_0)=(\sqrt{a}+\sqrt{b})^2$, this proves (6.8). The inequality (6.6) then follows by taking $a := \|f\|_{1,\infty}$ and $b := \|g\|_{1,\infty}$.

The last assertion follows from the fact that $\|f\|_{1,\infty}=0$ if and only if $\kappa_f(0)=0$ if and only if the set $A(0,f)=\{x\in X \mid f(x)\neq 0\}$ has measure zero. This proves Lemma 6.2. □

Example 6.3. This example shows that the weak triangle inequality (6.6) is sharp. Let $(\mathbb{R},\mathcal{A},m)$ be the Lebesgue measure space and define $f,g\colon\mathbb{R}\to\mathbb{R}$ by

$$f(x):=\frac{1}{x},\quad g(x):=\frac{1}{1-x}\quad\text{for } 0<x<1,$$

and $f(x):=g(x):=0$ for $x\leq 0$ and for $x\geq 1$. Then

$$\|f\|_{1,\infty}=\|g\|_{1,\infty}=1,\quad \|f+g\|_{1,\infty}=4.$$

Definition 6.4. Let $(X,\mathcal{A},\mu)$ be a measure space. A measurable function

$$f\colon X\longrightarrow\mathbb{R}$$

is called *weakly integrable* if

$$\|f\|_{1,\infty}<\infty.$$

The space of weakly integrable functions will be denoted by

$$\mathcal{L}^{1,\infty}(\mu):=\{f\colon X\to\mathbb{R}\mid f\text{ is measurable and }\|f\|_{1,\infty}<\infty\}.$$

The quotient space

$$L^{1,\infty}(\mu) := \mathcal{L}^{1,\infty}(\mu)/\overset{\mu}{\sim}$$

under the equivalence relation $f \overset{\mu}{\sim} g$ if $f = g$ μ-almost everywhere is called the *weak* L^1 *space*. It is not a normed vector space because the function $L^{1,\infty}(\mu) \to [0,\infty)\colon [f]_\mu \mapsto \|f\|_{1,\infty}$ does not satisfy the triangle inequality, in general, and hence *is not* a norm. However, it is a topological vector space and the topology is determined by the metric

$$d_{1,\infty}([f]_\mu, [g]_\mu) := \sqrt{\|f-g\|_{1,\infty}} \quad \text{for } f, g \in \mathcal{L}^{1,\infty}(\mu). \tag{6.9}$$

For the Lebesgue measure space $(\mathbb{R}^n, \mathcal{A}, m)$ we write

$$\mathcal{L}^{1,\infty}(\mathbb{R}^n) := \mathcal{L}^{1,\infty}(m) \quad \text{and} \quad L^{1,\infty}(\mathbb{R}^n) := L^{1,\infty}(m).$$

A subset of $L^{1,\infty}(\mu)$ is open in the topology determined by the metric (6.9) if and only if it is a union of sets of the form $\{[g]_\mu \in L^{1,\infty}(\mu) \mid \|f-g\|_{1,\infty} < r\}$ with $f \in \mathcal{L}^{1,\infty}(\mu)$ and $r > 0$. A sequence $[f_i]_\mu \in L^{1,\infty}(\mu)$ converges to $[f]_\mu$ in this topology if and only if $\lim_{i\to\infty} \|f_i - f\|_{1,\infty} = 0$. The inequality (6.3) in Lemma 6.2 shows that

$$L^1(\mu) \subset L^{1,\infty}(\mu)$$

for every measure space $(X, \mathcal{A}, \mu)$. In general, $L^{1,\infty}(\mu)$ is not equal to $L^1(\mu)$. For example, the function

$$f\colon \mathbb{R} \longrightarrow \mathbb{R}$$

defined by

$$f(x) := \begin{cases} \dfrac{1}{x} & \text{for } x > 0, \\ 0 & \text{for } x \le 0, \end{cases}$$

is weakly integrable, but not integrable.

Theorem 6.5. *The metric space $(L^{1,\infty}(\mu), d_{1,\infty})$ is complete.*

Proof. Choose a sequence of weakly integrable functions

$$f_i\colon X \longrightarrow \mathbb{R}$$

whose equivalence classes $[f_i]_\mu$ form a Cauchy sequence in $L^{1,\infty}(\mu)$ with respect to the metric (6.9). Then there is a subsequence

$$i_1 < i_2 < i_3 < \cdots$$

such that

$$\|f_{i_k} - f_{i_{k+1}}\|_{1,\infty} < 2^{-2k} \quad \text{for all } k \in \mathbb{N}.$$

For $k, \ell \in \mathbb{N}$ put

$$A_k := A(2^{-k}, f_{i_k} - f_{i_{k+1}})$$

and

$$E_\ell := \bigcup_{k=\ell}^{\infty} A_k, \quad E := \bigcap_{\ell=1}^{\infty} E_\ell.$$

Then

$$2^{-k}\mu(A_k) \le \left\| f_{i_k} - f_{i_{k+1}} \right\|_{1,\infty} < 2^{-2k} \quad \text{for all } k \in \mathbb{N},$$

hence

$$\mu(E_\ell) \le \sum_{k=\ell}^{\infty} \mu(A_k) \le \sum_{k=\ell}^{\infty} 2^{-k} = 2^{1-\ell}$$

for all $\ell \in \mathbb{N}$, and so $\mu(E) = 0$. If $x \in X \setminus E$, then there exists an $\ell \in \mathbb{N}$ such that $x \notin A_k$ for all $k \ge \ell$ and so

$$|f_{i_k}(x) - f_{i_{k+1}}(x)| \le 2^{-k} \quad \text{for all } k \ge \ell.$$

This shows that the limit

$$f(x) := \lim_{k\to\infty} f_{i_k}(x)$$

exists for all $x \in X \setminus E$. Extend f to a measurable function on X by setting

$$f(x) := 0 \quad \text{for } x \in E.$$

We prove that

$$\lim_{i\to\infty} \|f_i - f\|_{1,\infty} = 0$$

and hence also $f \in \mathcal{L}^{1,\infty}(\mu)$. To see this, fix a constant $\varepsilon > 0$ and choose an integer $i_0 \in \mathbb{N}$ such that

$$(i, j \in \mathbb{N}, i, j \ge i_0) \implies 4\left\|f_i - f_j\right\|_{1,\infty} < \varepsilon.$$

Now fix a constant $t > 0$ and choose $\ell \in \mathbb{N}$ such that

$$i_\ell \ge i_0, \quad 2^{2-\ell} t \le \varepsilon, \quad 2^{2-\ell} \le t.$$

If $x \notin E_\ell$, then $x \notin A_k$ for all $k \geq \ell$, hence

$$|f_{i_k}(x) - f_{i_{k+1}}(x)| \leq 2^{-k} \quad \text{for } k \geq \ell,$$

and so

$$|f_{i_\ell}(x) - f(x)| \leq \sum_{k=\ell}^{\infty} |f_{i_k}(x) - f_{i_{k+1}}(x)| \leq \sum_{k=\ell}^{\infty} 2^{-k} = 2^{1-\ell} \leq t/2.$$

This shows that $A(t/2, f_{i_\ell} - f) \subset E_\ell$ and hence

$$t\kappa_{f_{i_\ell}-f}(t/2) = t\mu(A(t/2, f_{i_\ell} - f)) \leq t\mu(E_\ell) \leq t2^{1-\ell} \leq \varepsilon/2.$$

With this understood, it follows from (6.7) with $\lambda = 1/2$ that

$$t\kappa_{f_i-f}(t) \leq t\kappa_{f_i-f_{i_\ell}}(t/2) + t\kappa_{f_{i_\ell}-f}(t/2) \leq 2\|f_i - f_{i_\ell}\|_{1,\infty} + \varepsilon/2 < \varepsilon$$

for all $i \in \mathbb{N}$ with $i \geq i_0$. Hence

$$\|f_i - f\|_{1,\infty} = \sup_{t>0} t\kappa_{f_i-f}(t) \leq \varepsilon$$

for every integer $i \geq i_0$ and this proves Theorem 6.5. □

6.2 Maximal functions

Let $(\mathbb{R}, \mathcal{A}, m)$ be the Lebesgue measure space on $\mathbb{R}$. In particular, the length of an interval $I \subset \mathbb{R}$ is $m(I)$. As a warmup we characterize the differentiability of a function that is obtained by integrating a signed measure.

Theorem 6.6. *Let $\lambda: \mathcal{A} \to \mathbb{R}$ be a signed measure and define*

$$f: \mathbb{R} \longrightarrow \mathbb{R}$$

by

$$f(x) := \lambda((-\infty, x)) \quad \text{for } x \in \mathbb{R}. \tag{6.10}$$

Fix two real numbers $x, A \in \mathbb{R}$. Then the following are equivalent:

(i) *f is differentiable at x and $f'(x) = A$.*

(ii) *For every $\varepsilon > 0$ there is a $\delta > 0$ such that, for every open interval $U \subset \mathbb{R}$,*

$$x \in U, \quad m(U) < \delta \implies \left|\frac{\lambda(U)}{m(U)} - A\right| \leq \varepsilon. \tag{6.11}$$

Proof. (i) $\Longrightarrow$ (ii). Fix a constant $\varepsilon > 0$. Since f is differentiable at x and we have $f'(x) = A$, there exists a constant $\delta > 0$ such that, for all $y \in \mathbb{R}$,

$$0 < |x - y| < \delta \implies \left| \frac{f(x) - f(y)}{x - y} - A \right| \leq \varepsilon. \tag{6.12}$$

Let $a, b \in \mathbb{R}$ such that $a < x < b$ and $b - a < \delta$. Then, by (6.12),

$$\left| \frac{f(x) - f(a)}{x - a} - A \right| \leq \varepsilon, \qquad \left| \frac{f(b) - f(x)}{b - x} - A \right| \leq \varepsilon,$$

or, equivalently,

$$-\varepsilon(x - a) \leq f(x) - f(a) - A(x - a) \leq \varepsilon(x - a),$$

$$-\varepsilon(b - x) \leq f(b) - f(x) - A(b - x) \leq \varepsilon(b - x).$$

Add these inequalities to obtain

$$-\varepsilon(b - a) \leq f(b) - f(a) - A(b - a) \leq \varepsilon(b - a).$$

Since $\lambda([a, b)) = f(b) - f(a)$ and $m([a, b)) = b - a$ it follows that

$$\left| \frac{\lambda([a, b))}{m([a, b))} - A \right| \leq \varepsilon.$$

Replace a by $a + 2^{-k}$ and take the limit $k \to \infty$ to obtain

$$\left| \frac{\lambda((a, b))}{m((a, b))} - A \right| \leq \varepsilon.$$

Thus we have proved that (i) $\Longrightarrow$ (ii).

(ii) $\Longrightarrow$ (i). Fix a constant $\varepsilon > 0$. Choose $\delta > 0$ such that (6.11) holds for every open interval $U \subset \mathbb{R}$. Choose $y \in \mathbb{R}$ such that $x < y < x + \delta$. Choose $k \in \mathbb{N}$ such that $y - x + 2^{-k} < \delta$. Then

$$U_k := (x - 2^{-k}, y)$$

is an open interval of length $m(U_k) < \delta$ containing x and hence

$$\left| \frac{\lambda(U_k)}{m(U_k)} - A \right| \leq \varepsilon$$

by (6.11). Take the limit $k \to \infty$ to obtain

$$\left| \frac{f(y) - f(x)}{y - x} - A \right| = \left| \frac{\lambda([x, y))}{m([x, y))} - A \right| = \lim_{k \to \infty} \left| \frac{\lambda(U_k)}{m(U_k)} - A \right| \leq \varepsilon.$$

Thus (6.12) holds for $x < y < x + \delta$ and an analogous argument proves the inequality for $x - \delta < y < x$. Thus (ii) implies (i) and this proves Theorem 6.6. $\square$

The main theorem of this chapter will imply that, when λ is absolutely continuous with respect to m, the derivative of the function f in (6.10) exists almost everywhere, defines a Lebesgue integrable function $f'\colon \mathbb{R} \to \mathbb{R}$, and that

$$\lambda(A) = \int_A f'\,dm$$

for all Lebesgue measurable sets $A \in \mathcal{A}$. It will then follow that an *absolutely continuous function* on $\mathbb{R}$ can be written as the integral of its derivative. This is the *fundamental theorem of calculus* in measure theory (Theorem 6.19).

The starting point for this program is the assertion of Theorem 6.6. It suggests the definition of the *derivative of a signed measure*

$$\lambda\colon \mathcal{A} \longrightarrow \mathbb{R}$$

at a point $x \in \mathbb{R}$ as the limit of the quotients $\lambda(U)/m(U)$ over all open intervals U containing x as $m(U)$ tends to zero, provided that the limit exists. This idea carries over to all dimensions and leads to the concept of a *maximal function,* which we explain next.

Notation. Fix a natural number $n \in \mathbb{N}$. Let $(\mathbb{R}^n, \mathcal{A}, m)$ denote the Lebesgue measure space and let

$$\mathcal{B} \subset 2^{\mathbb{R}^n}$$

denote the Borel σ-algebra of $\mathbb{R}^n$ with the standard topology. Thus $\mathcal{L}^1(\mathbb{R}^n)$ denotes the space of Lebesgue integrable functions $f\colon \mathbb{R}^n \to \mathbb{R}$. An element of $\mathcal{L}^1(\mathbb{R}^n)$ need not be Borel measurable, but differs from a Borel measurable function on a Lebesgue null set by Theorem 2.14 and part (v) of Theorem 1.55. For $x \in \mathbb{R}^n$ and $r > 0$, denote the open ball of radius r, centered at x, by

$$B_r(x) := \{y \in \mathbb{R}^n \mid |x - y| < r\}.$$

Here

$$|\xi| := \sqrt{\xi_1^2 + \cdots + \xi_n^2}$$

denotes the Euclidean norm of $\xi = (\xi_1, \ldots, \xi_n) \in \mathbb{R}^n$.

Definition 6.7 (Hardy–Littlewood maximal function). Let $\mu\colon \mathcal{B} \to [0, \infty)$ be a finite Borel measure. The *maximal function* of μ is the function

$$M\mu\colon \mathbb{R}^n \to \mathbb{R}$$

defined by

$$(M\mu)(x) := \sup_{r>0} \frac{\mu(B_r(x))}{m(B_r(x))}. \tag{6.13}$$

The *maximal function* of a signed measure $\lambda \colon \mathcal{B} \to \mathbb{R}$ is defined as the maximal function

$$M\lambda := M|\lambda| \colon \mathbb{R}^n \to \mathbb{R}$$

of its total variation $|\lambda| \colon \mathcal{B} \to [0, \infty)$.

Theorem 6.8 (Hardy–Littlewood maximal inequality). *Let $\lambda \colon \mathcal{B} \to \mathbb{R}$ be a signed Borel measure. Then the function $M\lambda \colon \mathbb{R}^n \to \mathbb{R}$ in Definition 6.7 is lower semi-continuous, i.e., the pre-image of the open interval (t, ∞) under $M\lambda$ is open for all $t \in \mathbb{R}$. Hence $M\lambda$ is Borel measurable. Moreover,*

$$\|M\lambda\|_{1,\infty} \leq 3^n |\lambda|(\mathbb{R}^n), \tag{6.14}$$

and so $M\lambda \in \mathcal{L}^{1,\infty}(\mathbb{R}^n)$.

Proof. See page 232. □

The proof of Theorem 6.8 relies on the following two lemmas.

Lemma 6.9. *Let $\mu \colon \mathcal{B} \to [0, \infty)$ be a finite Borel measure. Then the maximal function $M\mu \colon \mathbb{R}^n \to \mathbb{R}$ is lower semi-continuous and hence is Borel measurable.*

Proof. Let $t > 0$ and define

$$U_t := A(t, M\mu) = \{x \in \mathbb{R}^n \mid (M\mu)(x) > t\}. \tag{6.15}$$

We prove that U_t is open. Fix an element $x \in U_t$. Since $(M\mu)(x) > t$, there exists a number $r > 0$ such that

$$t < \frac{\mu(B_r(x))}{m(B_r(x))}.$$

Choose $\delta > 0$ such that

$$t\frac{(r+\delta)^n}{r^n} < \frac{\mu(B_r(x))}{m(B_r(x))}.$$

Further, choose $y \in \mathbb{R}^n$ such that $|y - x| < \delta$. Then $B_r(x) \subset B_{r+\delta}(y)$, and hence

$$\begin{aligned}
\mu(B_{r+\delta}(y)) &\geq \mu(B_r(x)) \\
&> t\frac{(r+\delta)^n}{r^n} m(B_r(x)) \\
&= t\frac{(r+\delta)^n}{r^n} m(B_r(y)) \\
&= t \cdot m(B_{r+\delta}(y)).
\end{aligned}$$

This implies

$$(M\mu)(y) \geq \frac{\mu(B_{r+\delta}(y))}{m(B_{r+\delta}(y))} > t,$$

and hence $y \in U_t$. This shows that U_t is open for all $t > 0$. It follows that $U_0 = \bigcup_{t>0} U_t$ is open and $U_t = \mathbb{R}^n$ is open for $t < 0$. Thus $M\mu$ is lower semi-continuous, as claimed. This proves Lemma 6.9. □

The Hardy–Littlewood estimate on the maximal function $M\mu$ is equivalent to an upper bound for the Lebesgue measure of the set U_t in (6.15). The proof relies on the next lemma about coverings by open balls.

Lemma 6.10 (Vitali's covering lemma). *Let $\ell \in \mathbb{N}$ and, for $i = 1, \dots, \ell$, let $x_i \in \mathbb{R}^n$ and $r_i > 0$. Define*

$$W := \bigcup_{i=1}^{\ell} B_{r_i}(x_i).$$

Then there exists a set

$$S \subset \{1, \dots, \ell\}$$

such that

$$B_{r_i}(x_i) \cap B_{r_j}(x_j) = \emptyset \quad \textit{for all } i, j \in S \textit{ with } i \neq j \tag{6.16}$$

and

$$W \subset \bigcup_{i \in S} B_{3r_i}(x_i). \tag{6.17}$$

Proof. Abbreviate $B_i := B_{r_i}(x_i)$ and choose the ordering such that

$$r_1 \geq r_2 \geq \cdots \geq r_\ell.$$

Choose $i_1 := 1$ and let $i_2 > 1$ be the smallest index such that $B_{i_2} \cap B_{i_1} = \emptyset$. Continue by induction to obtain a sequence

$$1 = i_1 < i_2 < \cdots < i_k \leq \ell$$

such that

$$B_{i_j} \cap B_{i_{j'}} = \emptyset \quad \text{for } j \neq j'$$

and

$$B_i \cap (B_{i_1} \cup \cdots \cup B_{i_j}) \neq \emptyset \quad \text{for } i_j < i < i_{j+1}$$

(respectively, for $i > i_k$ when $j = k$).

Then,

$$B_i \subset B_{3r_{i_1}}(x_{i_1}) \cup \cdots \cup B_{3r_{i_j}}(x_{i_j}) \quad \text{for } i_j < i < i_{j+1}$$

and hence

$$W = \bigcup_{i=1}^{\ell} B_i \subset \bigcup_{j=1}^{k} B_{3r_{i_j}}(x_{i_j}).$$

With $S := \{i_1, \dots, i_k\}$ this proves (6.17) and Lemma 6.10. □

Proof of Theorem 6.8. Fix a constant $t > 0$. Then the set

$$U_t := A(t, M\lambda)$$

is open by Lemma 6.9. Choose a compact set $K \subset U_t$. If $x \in K \subset U_t$, then $(M\lambda)(x) > t$ and so there exists a number $r(x) > 0$ such that

$$\frac{|\lambda|(B_{r(x)}(x))}{m(B_{r(x)}(x))} > t. \tag{6.18}$$

Since K is compact, there exist finitely many points

$$x_1, \dots, x_\ell \in K$$

such that

$$K \subset \bigcup_{i=1}^{\ell} B_{r_i}(x_i),$$

where $r_i := r(x_i)$. By Lemma 6.10, there is a subset $S \subset \{1, \dots, \ell\}$ such that the balls $B_{r_i}(x_i)$ for $i \in S$ are pairwise disjoint and $K \subset \bigcup_{i \in S} B_{3r_i}(x_i)$. Since $m(B_{3r}) = 3^n m(B_r)$ by Theorem 2.17, this gives

$$m(K) \le 3^n \sum_{i \in S} m(B_{r_i}(x_i)) < \frac{3^n}{t} \sum_{i \in S} |\lambda|(B_{r_i}(x_i)) \le \frac{3^n}{t} |\lambda|(\mathbb{R}^n).$$

Here the second step follows from (6.18) with $r_i = r(x_i)$ and the last step follows from the fact that the balls $B_{r_i}(x_i)$ for $i \in S$ are pairwise disjoint. Take the supremum over all compact sets $K \subset U_t$ to obtain

$$m(A(t, M\lambda)) = m(U_t) \le \frac{3^n}{t} |\lambda|(\mathbb{R}^n). \tag{6.19}$$

(See Theorem 2.13.) Multiply the inequality (6.19) by t and take the supremum over all real numbers $t > 0$ to obtain $\|M\lambda\|_{1,\infty} \le 3^n |\lambda|(\mathbb{R}^n)$. This proves Theorem 6.8. □

Definition 6.11. Let $f \in \mathcal{L}^1(\mathbb{R}^n)$. The *maximal function* of f is the function

$$Mf : \mathbb{R}^n \longrightarrow [0,\infty)$$

defined by

$$(Mf)(x) := \sup_{r>0} \frac{1}{m(B_r(x))} \int_{B_r(x)} |f|\, dm \quad \text{for } x \in \mathbb{R}^n. \tag{6.20}$$

Corollary 6.12. *Let $f \in \mathcal{L}^1(\mathbb{R}^n)$ and define the signed Borel measure μ_f on $\mathbb{R}^n$ by*

$$\mu_f(B) := \int_B f\, dm$$

for every Borel set $B \subset \mathbb{R}^n$. Then

$$Mf = M\mu_f \in \mathcal{L}^{1,\infty}(\mathbb{R}^n)$$

and

$$\|Mf\|_{1,\infty} \le 3^n \|f\|_1 .$$

Proof. The formula $|\mu_f|(B) = \int_B |f|\, dm$ for $B \in \mathcal{B}$ shows that $Mf = M\mu_f$. Hence the assertion follows from Theorem 6.8. □

Corollary 6.12 shows that the map $f \mapsto Mf$ descends to an operator (denoted by the same letter) from the Banach space $L^1(\mathbb{R}^n)$ to the topological vector space $L^{1,\infty}(\mathbb{R}^n)$. Corollary 6.12 also shows that the resulting operator

$$M : L^1(\mathbb{R}^n) \longrightarrow L^{1,\infty}(\mathbb{R}^n)$$

is continuous (because $|Mf - Mg| \le M(f-g)$). Note that it is not linear. By Theorem 6.8, M extends naturally to an operator $\lambda \mapsto M\lambda$ from the Banach space of signed Borel measures on $\mathbb{R}^n$ to $L^{1,\infty}(\mathbb{R}^n)$. (See Exercise 5.34.)

6.3 Lebesgue points

Definition 6.13. Let $f \in \mathcal{L}^1(\mathbb{R}^n)$. An element $x \in \mathbb{R}^n$ is a called a *Lebesgue point* of f if

$$\lim_{r\to 0} \frac{1}{m(B_r(x))} \int_{B_r(x)} |f - f(x)|\, dm = 0. \tag{6.21}$$

In particular, x is a Lebesgue point of f whenever f is continuous at x.

The next theorem is the main result of this chapter.

Theorem 6.14 (Lebesgue differentiation theorem). *Let $f \in \mathcal{L}^1(\mathbb{R}^n)$. Then there exists a Borel set $E \subset \mathbb{R}^n$ such that $m(E) = 0$ and every element of $\mathbb{R}^n \setminus E$ is a Lebesgue point of f.*

Proof. For $f \in \mathcal{L}^1(\mathbb{R}^n)$ and $r > 0$ define the function

$$T_r f : \mathbb{R}^n \longrightarrow [0, \infty)$$

by

$$(T_r f)(x) := \frac{1}{m(B_r(x))} \int_{B_r(x)} |f - f(x)| \, dm \quad \text{for } x \in \mathbb{R}^n. \tag{6.22}$$

One can prove via an approximation argument that $T_r f$ is Lebesgue measurable for every $r > 0$ and every $f \in \mathcal{L}^1(\mathbb{R}^n)$. However, we shall not use this fact in the proof. For $f \in \mathcal{L}^1(\mathbb{R}^n)$ define the function

$$Tf : \mathbb{R}^n \longrightarrow [0, \infty]$$

by

$$(Tf)(x) := \limsup_{r \to 0} (T_r f)(x) \quad \text{for } x \in \mathbb{R}^n. \tag{6.23}$$

We must prove that $Tf = 0$ almost everywhere for every $f \in \mathcal{L}^1(\mathbb{R}^n)$.

To see this, fix a function $f \in \mathcal{L}^1(\mathbb{R}^n)$ and assume without loss of generality that f is Borel measurable. (See Theorem 2.14 and part (v) of Theorem 1.55.) By Theorem 4.15, there exists a sequence of continuous functions $g_i : \mathbb{R}^n \to \mathbb{R}$ with compact support such that

$$\|f - g_i\|_1 < \frac{1}{2^i} \quad \text{for all } i \in \mathbb{N}.$$

Since g_i is continuous, we have $Tg_i = 0$. Moreover, the function

$$h_i := f - g_i$$

is Borel measurable and satisfies, for all $x \in \mathbb{R}^n$,

$$\begin{aligned}
(T_r h_i)(x) &= \frac{1}{m(B_r(x))} \int_{B_r(x)} |h_i - h_i(x)| \, dm \\
&\le \frac{1}{m(B_r(x))} \int_{B_r(x)} |h_i| \, dm + |h_i(x)| \\
&\le (M h_i)(x) + |h_i(x)|.
\end{aligned}$$

Thus

$$T_r h_i \leq M h_i + |h_i| \quad \text{for all } i \text{ and all } r > 0.$$

Take the limit superior as r tends to zero to obtain

$$T h_i \leq M h_i + |h_i| \quad \text{for all } i.$$

Moreover, it follows from the definition of T_r that

$$T_r f = T_r(g_i + h_i) \leq T_r g_i + T_r h_i \quad \text{for all } i \text{ and all } r > 0.$$

Take the limit superior as r tends to zero to obtain

$$Tf \leq Tg_i + Th_i = Th_i \leq Mh_i + |h_i| \quad \text{for all } i.$$

This implies

$$A(\varepsilon, Tf) \subset A(\varepsilon/2, Mh_i) \cup A(\varepsilon/2, h_i) \quad \text{for all } i \text{ and all } \varepsilon > 0. \tag{6.24}$$

(See (6.1) for the notation $A(\varepsilon, Tf)$, etc.) Since h_i and Mh_i are Borel measurable (see Theorem 6.8) the set

$$E_i(\varepsilon) := A(\varepsilon/2, Mh_i) \cup A(\varepsilon/2, h_i) \tag{6.25}$$

is a Borel set. Since $\|h_i\|_1 < 2^{-i}$, we have

$$m(A(\varepsilon/2, h_i)) \leq \frac{2}{\varepsilon} \|h_i\|_{1,\infty} \leq \frac{2}{\varepsilon} \|h_i\|_1 \leq \frac{1}{2^{i-1}\varepsilon}$$

and, by Theorem 6.8,

$$m(A(\varepsilon/2, Mh_i)) \leq \frac{2}{\varepsilon} \|Mh_i\|_{1,\infty} \leq \frac{2 \cdot 3^n}{\varepsilon} \|h_i\|_1 \leq \frac{3^n}{2^{i-1}\varepsilon}.$$

Thus

$$m(E_i(\varepsilon)) \leq \frac{3^n + 1}{2^{i-1}\varepsilon}.$$

Since this holds for all $i \in \mathbb{N}$, it follows that the Borel set

$$E(\varepsilon) := \bigcap_{i=1}^{\infty} E_i(\varepsilon)$$

has Lebesgue measure zero for all $\varepsilon > 0$.

Hence the Borel set

$$E := \bigcup_{k=1}^{\infty} E(1/k)$$

has Lebesgue measure zero. By (6.24) and (6.25), we have

$$A(1/k, Tf) \subset E(1/k)$$

for all $k \in \mathbb{N}$, and hence

$$\begin{aligned}\{x \in \mathbb{R}^n \mid (Tf)(x) \neq 0\} &= \bigcup_{k=1}^{\infty} \{x \in \mathbb{R}^n \mid (Tf)(x) > 1/k\} \\ &= \bigcup_{k=1}^{\infty} A(1/k, Tf) \\ &\subset \bigcup_{k=1}^{\infty} E(1/k) \\ &= E.\end{aligned}$$

This shows that

$$(Tf)(x) = 0 \quad \text{for all } x \in \mathbb{R}^n \setminus E,$$

and hence every element of $\mathbb{R}^n \setminus E$ is a Lebesgue point of f. This proves Theorem 6.14. □

The first consequence of Theorem 6.14 discussed here concerns a signed Borel measure λ on $\mathbb{R}^n$ that is absolutely continuous with respect to the Lebesgue measure. The following theorem provides a formula for the function f in Theorem 5.18 (also called the *Radon–Nikodým derivative of* λ).

Theorem 6.15. *Let $\lambda: \mathcal{B} \to \mathbb{R}$ be a signed Borel measure on $\mathbb{R}^n$ that is absolutely continuous with respect to the Lebesgue measure. Choose a Borel measurable function $f \in \mathcal{L}^1(\mathbb{R}^n)$ such that $\lambda(B) = \int_B f\, dm$ for all $B \in \mathcal{B}$. Then there exists a Borel set $E \subset \mathbb{R}^n$ such that $m(E) = 0$ and*

$$f(x) = \lim_{r \to 0} \frac{\lambda(B_r(x))}{m(B_r(x))} \quad \textit{for all } x \in \mathbb{R}^n \setminus E. \tag{6.26}$$

Proof. By Theorem 6.14, there exists a Borel set $E \subset \mathbb{R}^n$ of Lebesgue measure zero such that every element of $X \setminus E$ is a Lebesgue point. Since

$$\begin{aligned}\left|\frac{\lambda(B_r(x))}{m(B_r(x))} - f(x)\right| &= \frac{1}{m(B_r(x))}\left|\int_{B_r(x)} (f - f(x))\, dm\right| \\ &\leq \frac{1}{m(B_r(x))}\int_{B_r(x)} |f - f(x)|\, dm\end{aligned}$$

for all $r > 0$ and all $x \in \mathbb{R}^n$, it follows that (6.26) holds for all $x \in \mathbb{R}^n \setminus E$. This proves Theorem 6.15. □

Theorem 6.16. *Let $f \in \mathcal{L}^1(\mathbb{R}^n)$ and let $x \in \mathbb{R}^n$ be a Lebesgue point of f. Fix two constants $0 < \alpha < 1$ and $\varepsilon > 0$. Then there exists a $\delta > 0$ such that, for every Borel set $E \in \mathcal{B}$ and every $r > 0$,*

$$\begin{cases} r < \delta, \\ E \subset B_r(x), \\ m(E) > \alpha m(B_r(x)) \end{cases} \quad \Longrightarrow \quad \left|\frac{1}{m(E)}\int_E f\, dm - f(x)\right| < \varepsilon. \tag{6.27}$$

Proof. Since x is a Lebesgue point of f, there exists a constant $\delta > 0$ such that

$$m(B_r(x))^{-1}\int_{B_r(x)} |f - f(x)|\, dm < \alpha\varepsilon \quad \text{for } 0 < r < \delta.$$

Assume $0 < r < \delta$ and let $E \subset B_r(x)$ be a Borel set such that $m(E) \geq \alpha m(B_r(x))$. Then

$$\begin{aligned}\left|\frac{1}{m(E)}\int_E f\, dm - f(x)\right| &\leq \frac{1}{m(E)}\int_E |f - f(x)|\, dm \\ &\leq \frac{1}{\alpha m(B_r(x))}\int_{B_r(x)} |f - f(x)|\, dm \\ &< \varepsilon.\end{aligned}$$

This proves Theorem 6.16. □

The Lebesgue differentiation theorem (Theorem 6.14) can be viewed as a theorem about signed Borel measures that are absolutely continuous with respect to the Lebesgue measure. The next theorem is an analogous result for signed Borel measures that are singular with respect to the Lebesgue measure.

Theorem 6.17 (singular Lebesgue differentiation). *Let $\lambda\colon \mathcal{B} \to \mathbb{R}$ be a signed Borel measure on $\mathbb{R}^n$ such that $\lambda \perp m$. Then there exists a Borel set $E \subset \mathbb{R}^n$ such that $m(E) = 0 = |\lambda|(\mathbb{R}^n \setminus E)$ and*

$$\lim_{r\to 0} \frac{|\lambda|(B_r(x))}{m(B_r(x))} = 0 \quad \text{for all } x \in \mathbb{R}^n \setminus E. \tag{6.28}$$

Proof. The proof follows an argument in Heil [7, Section 3.4]. By assumption and Lemma 5.16, there exists a Borel set $A \subset \mathbb{R}^n$ such that

$$m(A) = 0, \quad |\lambda|(\mathbb{R}^n \setminus A) = 0.$$

For $\varepsilon > 0$, define the set

$$A_\varepsilon := \left\{ x \in \mathbb{R}^n \setminus A \;\middle|\; \limsup_{r\to 0} \frac{|\lambda|(B_r(x))}{m(B_r(x))} > \varepsilon \right\}.$$

We claim that A_ε is a Lebesgue null set for every $\varepsilon > 0$. To see this, fix two constants $\varepsilon > 0$ and $\delta > 0$. Since the Borel measure $|\lambda|$ is regular by Theorem 3.18, there exists an open set $U_\delta \subset \mathbb{R}^n$ such that

$$A \subset U_\delta, \quad |\lambda|(U_\delta) < \delta.$$

For $x \in A_\varepsilon$ choose a radius $r = r(x) > 0$ such that

$$\frac{|\lambda|(B_{r(x)}(x))}{m(B_{r(x)}(x))} > \varepsilon, \quad B_{r(x)}(x) \subset U_\delta,$$

and consider the open set

$$W_\delta := \bigcup_{x\in A_\varepsilon} B_{r(x)}(x) \subset U_\delta.$$

Fix a compact subset $K \subset W_\delta$ and cover K by finitely many of the balls $B_{r(x)}(x)$ with $x \in A_\varepsilon$. By Vitali's covering Lemma 6.10, there are elements $x_1, \dots, x_N \in A_\varepsilon$ such that the balls $B_{r(x_i)}(x_i)$ are pairwise disjoint and $K \subset \bigcup_{i=1}^N B_{3r(x_i)}(x_i)$. Thus

$$\begin{aligned} m(K) &\le \sum_{i=1}^N 3^n m(B_{r(x_i)}(x_i)) \\ &< \frac{3^n}{\varepsilon} \sum_{i=1}^N |\lambda|(B_{r(x_i)}(x_i)) \\ &\le \frac{3^n}{\varepsilon} \sum_{i=1}^N |\lambda|(U_\delta) \\ &\le \frac{3^n \delta}{\varepsilon}. \end{aligned}$$

This holds for every compact set $K \subset W_\delta$ and hence $m(W_\delta) \le 3^n\delta/\varepsilon$. Since $\delta > 0$ was arbitrary, the set A_ε is contained in an open set of arbitrarily small Lebesgue measure and so is a Lebesgue null set as claimed. This implies that

$$E := A \cup \bigcup_{k=1}^{\infty} A_{1/k}$$

is a Lebesgue null set. It satisfies $|\lambda|(\mathbb{R}^n \setminus E) = 0$ and (6.28) by definition, and this proves Theorem 6.17. □

6.4 Absolutely continuous functions

Definition 6.18. Let $I \subset \mathbb{R}$ be an interval. A function $f\colon I \to \mathbb{R}$ is called *absolutely continuous* if for every $\varepsilon > 0$ there exists a $\delta > 0$ such that, for every finite sequence $s_1 \le t_1 \le s_2 \le t_2 \le \cdots \le s_\ell \le t_\ell$ in I,

$$\sum_{i=1}^{\ell} |s_i - t_i| < \delta \implies \sum_{i=1}^{\ell} |f(s_i) - f(t_i)| < \varepsilon. \tag{6.29}$$

Every absolutely continuous function is continuous.

The equivalence of (i) and (iii) in the following result is the *fundamental theorem of calculus* for Lebesgue integrable functions. The equivalence of (i) and (ii) is known as the *Banach–Zarecki theorem.* For functions of bounded variation see Exercise 6.20 below.

Theorem 6.19 (fundamental theorem of calculus). *Let $I = [a,b] \subset \mathbb{R}$ be a compact interval, let $\mathcal{B} \subset 2^I$ be the Borel σ algebra, and let $m\colon \mathcal{B} \to [0,\infty]$ be the restriction of the Lebesgue measure to $\mathcal{B}$. Let $f\colon I \to \mathbb{R}$ be a function. Then the following are equivalent.*

(i) *f is absolutely continuous.*

(ii) *f is continuous, it has bounded variation, and if $E \subset I$ is a Lebesgue null set, then so is $f(E)$.*

(iii) *There is a Borel measurable function $g\colon I \to \mathbb{R}$ such that $\int_I |g|\,dm < \infty$ and, for all $x, y \in I$ with $x < y$,*

$$f(y) - f(x) = \int_x^y g(t)\,dt. \tag{6.30}$$

The right-hand side denotes the Lebesgue integral of g over the interval $[x,y]$.

If (iii) *holds, then there exists a Borel set $E \subset I$ such that $m(E) = 0$ and, for all $x \in I \setminus E$, f is differentiable at x and $f'(x) = g(x)$.*

Proof. We prove that (iii) implies the last assertion of the theorem. Thus assume that there exists a function $g \in \mathcal{L}^1(I)$ that satisfies (6.30) for all $x, y \in I$ with $x < y$. Then Theorem 6.14 asserts that there exists a Borel set $E \subset I$ of Lebesgue measure zero such that every element of $I \setminus E$ is a Lebesgue point of g. By Theorem 6.16 with $\alpha = 1/2$, every element $x \in I \setminus E$ satisfies condition (ii) in Theorem 6.6 with $A := g(x)$. Hence Theorem 6.6 asserts that the function f is differentiable at every point $x \in I \setminus E$ and satisfies $f'(x) = g(x)$ for $x \in I \setminus E$.

(iii) $\Longrightarrow$ (i). Thus assume f satisfies (iii) and define the signed measure

$$\lambda: \mathcal{B} \longrightarrow \mathbb{R}$$

by

$$\lambda(B) := \int_B g\, dm \tag{6.31}$$

for every Borel set $B \subset I$. Then λ is absolutely continuous with respect to the Lebesgue measure and

$$|\lambda|(B) = \int_B |g|\, dm$$

for ever Borel set $B \subset I$. Now let $\varepsilon > 0$. Since $|\lambda| \ll m$, it follows from Lemma 5.21 that there exists a constant $\delta > 0$ such that $|\lambda|(B) < \varepsilon$ for every Borel set $B \subset I$ with $m(B) < \delta$. Choose a sequence

$$s_1 \le t_1 \le \cdots \le s_\ell \le t_\ell$$

in I such that

$$\sum_{i=1}^{\ell} |t_i - s_i| < \delta$$

and define

$$U_i := (s_i, t_i) \quad \text{for } i = 1, \dots, \ell.$$

Then the Borel set

$$B := \bigcup_{i=1}^{\ell} U_i$$

has Lebesgue measure

$$m(B) = \sum_{i=1}^{\ell} |t_i - s_i| < \delta.$$

Hence

$$|\lambda|(B) < \varepsilon.$$

Since

$$|f(t_i) - f(s_i)| = \left| \int_{U_i} g \, dm \right| \le \int_{U_i} |g| \, dm = |\lambda|(U_i)$$

for all i, it follows that

$$\sum_{i=1}^{\ell} |f(t_i) - f(s_i)| \le \sum_{i=1}^{\ell} |\lambda|(U_i) = |\lambda|(B) < \varepsilon.$$

Hence f is absolutely continuous.

(i) $\Longrightarrow$ (ii). Assume f is absolutely continuous. It follows directly from the definition that f is continuous; that it has bounded variation is part (v) of Exercise 6.20. Now let $E \subset I$ be a Lebesgue null set and assume without loss of generality that $a, b \notin E$. Fix any constant $\varepsilon > 0$ and choose $\delta > 0$ such that (6.29) holds. Since the Lebesgue measure is outer regular by Theorem 2.13, there exists an open set $U \subset \mathrm{int}(I)$ such that $E \subset U$ and $m(U) < \delta$. Choose a (possibly finite) sequence of pairwise disjoint open intervals $U_i \subset I$ such that $U = \bigcup_i U_i$. Choose $s_i, t_i \in \overline{U_i}$ such that $f(s_i) = \inf_{U_i} f$ and $f(t_i) = \sup_{U_i} f$. Then it follows from (6.29) that $\sum_i m(f(U_i)) = \sum_i (f(t_i) - f(s_i)) < \varepsilon$ for every finite sum. Take the limit to obtain $m(f(U)) \le \sum_i m(f(U_i)) \le \varepsilon$. Since $\varepsilon > 0$ was chosen arbitrary and the Lebesgue measure is complete, it follows that $f(E)$ is a Lebesgue measurable set and $m(f(E)) = 0$.

(ii) $\Longrightarrow$ (iii). Assume f satisfies (ii). Then f has bounded variation and under this assumption Exercise 6.20 below outlines a proof that there exists a signed Borel measure $\lambda \colon \mathcal{B} \to \mathbb{R}$ such that $f(y) - f(x) = \lambda((x, y])$ for $x, y \in I$ with $x < y$. Since λ^+ and λ^- are regular by Theorem 3.18 and f is continuous, we have

$$f(y) - f(x) = \lambda([x, y]) = \lambda((x, y]) = \lambda([x, y)) = \lambda((x, y)) \tag{6.32}$$

for all $x, y \in I$ with $x < y$. By the Lebesgue decomposition theorem (Theorem 5.17), there exist two signed Borel measures $\lambda_{\mathrm{a}}, \lambda_{\mathrm{s}} \colon \mathcal{B} \to \mathbb{R}$ such that

$$\lambda = \lambda_{\mathrm{a}} + \lambda_{\mathrm{s}}, \quad \lambda_{\mathrm{a}} \ll m, \quad \lambda_{\mathrm{s}} \perp m. \tag{6.33}$$

Since $\lambda_{\mathrm{a}} \ll m$, it follows from Theorem 5.18 that there is an integrable function $g \in \mathcal{L}^1(I)$ such that

$$\lambda_{\mathrm{a}}(B) = \int_B g \, dm \tag{6.34}$$

for every Borel set $B \subset I$. Define the functions

$$f_{\mathrm{a}}, f_{\mathrm{s}} \colon I \longrightarrow \mathbb{R}$$

by

$$f_{\mathrm{a}}(x) := f(a) + \lambda_{\mathrm{a}}([a,x]) = f(a) + \int_a^x g(t)\,dt, \quad f_{\mathrm{s}}(x) := \lambda_{\mathrm{s}}([a,x]).$$

Then $f = f_{\mathrm{a}} + f_{\mathrm{s}}$ by (6.32) and (6.33). Since (iii) implies (i) and (i) implies (ii) (already proved), both functions f and f_{a} satisfy (ii) and f_{a} is absolutely continuous. Moreover $f_{\mathrm{s}} = f - f_{\mathrm{a}}$ is continuous.

It remains to prove that $f_{\mathrm{s}} \equiv 0$. The proof given below follows an argument in Heil [7, Section 3.5.4]. By Theorem 6.17, there exists a Lebesgue null set $E_{\mathrm{s}} \subset I$ such that $a, b \in E_{\mathrm{s}}$ (without loss of generality) and

$$|\lambda_{\mathrm{s}}|(I \setminus E_{\mathrm{s}}) = 0, \quad \lim_{r\to 0} \frac{|\lambda_{\mathrm{s}}((x-r,x+r))|}{r} = 0 \quad \text{for all } x \in I \setminus E_{\mathrm{s}}.$$

This implies that every element $x \in I \setminus E_{\mathrm{s}}$ satisfies condition (ii) in Theorem 6.6 with $A = 0$ and f replaced by f_{s}. Hence f_{s} is differentiable at every point $x \in I \setminus E_{\mathrm{s}}$ and $f_{\mathrm{s}}'(x) = 0$ for $x \in I \setminus E_{\mathrm{s}}$.

Since E_{s} is a Lebesgue null set, so are the sets $f(E_{\mathrm{s}})$ and $f_{\mathrm{a}}(E_{\mathrm{s}})$, because the functions f and f_{a} satisfy (ii). Since the difference of two Lebesgue null sets in $\mathbb{R}$ is a Lebesgue null set, by Exercise 2.27, it follows that

$$f_{\mathrm{s}}(E_{\mathrm{s}}) = \{f(x) - f_{\mathrm{a}}(x) \mid x \in E_{\mathrm{s}}\} \subset \{f(x) - f_{\mathrm{a}}(y) \mid x, y \in E_{\mathrm{s}}\}$$

is a Lebesgue null set.

We claim that $f_{\mathrm{s}}(I \setminus E_{\mathrm{s}})$ is also a Lebesgue null set. To see this, fix a constant $\varepsilon > 0$. For $n \in \mathbb{N}$ define the set

$$A_n := \left\{x \in I \setminus E_{\mathrm{s}} \,\middle|\, y \in I,\ |x-y| < \frac{1}{n} \implies |f_{\mathrm{s}}(x) - f_{\mathrm{s}}(y)| \le \varepsilon|x-y|\right\}.$$

Then

$$A_n \subset A_{n+1} \quad \text{for all } n \in \mathbb{N}$$

and

$$I \setminus E_{\mathrm{s}} = \bigcup_{n\in\mathbb{N}} A_n.$$

Here the last assertion follows from the fact that f_{s} is differentiable on $I \setminus E_{\mathrm{s}}$ with derivative zero. We prove that the Lebesgue outer measure of the set $f_{\mathrm{s}}(I \setminus E_{\mathrm{s}})$ satisfies the estimate

$$\nu(f_{\mathrm{s}}(I \setminus E_{\mathrm{s}})) \le \varepsilon(b - a + \varepsilon). \tag{6.35}$$

To see this, cover the set A_n by at most countably many open intervals U_i, each of length less than $1/n$, such that

$$\sum_i m(U_i) \le \nu(A_n) + \varepsilon$$

and each interval U_i contains an element of A_n. Then $f(U_i)$ is contained in an interval of length at most $\varepsilon m(U_i)$, by the definition of A_n. Hence

$$\nu(f(A_n)) \leq \varepsilon \sum_i m(U_i) \leq \varepsilon(\nu(A_n) + \varepsilon) \leq \varepsilon(b - a + \varepsilon).$$

Since the Lebesgue outer measure is continuous from below by part (iii) of Theorem 2.13 and

$$f_{\mathrm{s}}(I \setminus E_{\mathrm{s}}) = \bigcup_{n \in \mathbb{N}} f_{\mathrm{s}}(A_n),$$

it follows that

$$\nu(f_{\mathrm{s}}(I \setminus E_{\mathrm{s}})) = \lim_{n \to \infty} \nu(f(A_n)) \leq \varepsilon(b - a + \varepsilon).$$

This proves (6.35). Since $\varepsilon > 0$ was chosen arbitrary, this implies that $f_{\mathrm{s}}(I \setminus E_{\mathrm{s}})$ is a Lebesgue null set, as claimed. Since $f_{\mathrm{s}}(E_{\mathrm{s}})$ is also a Lebesgue null set, as noted above, it follows that $f_{\mathrm{s}}(I)$ is a Lebesgue null set. Finally, since f_{s} is continuous and $f_{\mathrm{s}}(0) = 0$ by definition, this implies $f_{\mathrm{s}} \equiv 0$. Hence $f = f_{\mathrm{a}}$ is absolutely continuous and this proves Theorem 6.19. □

6.5 Exercises

Exercise 6.20. Let $I = [a, b] \subset \mathbb{R}$ be a compact interval and let $\mathcal{B} \subset 2^I$ be the Borel σ-algebra. A function $f : I \to \mathbb{R}$ is said to be of *bounded variation* if

$$V(f) := \sup_{a = t_0 < t_1 < \cdots < t_\ell = b} \sum_{i=1}^{\ell} |f(t_i) - f(t_{i-1})| < \infty. \tag{6.36}$$

Denote by $\mathrm{BV}(I)$ the set of all functions $f : I \to \mathbb{R}$ of bounded variation. This is a real vector space. Functions of bounded variation have at most countably many discontinuities and the left and right limits exist everywhere. Prove the following.

(i) Every monotone function $f : I \to \mathbb{R}$ has bounded variation.

(ii) Let $f \in \mathrm{BV}(I)$ be right continuous. Then there exist right continuous monotone functions $f^{\pm} : I \to \mathbb{R}$ such that $f = f^+ - f^-$.

Hint. Define

$$F(x) := V(f|_{[a,x]}) = \sup_{a = t_0 < t_1 < \cdots < t_\ell = x} \sum_{i=1}^{\ell} |f(t_i) - f(t_{i-1})| \tag{6.37}$$

for $a \leq x \leq b$. Prove that F is right continuous and $F \pm f$ are monotone.

(iii) If f is continuous, then the function F in (6.37) is continuous.

(iv) Let $f: I \to \mathbb{R}$ be right continuous. Then $f \in \mathrm{BV}(I)$ if and only if there exists a signed Borel measure $\lambda = \lambda_f: \mathcal{B} \to \mathbb{R}$ such that $\lambda(\{a\}) = 0$ and

$$f(x) - f(a) = \lambda([a, x]) \quad \text{for } a \leq x \leq b. \tag{6.38}$$

Hint. Assume f is monotone. For $h \in C(I)$ define

$$\begin{aligned} \Lambda_f(h) &:= \int_a^b h\, df \\ &:= \sup_{a=t_0<t_1<\cdots<t_\ell=b} \sum_{i=1}^{\ell} \Big(\inf_{[t_{i-1},t_i]} h\Big) \cdot (f(t_i) - f(t_{i-1})). \end{aligned} \tag{6.39}$$

(This is the *Riemann–Stieltjes integral.* See Körner [9] and compare it with the Riemann integral [9, 18, 21].) Prove that $\Lambda_f: C(I) \to \mathbb{R}$ is a positive linear functional. Use the Riesz representation theorem (Theorem 3.15) to find a Borel measure $\lambda_f: \mathcal{B} \to [0, \infty)$ such that $\Lambda_f(h) = \int_I h\, d\lambda_f$ for all $h \in C(I)$. Use the fact that f is right continuous to prove that λ_f satisfies (6.38).

(v) If $f \in \mathrm{BV}(I)$ is right continuous and λ_f is as in (iii), then $V(f) = |\lambda_f|(I)$.

(vi) Every absolutely continuous function $f: I \to \mathbb{R}$ has bounded variation.

Exercise 6.21. Let $(\mathbb{R}, \mathcal{A}, m)$ be the Lebesgue measure space and fix a constant $0 < \varepsilon < 1/2$. Prove that there does not exist a Lebesgue measurable set $E \subset \mathbb{R}$ such that

$$\varepsilon < \frac{m(E \cap I)}{m(I)} < 1 - \varepsilon$$

for every nonempty bounded open interval $I \subset \mathbb{R}$.

Hint. Consider the function

$$f := \chi_{E \cap [-1,1]}$$

and define the measure $\mu_f: \mathcal{B} \to \mathbb{R}$ by

$$\mu_f(B) := \int_B f\, dm = m(B \cap E \cap [-1, 1]).$$

Examine the Lebesgue points of f. (Compare this with Exercise 2.32.)

Exercise 6.22. Prove the VITALI–CARATHÉODORY THEOREM. *Let $(X, \mathcal{U})$ be a locally compact Hausdorff space and let $\mathcal{B} \subset 2^X$ be its Borel σ-algebra. Let $\mu: \mathcal{B} \to [0, \infty]$ be an outer regular Borel measure that is inner regular on open sets. Let $f \in \mathcal{L}^1(\mu)$ and let $\varepsilon > 0$. Then there exists an upper semi-continuous function $u: X \to \mathbb{R}$ that is bounded above and a lower semi-continuous function $v: X \to \mathbb{R}$ that is bounded below such that*

$$u \leq f \leq v, \qquad \int_X (v - u)\, d\mu < \varepsilon. \tag{6.40}$$

Hint. Assume first that $f \geq 0$. Use Theorem 1.26 to find a sequence of measurable sets $E_i \in \mathcal{A}$, not necessarily disjoint, and a sequence of real numbers $c_i > 0$ such that $\mu(E_i) < \infty$ for all i and

$$f = \sum_{i=1}^{\infty} c_i \chi_{E_i}.$$

Thus

$$\sum_{i=1}^{\infty} c_i \mu(E_i) = \int_X f\, d\mu < \infty.$$

Choose a sequence of compact sets $K_i \subset X$ and a sequence of open sets $U_i \subset X$ such that $K_i \subset E_i \subset U_i$ and $c_i \mu(U_i \setminus K_i) < \varepsilon 2^{-i-1}$ for all i. Choose $n \in \mathbb{N}$ such that $\sum_{i=n+1}^{\infty} c_i \mu(E_i) < \varepsilon/2$ and define

$$u := \sum_{i=1}^{n} c_i \chi_{K_i}, \qquad v := \sum_{i=1}^{\infty} c_i \chi_{U_i}.$$

Show that $\int_X (v - u)\, d\mu < \varepsilon$, v is *lower semi-continuous* (i.e., $v^{-1}((t, \infty))$ is open for all $t \in \mathbb{R}$), and u is *upper semi-continuous* (i.e., $u^{-1}((-\infty, t))$ is open for all $t \in \mathbb{R}$).

Exercise 6.23. Fix two real numbers $a < b$ and prove the following.

(i) If $f: [a, b] \to \mathbb{R}$ is everywhere differentiable, then $f': [a, b] \to \mathbb{R}$ is Borel measurable.

(ii) If $f: [a, b] \to \mathbb{R}$ is everywhere differentiable and $\int_a^b |f'(t)|\, dt < \infty$, then f is absolutely continuous.

Hint. Fix a constant $\varepsilon > 0$. By the Vitali–Carathéodory theorem in Exercise 6.22, there is a lower semi-continuous function $g: [a, b] \to \mathbb{R}$ such that

$$g > f', \qquad \int_a^b g(t)\, dt < \int_a^b f'(t)\, dt + \varepsilon.$$

For $\eta > 0$ define the function

$$F_\eta \colon [a,b] \longrightarrow \mathbb{R}$$

by

$$F_\eta(x) := \int_a^x g(t)\,dt - f(x) + f(a) + \eta(x-a)$$

for $a \le x \le b$. Consider a point $a \le x < b$. Since $g(x) > f'(x)$ and g is lower semi-continuous, find a number $\delta_x > 0$ such that

$$g(t) > f'(x), \qquad \frac{f(t)-f(x)}{t-x} < f'(x) + \eta \quad \text{for } x < t < x + \delta_x.$$

Deduce that

$$F_\eta(t) > F_\eta(x) \quad \text{for } x < t < x + \delta_x.$$

Since $F_\eta(a) = 0$, there exists a maximal element $x \in [a,b]$ such that $F_\eta(x) = 0$. If $x < b$, it follows from the previous discussion that $F_\eta(t) > 0$ for $x < t \le b$. In either case $F_\eta(b) \ge 0$ and hence

$$f(b) - f(a) \le \int_a^b g(t)\,dt + \eta(b-a) < \int_a^b f'(t)\,dt + \varepsilon + \eta(b-a).$$

Since this holds for all $\eta > 0$ and all $\varepsilon > 0$, it follows that

$$f(b) - f(a) \le \int_a^b f'(t)\,dt.$$

Replace f by $-f$ to obtain the relation $f(b) - f(a) = \int_a^b f'(t)\,dt$. Now deduce that

$$f(x) - f(a) = \int_a^x f'(t)\,dt$$

for all $x \in [a,b]$.

Example 6.24. (i) The *Cantor function* is the unique monotone function

$$f \colon [0,1] \longrightarrow [0,1]$$

that satisfies

$$f\Bigl(2\sum_{i=1}^\infty \frac{a_i}{3^i}\Bigr) = \sum_{i=1}^\infty \frac{a_i}{2^i}$$

for all sequences $a_i \in \{0,1\}$. It is continuous and nonconstant, and its derivative exists and vanishes on the complement of the standard Cantor set

$$C := \bigcap_{n=1}^{\infty} \bigcup_{a_i \in \{0,1\}} \Big[2\sum_{i=1}^{n} \frac{a_i}{3^i}, 2\sum_{i=1}^{n} \frac{a_i}{3^i} + \frac{1}{3^n} \Big].$$

This Cantor set has Lebesgue measure zero. Hence f is almost everywhere differentiable and its derivative is integrable. However, f is not equal to the integral of its derivative, and therefore is not absolutely continuous.

(ii) The following construction was explained to me by Theo Buehler. Define the homeomorphisms

$$g\colon [0,1] \longrightarrow [0,2] \quad \text{and} \quad h\colon [0,2] \longrightarrow [0,1]$$

by

$$g(x) := f(x) + x, \quad h := g^{-1}.$$

The image $g([0,1] \setminus C)$ is a countable union of disjoint open intervals of total length one and hence has Lebesgue measure one. Thus its complement

$$K := g(C) \subset [0,2]$$

is a modified Cantor set of Lebesgue measure one. Hence, by Theorem 6.19, g is not absolutely continuous. Moreover, by Lemma 2.15, there exists a set $E \subset K$ which is not Lebesgue measurable. However, its image

$$F := h(E) \subset [0,1]$$

under h is a subset of the Lebesgue null set C and hence is a Lebesgue null set. Thus F is a Lebesgue measurable set and $E = h^{-1}(F)$ is not Lebesgue measurable. This shows that the function $h\colon [0,2] \to [0,1]$ is not measurable with respect to the Lebesgue σ-algebras on both domain and target (i.e., it is not *Lebesgue-Lebesgue measurable*).

(iii) Let $I, J \subset \mathbb{R}$ be intervals. Then it follows from Lemma 2.15 and Theorem 6.19 that every Lebesgue-Lebesgue measurable homeomorphism $h\colon I \to J$ has an absolutely continuous inverse.

(iv) Let $h\colon [0,2] \to [0,1]$ and $F \subset C \subset [0,1]$ be as in part (ii). Then the characteristic function $\chi_F\colon \mathbb{R} \to \mathbb{R}$ is Lebesgue measurable and $h\colon [0,2] \to \mathbb{R}$ is continuous. However, the composition $\chi_F \circ h\colon [0,2] \to \mathbb{R}$ is not Lebesgue measurable because the set $(\chi_F \circ h)^{-1}(1) = E$ is not Lebesgue measurable.

(v) By contrast, if $I, J \subset \mathbb{R}$ are intervals, $f\colon J \to \mathbb{R}$ is Lebesgue measurable, and $h\colon I \to J$ is a C^1 diffeomorphism, then $f \circ h\colon I \to \mathbb{R}$ is again Lebesgue measurable by Theorem 2.17.

Chapter 7

Product measures

The purpose of this chapter is to study products of two measurable spaces (Section 7.1), introduce the product measure (Section 7.2), and prove Fubini's Theorem (Section 7.3). The archetypal example is the Lebesgue measure on $\mathbb{R}^{k+\ell} = \mathbb{R}^k \times \mathbb{R}^\ell$; it is the completion of the product measure associated to the Lebesgue measures on $\mathbb{R}^k$ and $\mathbb{R}^\ell$ (Section 7.4). Applications include the convolution (Section 7.5), Marcinkiewicz interpolation (Section 7.6), and the Calderón–Zygmund inequality (Section 7.7).

7.1 The product σ-algebra

Assume throughout that $(X, \mathcal{A})$ and $(Y, \mathcal{B})$ are measurable spaces.

Definition 7.1. The *product σ-algebra* of $\mathcal{A}$ and $\mathcal{B}$ is defined as the smallest σ-algebra on the product space

$$X \times Y := \{(x, y) \mid x \in X,\ y \in Y\}$$

that contains all subsets of the form $A \times B$, where $A \in \mathcal{A}$ and $B \in \mathcal{B}$. It will be denoted by $\mathcal{A} \otimes \mathcal{B} \subset 2^{X \times Y}$.

Lemma 7.2. *Let $E \in \mathcal{A} \otimes \mathcal{B}$ and let $f: X \times Y \to \overline{\mathbb{R}}$ be an $(\mathcal{A} \otimes \mathcal{B})$-measurable function. Then the following holds.*

(i) *For every $x \in X$, the function $f_x: Y \to \overline{\mathbb{R}}$, defined by $f_x(y) := f(x, y)$ for $y \in Y$, is $\mathcal{B}$-measurable and*

$$E_x := \{y \in Y \mid (x, y) \in E\} \in \mathcal{B}. \tag{7.1}$$

(ii) *For every $y \in Y$, the function $f^y: X \to \overline{\mathbb{R}}$, defined by $f^y(x) := f(x, y)$ for $x \in X$, is $\mathcal{A}$-measurable and*

$$E^y := \{x \in X \mid (x, y) \in E\} \in \mathcal{A}. \tag{7.2}$$

Proof. Define $\Omega \subset 2^{X\times Y}$ by

$$\Omega := \{E \subset X \times Y \mid E_x \in \mathcal{B} \text{ for all } x \in X\}.$$

We prove that Ω is a σ-algebra. To see this, note first that $X \times Y \in \Omega$. Second, if $E \in \Omega$, then $E_x \in \mathcal{B}$ for all $x \in X$, hence

$$(E^{c})_x = \{y \in Y \mid (x, y) \notin E\} = (E_x)^{c} \in \mathcal{B}$$

for all $x \in X$, and hence $E^{c} \in \Omega$. Third, if $E_i \in \Omega$ is a sequence and $E := \bigcup_{i=1}^{\infty} E_i$, then $E_x = \bigcup_{i=1}^{\infty}(E_i)_x \in \mathcal{B}$ for all $x \in X$, and hence $E \in \Omega$. This shows that Ω is a σ-algebra. Since $A \times B \in \Omega$ for all $A \in \mathcal{A}$ and all $B \in \mathcal{B}$, it follows that $\mathcal{A} \otimes \mathcal{B} \subset \Omega$. This proves (7.1) for all $x \in X$.

Now fix an element $x \in X$. If $V \subset \overline{\mathbb{R}}$ is open, then $E := f^{-1}(V) \in \mathcal{A} \otimes \mathcal{B}$ and hence $(f_x)^{-1}(V) = E_x \in \mathcal{B}$ by (7.1). Thus f_x is $\mathcal{B}$-measurable. This proves (i).

The proof of (ii) is analogous and this proves Lemma 7.2. □

Definition 7.3. Let Z be a set. A collection of subsets $\mathcal{M} \subset 2^{Z}$ is called a *monotone class* if it satisfies the following two axioms:

(a) if $A_i \in \mathcal{M}$ for $i \in \mathbb{N}$ such that $A_i \subset A_{i+1}$ for all i, then $\bigcup_{i=1}^{\infty} A_i \in \mathcal{M}$;

(b) if $B_i \in \mathcal{M}$ for $i \in \mathbb{N}$ such that $B_i \supset B_{i+1}$ for all i, then $\bigcap_{i=1}^{\infty} B_i \in \mathcal{M}$.

Definition 7.4. A subset $Q \subset X \times Y$ is called *elementary* if it is the union of finitely many pairwise disjoint subsets of the form $A \times B$ with $A \in \mathcal{A}$ and $B \in \mathcal{B}$.

The next lemma is a useful characterization of the product σ-algebra.

Lemma 7.5. *The product σ-algebra $\mathcal{A} \otimes \mathcal{B}$ is the smallest monotone class in $X \times Y$ that contains all elementary subsets.*

Proof. Let $\mathcal{E} \subset 2^{X\times Y}$ denote the collection of all elementary subsets and define $\mathcal{M} \subset 2^{X\times Y}$ as the smallest monotone class that contains $\mathcal{E}$. This is well defined because the intersection of any collection of monotone classes is again a monotone class. Since every σ-algebra is a monotone class and every elementary set is an element of $\mathcal{A} \otimes \mathcal{B}$, it follows that

$$\mathcal{M} \subset \mathcal{A} \otimes \mathcal{B}.$$

Since $\mathcal{E} \subset \mathcal{M}$ by definition, the converse inclusion follows once we know that $\mathcal{M}$ is a σ-algebra. We prove this in seven steps.

Step 1. *For every set $P \subset X \times Y$ the collection*

$$\Omega(P) := \{Q \subset X \times Y \mid P \setminus Q,\ Q \setminus P,\ P \cup Q \in \mathcal{M}\}$$

is a monotone class.

This follows immediately from the definition of monotone class.

Step 2. *If $P, Q \subset X \times Y$, then $Q \in \Omega(P)$ if and only if $P \in \Omega(Q)$.*

This follows immediately from the definition of $\Omega(P)$ in Step 1.

Step 3. *If $P, Q \in \mathcal{E}$, then $P \cap Q, P \setminus Q, P \cup Q \in \mathcal{E}$.*

For the intersection this follows from the fact that

$$(A_1 \times B_1) \cap (A_2 \times B_2) = (A_1 \cap A_2) \times (B_1 \cap B_2).$$

For the complement it follows from the fact that

$$(A_1 \times B_1) \setminus (A_2 \times B_2) = ((A_1 \setminus A_2) \times B_1) \cup ((A_1 \cap A_2) \times (B_1 \setminus B_2)).$$

For the union this follows from the fact that $P \cup Q = (P \setminus Q) \cup Q$.

Step 4. *If $P \in \mathcal{E}$, then $\mathcal{M} \subset \Omega(P)$.*

Let $P \in \mathcal{E}$. Then $P \setminus Q, Q \setminus P, P \cup Q \in \mathcal{E} \subset \mathcal{M}$ for all $Q \in \mathcal{E}$ by Step 3. Hence $Q \in \Omega(P)$ for all $Q \in \mathcal{E}$ by the definition of $\Omega(P)$ in Step 1. Thus we have proved that $\mathcal{E} \subset \Omega(P)$. Since $\Omega(P)$ is a monotone class by Step 1, it follows that $\mathcal{M} \subset \Omega(P)$. This proves Step 4.

Step 5. *If $P \in \mathcal{M}$, then $\mathcal{M} \subset \Omega(P)$.*

Fix a set $P \in \mathcal{M}$. Then $P \in \Omega(Q)$ for all $Q \in \mathcal{E}$ by Step 4. Hence $Q \in \Omega(P)$ for all $Q \in \mathcal{E}$ by Step 2. Thus $\mathcal{E} \subset \Omega(P)$ and hence it follows from Step 1 that $\mathcal{M} \subset \Omega(P)$. This proves Step 5.

Step 6. *If $P, Q \in \mathcal{M}$, then $P \setminus Q, P \cup Q \in \mathcal{M}$.*

If $P, Q \in \mathcal{M}$, then $Q \in \mathcal{M} \subset \Omega(P)$ by Step 5, and hence $P \setminus Q, P \cup Q \in \mathcal{M}$ by the definition of $\Omega(P)$ in Step 1.

Step 7. *$\mathcal{M}$ is a σ-algebra.*

By definition, $X \times Y \in \mathcal{E} \subset \mathcal{M}$. If $P \in \mathcal{M}$, then $P^c = (X \times Y) \setminus P \in \mathcal{M}$ by Step 6. If $P_i \in \mathcal{M}$ for $i \in \mathbb{N}$, then $Q_n := \bigcup_{i=1}^n P_i \in \mathcal{M}$ for all $n \in \mathbb{N}$ by Step 6 and hence $\bigcup_{i=1}^\infty P_i = \bigcup_{n=1}^\infty Q_n \in \mathcal{M}$ because $\mathcal{M}$ is a monotone class. This proves Step 7 and Lemma 7.5. □

Lemma 7.6. *Let $(X, \mathcal{U}_X)$ and $(Y, \mathcal{U}_Y)$ be topological spaces, let $\mathcal{U}_{X\times Y}$ be the product topology on $X \times Y$ (see Appendix* B*), and let $\mathcal{B}_X$, $\mathcal{B}_Y$, $\mathcal{B}_{X\times Y}$ be the associated Borel σ-algebras. Then*

$$\mathcal{B}_X \otimes \mathcal{B}_Y \subset \mathcal{B}_{X\times Y}. \tag{7.3}$$

If $(X, \mathcal{U}_X)$ is a second countable locally compact Hausdorff space, then

$$\mathcal{B}_X \otimes \mathcal{B}_Y = \mathcal{B}_{X\times Y}. \tag{7.4}$$

Proof. The projections $\pi_X : X \times Y \to X$ and $\pi_Y : X \times Y \to Y$ are continuous, and hence Borel measurable by Theorem 1.20. Thus $\pi_X^{-1}(A) = A \times Y \in \mathcal{B}_{X\times Y}$ for all $A \in \mathcal{B}_X$ and $\pi_Y^{-1}(B) = X \times B \in \mathcal{B}_{X\times Y}$ for all $B \in \mathcal{B}_Y$. Hence $A \times B \in \mathcal{B}_{X\times Y}$ for all $A \in \mathcal{B}_X$ and all $B \in \mathcal{B}_Y$, and this implies (7.3).

Now assume $(X, \mathcal{U}_X)$ is a second countable locally compact Hausdorff space and choose a countable basis $\{U_i \mid i \in \mathbb{N}\}$ of $\mathcal{U}_X$ such that $\bar{U}_i$ is compact for all $i \in \mathbb{N}$. Fix an open set $W \in \mathcal{U}_{X\times Y}$ and, for $i \in \mathbb{N}$, define

$$V_i := \{y \in Y \mid (x, y) \in W \text{ for all } x \in \bar{U}_i\}.$$

We prove that V_i is open. Let $y_0 \in V_i$. Then $(x, y_0) \in W$ for all $x \in \bar{U}_i$. Hence, for every $x \in \bar{U}_i$, there exist open sets $U(x) \in \mathcal{U}_X$ and $V(x) \in \mathcal{U}_Y$ such that $(x, y_0) \in U(x) \times V(x) \subset W$. Since $\bar{U}_i$ is compact, there are finitely many elements $x_1, \dots, x_\ell \in \bar{U}_i$ such that $\bar{U}_i \subset U(x_1) \cup \dots \cup U(x_\ell)$. Define

$$V := V(x_1) \cap \dots \cap V(x_\ell).$$

Then V is open and $\bar{U}_i \times V \subset W$, so $y_0 \in V \subset V_i$. This shows that V_i is open for all $i \in \mathbb{N}$. Next we prove that

$$W = \bigcup_{i=1}^{\infty} (U_i \times V_i). \tag{7.5}$$

Let $(x_0, y_0) \in W$. Then there exist open sets $U \in \mathcal{U}_X$ and $V \in \mathcal{U}_Y$ such that

$$(x_0, y_0) \in U \times V \subset W.$$

Since $(X, \mathcal{U}_X)$ is a locally compact Hausdorff space, Lemma A.3 asserts that there exists an open set $U' \subset X$ such that $x_0 \in U' \subset \overline{U'} \subset U$. Since the sets U_i form a basis of the topology, there exists an integer $i \in \mathbb{N}$ such that $x_0 \in U_i \subset U'$ and hence $x_0 \in \bar{U}_i \subset \overline{U'} \subset U$. Thus

$$\bar{U}_i \times \{y_0\} \subset U \times V \subset W,$$

hence $y_0 \in V_i$, and so $(x_0, y_0) \in U_i \times V_i \subset W$. Since the element $(x_0, y_0) \in W$ was chosen arbitrarily, this proves (7.5). Thus we have shown that

$$\mathcal{U}_{X\times Y} \subset \mathcal{B}_X \otimes \mathcal{B}_Y$$

and this implies

$$\mathcal{B}_{X\times Y} \subset \mathcal{B}_X \otimes \mathcal{B}_Y.$$

Hence (7.4) follows from (7.3). This proves Lemma 7.6. □

Lemma 7.7. *Let $(X, \mathcal{A})$ be a measurable space such that the cardinality of X is greater than that of $2^{\mathbb{N}}$. Then the diagonal $\Delta := \{(x, x) \mid x \in X\}$ is not an element of $\mathcal{A} \otimes \mathcal{A}$.*

Proof. The proof has three steps.

Step 1. *Let Y be a set. For $\mathcal{E} \subset 2^Y$ denote by $\sigma(\mathcal{E}) \subset 2^Y$ the smallest σ-algebra containing $\mathcal{E}$. If $D \in \sigma(\mathcal{E})$, then there exists a sequence $E_i \in \mathcal{E}$, $i \in \mathbb{N}$, such that*

$$D \in \sigma(\{E_i \mid i \in \mathbb{N}\}).$$

The union of the sets $\sigma(\mathcal{E}')$ over all countable subsets $\mathcal{E}' \subset \mathcal{E}$ is a σ-algebra that contains $\mathcal{E}$ and is contained in $\sigma(\mathcal{E})$. Hence it is equal to $\sigma(\mathcal{E})$.

Step 2. *Let Y be a set, let $\mathcal{E} \subset 2^Y$, and let $D \in \sigma(\mathcal{E})$. Then there are a sequence $E_i \in \mathcal{E}$ and a set $\mathcal{J} \subset 2^{\mathbb{N}}$ such that*

$$D = \bigcup_{I\in\mathcal{J}} \Big(\bigcap_{i\in I} E_i \cap \bigcap_{i\in\mathbb{N}\setminus I} E_i^c \Big).$$

By Step 1 there exists a sequence $E_i \in \mathcal{E}$ such that $D \in \sigma(\{E_i \mid i \in \mathbb{N}\})$. For $I \subset \mathbb{N}$ define

$$E_I := \Big(\bigcap_{i\in I} E_i\Big) \cap \Big(\bigcap_{i\in\mathbb{N}\setminus I} E_i^c\Big).$$

These sets form a partition of Y. Hence the collection

$$\mathcal{F} := \Big\{ \bigcup_{I\in\mathcal{J}} E_I \,\Big|\, \mathcal{J} \subset 2^{\mathbb{N}} \Big\}$$

is a σ-algebra on Y. Since $E_i \in \mathcal{F}$ for each $i \in \mathbb{N}$, it follows that $D \in \mathcal{F}$. This proves Step 2.

Step 3. $\Delta \notin \mathcal{A} \otimes \mathcal{A}$.

Let $\mathcal{E} \subset 2^{X \times X}$ be the collection of all sets of the form $A \times B$ with $A, B \in \mathcal{A}$. Let $D \in \mathcal{A} \otimes \mathcal{A}$. By Step 2, there are sequences $A_i, B_i \in \mathcal{A}$ and a set $\mathcal{I} \subset 2^{\mathbb{N}}$ such that

$$D = \bigcup_{I \in \mathcal{I}} E_I,$$

where

$$E_I := \Big(\bigcap_{i \in I} (A_i \times B_i) \cap \bigcap_{i \in \mathbb{N} \setminus I} (A_i \times B_i)^{c} \Big).$$

Thus

$$E_I = \bigcup_{J \subset \mathbb{N} \setminus I} A_{IJ} \times B_{IJ},$$

where

$$A_{IJ} := \Big(\bigcap_{i \in I} A_i \Big) \cap \Big(\bigcap_{j \in J} (X \setminus A_j) \Big)$$

and

$$B_{IJ} := \Big(\bigcap_{i \in I} B_i \Big) \cap \Big(\bigcap_{j \in \mathbb{N} \setminus (I \cup J)} (X \setminus B_j) \Big).$$

If $D \subset \Delta$, then for all I and J, we have $A_{IJ} \times B_{IJ} \subset \Delta$ and so $A_{IJ} \times B_{IJ}$ is either empty or a singleton. Thus the cardinality of D is at most the cardinality of the set of pairs of disjoint subsets of $\mathbb{N}$, which is equal to the cardinality of $2^{\mathbb{N}}$. Since the cardinality of the diagonal is bigger than that of $2^{\mathbb{N}}$, it follows that $\Delta \notin \mathcal{A} \otimes \mathcal{A}$, as claimed. This proves Lemma 7.7. □

Example 7.8. Let X be an uncountable set, of cardinality greater than that of $2^{\mathbb{N}}$, and equipped with the discrete topology so that $\mathcal{B}_X = \mathcal{U}_X = 2^X$. Then Δ is an open subset of $X \times X$ with respect to the product topology (which is also discrete because points are open). Hence $\Delta \in \mathcal{B}_{X \times X} = 2^{X \times X}$. However, $\Delta \notin \mathcal{B}_X \otimes \mathcal{B}_X$ by Lemma 7.7. Thus the product $\mathcal{B}_X \otimes \mathcal{B}_X$ of the Borel σ-algebras is not the Borel σ-algebra of the product. In other words, the inclusion (7.3) in Lemma 7.6 is strict in this example. Note also that the distance function $d\colon X \times X \to \mathbb{R}$ defined by $d(x, y) := 1$ for $x \neq y$ and $d(x, x) := 0$ is continuous with respect to the product topology, but is not measurable with respect to the product of the Borel σ-algebras.

7.2 The product measure

The definition of the product measure on the product σ-algebra is based on the following theorem. For a measure space $(X, \mathcal{A}, \mu)$ and a measurable function $\phi: X \to [0, \infty]$ we use the notation

$$\int_X \phi(x)\, d\mu(x) := \int_X \phi\, d\mu.$$

Theorem 7.9. *Let $(X, \mathcal{A}, \mu)$ and $(Y, \mathcal{B}, \nu)$ be σ-finite measure spaces and let $Q \in \mathcal{A} \otimes \mathcal{B}$. Then the functions*

$$X \longrightarrow [0, \infty]: x \longmapsto \nu(Q_x) \quad \text{and} \quad Y \longrightarrow [0, \infty]: y \longmapsto \mu(Q^y) \tag{7.6}$$

are measurable and

$$\int_X \nu(Q_x)\, d\mu(x) = \int_Y \mu(Q^y)\, d\nu(y). \tag{7.7}$$

Definition 7.10. Let $(X, \mathcal{A}, \mu)$ and $(Y, \mathcal{B}, \nu)$ be σ-finite measure spaces. The *product measure* of μ and ν is the map

$$\mu \otimes \nu: \mathcal{A} \otimes \mathcal{B} \longrightarrow [0, \infty]$$

defined by

$$(\mu \otimes \nu)(Q) := \int_X \nu(Q_x)\, d\mu(x) = \int_Y \mu(Q^y)\, d\nu(y) \tag{7.8}$$

for $Q \in \mathcal{A} \otimes \mathcal{B}$. That $\mu \otimes \nu$ is σ-additive, and hence is a measure, follows from Theorem 1.38 and the fact that $\nu(Q_x) = \sum_{i=1}^{\infty} \nu((Q_i)_x)$ for every sequence of pairwise disjoint sets $Q_i \in \mathcal{A} \otimes \mathcal{B}$. The product measure satisfies

$$(\mu \otimes \nu)(A \times B) = \mu(A) \cdot \nu(B) \tag{7.9}$$

for $A \in \mathcal{A}$ and $B \in \mathcal{B}$ and hence is σ-finite.

Proof of Theorem 7.9. Define

$$\Omega := \left\{ Q \in \mathcal{A} \otimes \mathcal{B} \;\middle|\; \begin{array}{l} \text{the functions (7.6) are measurable} \\ \text{and equality (7.7) holds} \end{array} \right\}.$$

We prove in five steps that $\Omega = \mathcal{A} \otimes \mathcal{B}$.

Step 1. *If $A \in \mathcal{A}$ and $B \in \mathcal{B}$, then $Q := A \times B \in \Omega$.*

By assumption

$$Q_x = \begin{cases} B & \text{if } x \in A, \\ \emptyset & \text{if } x \notin A, \end{cases} \qquad Q^y = \begin{cases} A & \text{if } y \in B, \\ \emptyset & \text{if } y \notin B. \end{cases}$$

Define the function

$$\phi \colon X \longrightarrow [0, \infty]$$

by

$$\phi(x) := \nu(Q_x) = \nu(B)\chi_A(x) \quad \text{for } x \in X$$

and the function

$$\psi \colon Y \longrightarrow [0, \infty]$$

by

$$\psi(y) := \mu(Q^y) = \mu(A)\chi_B(y) \quad \text{for } y \in Y.$$

Then ϕ, ψ are measurable and

$$\int_X \phi \, d\mu = \mu(A)\nu(B) = \int_Y \psi \, d\nu.$$

Thus $Q \in \Omega$.

Step 2. *If $Q_1, Q_1 \in \Omega$ and $Q_1 \cap Q_2 = \emptyset$, then $Q := Q_1 \cup Q_2 \in \Omega$.*

Define

$$\phi_i(x) := \nu((Q_i)_x), \quad \phi(x) := \nu(Q_x), \tag{7.10a}$$

$$\psi_i(y) := \nu((Q_i)^y), \quad \psi(y) := \nu(Q^y) \tag{7.10b}$$

for $x \in X$, $y \in Y$ and $i = 1, 2$. Then $\phi = \phi_1 + \phi_2$ and $\psi = \psi_1 + \psi_2$. Moreover,

$$\int_X \phi_i \, d\mu = \int_Y \psi_i \, d\nu$$

for $i = 1, 2$ because $Q_i \in \Omega$. Hence

$$\int_X \phi \, d\mu = \int_Y \psi \, d\nu$$

and so $Q \in \Omega$.

Step 3. *If $Q_i \in \Omega$ for $i \in \mathbb{N}$ and $Q_i \subset Q_{i+1}$ for all i, then $Q := \bigcup_{i=1}^\infty Q_i \in \Omega$.*

Define $\phi_i, \phi\colon X \to [0,\infty]$ and $\psi_i, \psi\colon Y \to [0,\infty]$ by (7.10) for $i \in \mathbb{N}$. Since

$$Q_x = \bigcup_{i=1}^\infty (Q_i)_x, \quad Q^y = \bigcup_{i=1}^\infty (Q_i)^y$$

and $(Q_i)_x \in \mathcal{B}$ and $(Q_i)^y \in \mathcal{A}$ for all i, it follows from Theorem 1.28 (iv) that

$$\phi(x) = \nu(Q_x) = \lim_{i\to\infty} \nu((Q_i)_x) = \lim_{i\to\infty} \phi_i(x) \quad \text{for all } x \in X,$$

$$\psi(y) = \nu(Q^y) = \lim_{i\to\infty} \nu((Q_i)^y) = \lim_{i\to\infty} \psi_i(y) \quad \text{for all } y \in Y.$$

By the Lebesgue monotone convergence theorem (Theorem 1.37) this implies

$$\int_X \phi\, d\mu = \lim_{i\to\infty} \int_X \phi_i\, d\mu = \lim_{i\to\infty} \int_Y \psi_i\, d\mu = \int_Y \psi\, d\mu.$$

Thus $Q \in \Omega$ and this proves Step 3.

Step 4. *Let $A \in \mathcal{A}$ and $B \in \mathcal{B}$ such that $\mu(A) < \infty$ and $\nu(B) < \infty$. If $Q_i \in \Omega$ for $i \in \mathbb{N}$ such that $A \times B \supset Q_1 \supset Q_2 \supset \cdots$, then $Q := \bigcap_{i=1}^\infty Q_i \in \Omega$.*

Let $\phi_i, \phi, \psi_i, \psi$ be as in the proof of Step 3. Since $(Q_i)_x \subset B$ and $\nu(B) < \infty$, it follows from part (v) of Theorem 1.28 that ϕ_i converges pointwise to ϕ. Moreover, $\phi_i \le \nu(B)\chi_A$ for all i and the function $\nu(B)\chi_A\colon X \to [0,\infty)$ is integrable because $\mu(A) < \infty$ and $\nu(B) < \infty$. Hence it follows from the Lebesgue dominated convergence theorem (Theorem 1.45) that

$$\int_X \phi\, d\mu = \lim_{i\to\infty} \int_X \phi_i\, d\mu.$$

The same argument shows that

$$\int_Y \psi\, d\mu = \lim_{i\to\infty} \int_Y \psi_i\, d\mu.$$

Since $Q_i \in \Omega$ for all i, this implies

$$\int_X \phi\, d\mu = \int_Y \psi\, d\mu$$

and hence $Q \in \Omega$. This proves Step 4.

Step 5. $\Omega = \mathcal{A} \otimes \mathcal{B}$.

Since $(X, \mathcal{A}, \mu)$ and $(Y, \mathcal{B}, \nu)$ are σ-finite, there exist sequences of measurable sets $X_n \in \mathcal{A}$ and $Y_n \in \mathcal{B}$ such that

$$X_n \subset X_{n+1}, \quad Y_n \subset Y_{n+1}, \quad \mu(X_n) < \infty, \quad \nu(Y_n) < \infty$$

for all $n \in \mathbb{N}$ and $X = \bigcup_{n=1}^{\infty} X_n$ and $Y = \bigcup_{n=1}^{\infty} Y_n$. Define

$$\mathcal{M} := \{Q \in \mathcal{A} \otimes \mathcal{B} \mid Q \cap (X_n \times Y_n) \in \Omega \text{ for all } n \in \mathbb{N}\}.$$

Then $\mathcal{M}$ is a monotone class by Steps 3 and 4, $\mathcal{E} \subset \mathcal{M}$ by Steps 1 and 2, and $\mathcal{M} \subset \mathcal{A} \otimes \mathcal{B}$ by definition. Hence it follows from Lemma 7.5 that $\mathcal{M} = \mathcal{A} \otimes \mathcal{B}$. In other words, $Q \cap (X_n \cap Y_n) \in \Omega$ for all $Q \in \mathcal{A} \otimes \mathcal{B}$. By Step 3, this implies

$$Q = \bigcup_{n=1}^{\infty} \big(Q \cap (X_n \times Y_n)\big) \in \Omega \quad \text{for all } Q \in \mathcal{A} \otimes \mathcal{B}.$$

Thus $\mathcal{A} \otimes \mathcal{B} \subset \Omega \subset \mathcal{A} \otimes \mathcal{B}$ and so $\Omega = \mathcal{A} \otimes \mathcal{B}$, as claimed. This proves Step 5 and theorem (Theorem 7.9). □

Examples and exercises

Example 7.11. Let $X = Y = [0, 1]$, let $\mathcal{A} \subset 2^X$ be the σ-algebra of Lebesgue measurable sets, let $\mathcal{B} := 2^Y$, let $\mu: \mathcal{A} \to [0, 1]$ be the Lebesgue measure, and let $\nu: \mathcal{B} \to [0, \infty]$ be the counting measure. Consider the diagonal

$$\Delta := \{(x, x) \mid 0 \le x \le 1\} = \bigcap_{n=1}^{\infty} \bigcup_{i=1}^{n} \Big[\frac{i-1}{n}, \frac{i}{n}\Big]^2 \in \mathcal{A} \otimes \mathcal{B}.$$

Its characteristic function

$$f := \chi_\Delta : X \times Y \longrightarrow \mathbb{R}$$

is given by

$$f(x, y) := \begin{cases} 1 & \text{if } x = y, \\ 0 & \text{if } x \ne y. \end{cases}$$

Hence

$$\mu(\Delta^y) = \int_X f(x, y)\, d\mu(x) = 0 \quad \text{for } 0 \le y \le 1,$$

$$\nu(\Delta_x) = \int_Y f(x, y)\, d\nu(y) = 1 \quad \text{for } 0 \le x \le 1,$$

and so

$$\int_X \mu(\Delta^y)\, d\nu = 0 \neq 1 = \int_Y \mu(\Delta_x)\, d\mu(x).$$

Thus the hypothesis that $(X, \mathcal{A}, \mu)$ and $(Y, \mathcal{B}, \nu)$ are σ-finite cannot be removed in Theorem 7.9.

Example 7.12. Let $X := Y := [0, 1]$, let $\mathcal{A} = \mathcal{B} \subset 2^{[0,1]}$ be the σ-algebra of Lebesgue measurable sets, and let $\mu = \nu$ be the Lebesgue measure.

Claim 1. *Assume the continuum hypothesis. Then there is a set $Q \subset [0, 1]^2$ such that $[0, 1] \setminus Q_x$ is countable for all x and Q^y is countable for all y.*

Let Q be as in Claim 1 and define

$$f := \chi_Q : [0, 1]^2 \longrightarrow \mathbb{R}.$$

Then

$$\mu(Q^y) = \int_X f(x, y)\, d\mu(x) = 0 \quad \text{for } 0 \le y \le 1,$$

$$\nu(Q_x) = \int_Y f(x, y)\, d\nu(y) = 1 \quad \text{for } 0 \le x \le 1,$$

and hence

$$\int_X \mu(Q^y)\, d\nu = 0 \neq 1 = \int_Y \mu(Q)\, d\mu(x).$$

The sets Q_x and Q^y are all measurable and the integrals are finite, but the set Q is not $\mathcal{A} \otimes \mathcal{B}$-measurable. This shows that the hypothesis $Q \in \mathcal{A} \otimes \mathcal{B}$ in Theorem 7.9 cannot be replaced by the weaker hypothesis that the sets Q_x and Q^y are all measurable, even when the integrals are finite. It also shows that Lemma 7.2 does not have a converse. Namely, f_x and f^y are measurable for all x and y, but f is not $\mathcal{A} \otimes \mathcal{B}$-measurable.

Claim 2. *Assume the continuum hypothesis. Then there exists a bijection $j : [0, 1] \to W$ with values in a well-ordered set $(W, \prec)$ such that the set $\{w \in W \mid w \prec z\}$ is countable for all $z \in W$.*

Claim 2 $\Longrightarrow$ *Claim* 1. Let j be as in Claim 2 and put

$$Q := \{(x, y) \in [0, 1]^2 \mid j(x) \prec j(y)\}.$$

Then the set $Q^y = \{x \in [0, 1] \mid j(x) \prec j(y)\}$ is countable for all $y \in [0, 1]$ and the set $[0, 1] \setminus Q_x = \{y \in [0, 1] \mid j(y) \preccurlyeq j(x)\}$ is countable for all $x \in [0, 1]$. $\square$

Proof of Claim 2. By Zorn's lemma, every set admits a well ordering. Choose any well ordering $\prec$ on $A := [0, 1]$ and define

$$B := \{b \in A \mid \text{ the set } \{a \in A \mid a \prec b\} \text{ is uncountable}\}.$$

If $B = \emptyset$ choose $W := A = [0, 1]$ and $j = \mathrm{id}$. If $B \neq \emptyset$, then by the well ordering axiom, B contains a smallest element b_0. Since $b_0 \in B$, the set

$$W := B \setminus A = \{w \in A \mid w \prec b_0\}$$

is uncountable. Since $W \cap B = \emptyset$ the set $\{w \in W \mid w \prec z\}$ is countable for all $z \in W$. Since W is an uncountable subset of $[0, 1]$, the continuum hypothesis asserts that there exists a bijection $j : [0, 1] \to W$. This proves Claim 2. $\square$

Example 7.13. Let X and Y be countable sets, let $\mathcal{A} = 2^X$ and $\mathcal{B} = 2^Y$, and let $\mu : 2^X \to [0, \infty]$ and $\nu : 2^Y \to [0, \infty]$ be the counting measures. Then $\mathcal{A} \otimes \mathcal{B} = 2^{X \times Y}$ and $\mu \otimes \nu : 2^{X \times Y} \to [0, \infty]$ is the counting measure.

Example 7.14. Let $(X, \mathcal{A}, \mu)$ and $(Y, \mathcal{B}, \nu)$ be probability measure spaces, so that $\mu(X) = \nu(Y) = 1$. Then $\mu \otimes \nu : \mathcal{A} \otimes \mathcal{B} \to [0, 1]$ is also a probability measure. A trivial example is $\mathcal{A} = \{\emptyset, X\}$ and $\mathcal{B} = \{\emptyset, Y\}$. In this case the product σ-algebra is $\mathcal{A} \otimes \mathcal{B} = \{\emptyset, X \times Y\}$ and the product measure is given by $(\mu \otimes \nu)(\emptyset) = 0$ and $(\mu \otimes \nu)(X \times Y) = 1$.

Exercise 7.15. Let $(X, \mathcal{A}, \mu)$, $(Y, \mathcal{B}, \nu)$ be σ-finite measure spaces and let

$$\phi : X \longrightarrow X, \quad \psi : Y \longrightarrow Y$$

be bijections. Define the bijection $\phi \times \psi : X \times Y \to X \times Y$ by

$$(\phi \times \psi)(x, y) := (\phi(x), \psi(y))$$

for $x \in X$ and $y \in Y$. Prove that

$$(\phi \times \psi)_*(\mathcal{A} \otimes \mathcal{B}) = \phi_*\mathcal{A} \otimes \psi_*\mathcal{B}, \quad (\phi \times \psi)_*(\mu \otimes \nu) = \phi_*\mu \otimes \psi_*\nu.$$

Hint. Use Theorem 1.19 to show that $\phi_*\mathcal{A} \otimes \psi_*\mathcal{B} \subset (\phi \times \psi)_*(\mathcal{A} \otimes \mathcal{B})$. See also Exercise 2.34.

Exercise 7.16. For $n \in \mathbb{N}$ let $\mathcal{B}_n \subset \mathbb{R}^n$ be the Borel σ-algebra and let

$$\mu_n : \mathcal{B}_n \longrightarrow [0, \infty]$$

be the restriction of the Lebesgue measure to $\mathcal{B}_n$. Let $k, \ell \in \mathbb{N}$ and $n := k + \ell$. Identify $\mathbb{R}^k \times \mathbb{R}^\ell$ with $\mathbb{R}^n$ in the obvious manner. Then

$$\mathcal{B}_k \otimes \mathcal{B}_\ell = \mathcal{B}_n$$

by Lemma 7.6. Prove that the product measure $\mu_k \otimes \mu_\ell$ is translation-invariant and satisfies $(\mu_k \otimes \mu_\ell)([0, 1]^n) = 1$. Deduce that

$$\mu_k \otimes \mu_\ell = \mu_n.$$

Hint. Use Exercise 7.15. We return to this example in Section 7.4.

7.3 Fubini's Theorem

There are three versions of Fubini's Theorem. The first concerns nonegative functions that are measurable with respect to the product σ-algebra (Theorem 7.17), the second concerns real-valued functions that are integrable with respect to the product measure (Theorem 7.20), and the third concerns real valued functions that are integrable with respect to the completion of the product measure (Theorem 7.23).

Theorem 7.17 (Fubini for positive functions). *Let $(X, \mathcal{A}, \mu)$, $(Y, \mathcal{B}, \nu)$ be σ-finite measure spaces and let $\mu \otimes \nu : \mathcal{A} \otimes \mathcal{B} \to [0, \infty]$ be the product measure in Definition 7.10. Let $f : X \times Y \to [0, \infty]$ be an $\mathcal{A} \otimes \mathcal{B}$-measurable function. Then the function $X \to [0, \infty] : x \mapsto \int_Y f(x, y)\, d\nu(y)$ is $\mathcal{A}$-measurable, the function $Y \to [0, \infty] : y \mapsto \int_X f(x, y)\, d\mu(x)$ is $\mathcal{B}$-measurable, and*

$$\begin{aligned} \int_{X \times Y} f \, d(\mu \otimes \nu) &= \int_X \left(\int_Y f(x, y)\, d\nu(y) \right) d\mu(x) \\ &= \int_Y \left(\int_X f(x, y)\, d\mu(x) \right) d\nu(y). \end{aligned} \tag{7.11}$$

Example 7.18. equation (1.20) is equivalent to equation (7.11) for the counting measure on $X = Y = \mathbb{N}$.

Proof of Theorem 7.17. Let

$$f_x(y) := f^y(x) := f(x,y) \quad \text{for } (x,y) \in X \times Y$$

and define the functions

$$\phi: X \to [0,\infty] \quad \text{and} \quad \psi: Y \to [0,\infty]$$

by

$$\phi(x) := \int_Y f_x \, d\nu, \quad \psi(y) := \int_X f^y \, d\mu \tag{7.12}$$

for $x \in X$ and $y \in Y$. We prove in three steps that ϕ is $\mathcal{A}$-measurable, ψ is $\mathcal{B}$-measurable, and ϕ and ψ satisfy (7.11).

Step 1. *The assertion holds when $f: X \times Y \to [0,\infty)$ is the characteristic function of an $\mathcal{A} \otimes \mathcal{B}$-measurable set.*

Let $Q \in \mathcal{A} \otimes \mathcal{B}$ and $f = \chi_Q$. Then $f_x = \chi_{Q_x}$ and $f^y = \chi_{Q^y}$, and so

$$\phi(x) = \nu(Q_x), \quad \psi(y) = \mu(Q^y)$$

for all $x \in X$ and all $y \in Y$. Hence it follows from Theorem 7.9 that

$$\int_X \phi \, d\mu = \int_Y \psi \, d\mu = (\mu \otimes \nu)(Q) = \int_X f \, d(\mu \otimes \nu).$$

Here the third equality follows from the definition of the measure $\mu \otimes \nu$. This proves Step 1.

Step 2. *The assertion holds when $f: X \times Y \to [0,\infty)$ is an $\mathcal{A} \otimes \mathcal{B}$-measurable step-function.*

This follows immediately from Step 1 and the linearity of the integral.

Step 3. *The assertion holds when $f: X \times Y \to [0,\infty]$ is $\mathcal{A} \otimes \mathcal{B}$-measurable.*

By Theorem 1.26, there exists a sequence of $\mathcal{A} \otimes \mathcal{B}$-measurable step-functions

$$s_n: X \times Y \to [0,\infty)$$

such that $s_n \le s_{n+1}$ for all $n \in \mathbb{N}$ and s_n converges pointwise to f. Define

$$\phi_n(x) := \int_Y s_n(x,y) \, d\nu(y) \quad \text{for } x \in X,$$

$$\psi_n(x) := \int_X s_n(x,y) \, d\mu(x) \quad \text{for } y \in Y.$$

Then

$$\phi_n \le \phi_{n+1}, \quad \psi_n \le \psi_{n+1} \quad \text{for all } n \in \mathbb{N}$$

by part (i) of Theorem 1.35. Moreover, it follows from the Lebesgue monotone convergence theorem (Theorem 1.37) that

$$\phi(x) = \lim_{n\to\infty} \phi_n(x), \quad \psi(y) = \lim_{n\to\infty} \psi_n(y)$$

for all $x \in X$ and all $y \in Y$. Use the Lebesgue monotone convergence theorem (Theorem 1.37) again as well as Step 2 to obtain

$$\begin{aligned}\int_X \phi\, d\mu &= \lim_{n\to\infty} \int_X \phi_n\, d\mu \\ &= \lim_{n\to\infty} \int_{X\times Y} s_n\, d(\mu \otimes \nu) \\ &= \int_{X\times Y} f\, d(\mu \otimes \nu) \\ &= \lim_{n\to\infty} \int_Y \psi_n\, d\nu \\ &= \int_Y \psi\, d\nu.\end{aligned}$$

This proves Step 3 and Theorem 7.17. □

A first application of Fubini's Theorem 7.17 is *Minkowski's inequality* for a measurable function on a product space that is p-integrable with respect to one variable such that the resulting L^p norms define an integrable function of the other variable.

Theorem 7.19 (Minkowski). *Fix a constant* $1 \le p < \infty$. *Let* $(X, \mathcal{A}, \mu)$ *and* $(Y, \mathcal{B}, \nu)$ *be σ-finite measure spaces and let*

$$f : X \times Y \longrightarrow [0, \infty]$$

be $\mathcal{A} \otimes \mathcal{B}$*-measurable. Then*

$$\left(\int_X \left(\int_Y f(x,y)\, d\nu(y)\right)^p d\mu(x)\right)^{1/p} \le \int_Y \left(\int_X f(x,y)^p\, d\mu(x)\right)^{1/p} d\nu(y).$$

In the notation

$$f_x(y) := f^y(x) := f(x,y),$$

Minkowski's inequality has the form

$$\left(\int_X \|f_x\|^p_{L^1(\nu)}\, d\mu(x)\right)^{1/p} \le \int_Y \|f^y\|_{L^p(\mu)}\, d\nu(y). \tag{7.13}$$

Proof. By Lemma 7.2, $f_x\colon Y \to [0,\infty]$ is $\mathcal{B}$-measurable for all $x \in X$ and $f_y\colon X \to [0,\infty]$ is $\mathcal{A}$-measurable for all $y \in Y$. Moreover, by Theorem 7.17, the function $X \to [0,\infty]\colon x \mapsto \|f_x\|_{L^1(\nu)}^p$ is $\mathcal{A}$-measurable and the function $Y \to [0,\infty]\colon y \mapsto \|f^y\|_{L^p(\mu)}$ is $\mathcal{B}$-measurable. Hence both sides of the inequality (7.13) are well defined. Theorem 7.17 also shows that for $p = 1$ equality holds in (7.13). Hence assume $1 < p < \infty$ and a choose $1 < q < \infty$ such that $1/p + 1/q = 1$. It suffices to assume

$$c := \int_Y \|f^y\|_{L^p(\mu)}\, d\nu(y) < \infty.$$

Define $\phi\colon X \to [0,\infty]$ by

$$\phi(x) := \int_Y f_x\, d\nu \quad \text{for } x \in X$$

and let $g \in \mathcal{L}^q(\mu)$. Then the function

$$X \times Y \to [0,\infty]\colon (x,y) \mapsto f(x,y)|g(x)|$$

is $\mathcal{A} \otimes \mathcal{B}$-measurable. Hence it follows from Theorem 7.17 that

$$\begin{aligned}
\int_X \phi|g|\, d\mu &= \int_X \left(\int_Y f(x,y)|g(x)|\, d\nu(y)\right) d\mu(x) \\
&= \int_Y \left(\int_X f(x,y)|g(x)|\, d\mu(x)\right) d\nu(y) \\
&\le \int_Y \|f^y\|_{L^p(\mu)} \|g\|_{L^q(\mu)}\, d\nu(y) \\
&= c\, \|g\|_{L^q(\mu)}\,.
\end{aligned}$$

Here the third step follows from Hölder's inequality in Theorem 4.1. Since $(X,\mathcal{A},\mu)$ is semi-finite by part (ii) of Lemma 4.30, it follows from Lemma 4.34 that

$$\|\phi\|_{L^p(\mu)} \le c.$$

This proves Theorem 7.19. □

Theorem 7.20 (Fubini for integrable functions). *Let $(X, \mathcal{A}, \mu)$ and $(Y, \mathcal{B}, \nu)$ be σ-finite measure spaces, let $\mu \otimes \nu: \mathcal{A} \otimes \mathcal{B} \to [0, \infty]$ be the product measure, and let $f \in \mathcal{L}^1(\mu \otimes \nu)$. Define*

$$f_x(y) := f^y(x) := f(x, y) \quad \text{for } x \in X \text{ and } y \in Y.$$

Then the following holds.

(i) *$f_x \in \mathcal{L}^1(\nu)$ for μ-almost every $x \in X$ and the map*

$$\phi: X \longrightarrow \mathbb{R}$$

defined by

$$\phi(x) := \begin{cases} \int_Y f_x \, d\nu & \text{if } f_x \in \mathcal{L}^1(\nu), \\ 0 & \text{if } f_x \notin \mathcal{L}^1(\nu), \end{cases} \tag{7.14}$$

is μ-integrable.

(ii) *$f^y \in \mathcal{L}^1(\mu)$ for ν-almost every $y \in Y$ and the map*

$$\psi: Y \longrightarrow \mathbb{R}$$

defined by

$$\psi(y) := \begin{cases} \int_X f^y \, d\mu & \text{if } f^y \in \mathcal{L}^1(\mu), \\ 0 & \text{if } f^y \notin \mathcal{L}^1(\mu), \end{cases} \tag{7.15}$$

is ν-integrable.

(iii) *Let $\phi \in \mathcal{L}^1(\mu)$ and $\psi \in \mathcal{L}^1(\nu)$ be as in* (i) *and* (ii). *Then*

$$\int_X \phi \, d\mu = \int_{X \times Y} f \, d(\mu \otimes \nu) = \int_Y \psi \, d\nu. \tag{7.16}$$

Proof. We prove part (i) and the first equality in (7.16). The functions

$$f^\pm := \max\{\pm f, 0\}: X \times Y \longrightarrow [0, \infty)$$

are $\mathcal{A} \times \mathcal{B}$-measurable by Theorem 1.24. Hence the functions

$$f_x^\pm := \max\{\pm f_x, 0\} = f^\pm(x, \cdot): Y \longrightarrow [0, \infty)$$

are $\mathcal{B}$-measurable by Lemma 7.2.

Define

$$\Phi^{\pm} : Y \longrightarrow [0, \infty]$$

by

$$\Phi^{\pm}(x) := \int_Y f_x^{\pm} \, d\nu \quad \text{for } x \in X.$$

By Theorem 7.17, the functions $\Phi^{\pm} : X \to [0, \infty]$ are $\mathcal{A}$-measurable and

$$\int_X \Phi^{\pm} \, d\mu = \int_{X \times Y} f^{\pm} \, d(\mu \otimes \nu) \leq \int_{X \times Y} |f| \, d(\mu \otimes \nu) < \infty. \tag{7.17}$$

Now Lemma 1.47 asserts that the $\mathcal{A}$-measurable set

$$E := \left\{ x \in X \;\middle|\; \int_Y |f_x| \, d\nu = \infty \right\} = \{x \in X \mid \Phi^+(x) = \infty \text{ or } \Phi^-(x) = \infty\}$$

has measure $\mu(E) = 0$. Moreover, for all $x \in X$,

$$x \in E \iff f_x \notin \mathcal{L}^1(\nu).$$

Define

$$\phi^{\pm} : X \to [0, \infty)$$

by

$$\phi^{\pm}(x) := \begin{cases} \Phi^{\pm}(x) & \text{if } x \notin E, \\ 0 & \text{if } x \in E, \end{cases} \qquad \text{for } x \in X.$$

Then it follows from (7.17) that $\phi^{\pm} \in \mathcal{L}^1(\mu)$ and

$$\int_X \phi^{\pm} \, d\mu = \int_{X \times Y} f^{\pm} \, d(\mu \otimes \nu).$$

Hence $\phi = \phi^+ - \phi^- \in \mathcal{L}^1(\mu)$ and

$$\begin{aligned} \int_X \phi \, d\mu &= \int_X \phi^+ \, d\mu - \int_X \phi^- \, d\mu \\ &= \int_{X \times Y} f^+ \, d(\mu \otimes \nu) - \int_{X \times Y} f^- \, d(\mu \otimes \nu) \\ &= \int_{X \times Y} f \, d(\mu \otimes \nu). \end{aligned}$$

This proves (i) and the first equality in (7.16). An analogous argument proves (ii) and the second equality in (7.16). This proves Theorem 7.20. □

Example 7.21. Let $(X, \mathcal{A}, \mu) = (Y, \mathcal{B}, \nu)$ be the Lebesgue measure space in the unit interval $[0, 1]$ as in Example 7.12. Let $g_n: [0, 1] \to [0, \infty)$ be a sequence of smooth functions such that

$$\int_0^1 g_n(x)\,dx = 1,$$

and

$$g_n(x) = 0 \quad \text{for } x \in [0, 1] \setminus [2^{-n-1}, 2^{-n}]$$

for all $n \in \mathbb{N}$. Define

$$f: [0, 1]^2 \longrightarrow \mathbb{R}$$

by

$$f(x, y) := \sum_{n=1}^{\infty} (g_n(x) - g_{n+1}(x))g_n(y).$$

The sum on the right is finite for every pair $(x, y) \in [0, 1]^2$. Then

$$\int_X f(x, y)\,dx = 0,$$

$$\int_Y f(x, y)\,dy = \sum_{n=1}^{\infty} (g_n(x) - g_{n+1}(x)) = g_1(x),$$

and hence

$$\int_0^1 \left(\int_0^1 f(x, y)\,dx \right) dy = 0 \neq 1 = \int_0^1 \left(\int_0^1 f(x, y)\,dy \right) dx.$$

Thus the hypothesis $f \in \mathcal{L}^1(\mu \otimes \nu)$ cannot be removed in Theorem 7.20.

Example 7.22. This example shows that the product measure is typically not complete. Let $(X, \mathcal{A}, \mu)$ and $(Y, \mathcal{B}, \nu)$ be two complete σ-finite measure spaces. Suppose $(X, \mathcal{A}, \mu)$ admits a nonempty null set $A \in \mathcal{A}$ and $\mathcal{B} \neq 2^Y$. Choose $B \in 2^Y \setminus \mathcal{B}$. Then $A \times B \notin \mathcal{A} \otimes \mathcal{B}$. However, $A \times B$ is contained in the $\mu \otimes \nu$-null set $A \times Y$ and so belongs to the completion $(\mathcal{A} \otimes \mathcal{B})^*$.

In the first version of Fubini's Theorem integrability was not an issue. In the second version integrability of f_x was only guaranteed for *almost all* x. In the third version the function f_x may not even be measurable for all x.

Theorem 7.23 (Fubini for the completion). *Let $(X, \mathcal{A}, \mu)$ and $(Y, \mathcal{B}, \nu)$ be complete σ-finite measure spaces, let $(X \times Y, (\mathcal{A} \otimes \mathcal{B})^*, (\mu \otimes \nu)^*)$ denote the completion of the product space, and let $f \in \mathcal{L}^1((\mu \otimes \nu)^*)$. Define*

$$f_x(y) := f^y(x) := f(x, y) \quad \text{for } x \in X \text{ and } y \in Y.$$

Then the following holds.

(i) *$f_x \in \mathcal{L}^1(\nu)$ for μ-almost every $x \in X$ and the map*

$$\phi \colon X \longrightarrow \mathbb{R}$$

defined by

$$\phi(x) := \begin{cases} \int_Y f_x \, d\nu & \text{if } f_x \in \mathcal{L}^1(\nu), \\ 0 & \text{if } f_x \notin \mathcal{L}^1(\nu), \end{cases} \tag{7.18}$$

is μ-integrable.

(ii) *$f^y \in \mathcal{L}^1(\mu)$ for ν-almost every $y \in Y$ and the map*

$$\psi \colon Y \longrightarrow \mathbb{R}$$

defined by

$$\psi(y) := \begin{cases} \int_X f^y \, d\mu & \text{if } f^y \in \mathcal{L}^1(\mu), \\ 0 & \text{if } f^y \notin \mathcal{L}^1(\mu), \end{cases} \tag{7.19}$$

is ν-integrable.

(iii) *Let $\phi \in \mathcal{L}^1(\mu)$ and $\psi \in \mathcal{L}^1(\nu)$ be as in* (i) *and* (ii)*. Then*

$$\int_X \phi \, d\mu = \int_{X \times Y} f \, d(\mu \otimes \nu)^* = \int_Y \psi \, d\nu. \tag{7.20}$$

Proof. By part (v) of Theorem 1.55, there exists a function $g \in \mathcal{L}^1(\mu \otimes \nu)$ such that the set $N := \{(x, y) \in X \times Y \mid f(x, y) \neq g(x, y)\} \in (\mathcal{A} \otimes \mathcal{B})^*$ has measure zero, i.e., $(\mu \otimes \nu)^*(N) = 0$. By, the definition of the completion, there exists a set $Q \in \mathcal{A} \otimes \mathcal{B}$ such that $N \subset Q$ and $(\mu \otimes \nu)(Q) = 0$. Thus

$$\int_X \nu(Q_x) \, d\mu(x) = \int_Y \mu(Q^y) \, d\nu(y) = 0.$$

Hence, by Lemma 1.49,

$$\mu(E) = 0, \quad E := \{x \in X \mid \nu(Q_x) \neq 0\},$$

$$\nu(F) = 0, \quad F := \{y \in Y \mid \mu(Q^y) \neq 0\}.$$

Since $f = g$ on $(X \times Y) \setminus Q$, we have $f_x = g_x$ on $Y \setminus Q_x$ for all $x \in X$ and $f^y = g^y$ on $X \setminus Q^y$ for all $y \in Y$. By Theorem 7.20, for $g \in \mathcal{L}^1(\mu \otimes \nu)$ there are measurable sets $E' \in \mathcal{A}$ and $F' \in \mathcal{B}$ such that $\mu(E') = \nu(F') = 0$ and

$$g_x \in \mathcal{L}^1(\nu) \quad \text{for all } x \in X \setminus E',$$

$$g^y \in \mathcal{L}^1(\mu) \quad \text{for all } y \in Y \setminus F'.$$

If $x \in X \setminus (E \cup E')$, then $\nu(Q_x) = 0$ and $f_x = g_x$ on $Y \setminus Q_x$. Since $(Y, \mathcal{B}, \nu)$ is complete and $g_x \in \mathcal{L}^1(\nu)$, every function that differs from g_x on a set of measure zero is also $\mathcal{B}$-measurable and ν-integrable. Hence $f_x \in \mathcal{L}^1(\nu)$ for all $x \in X \setminus (E \cup E')$. The same argument shows that $f^y \in \mathcal{L}^1(\mu)$ for all $y \in Y \setminus (F \cup F')$. Define the

$$\phi: X \longrightarrow \mathbb{R}$$

by

$$\phi(x) := \begin{cases} \int_Y f_x \, d\nu & \text{for } x \in X \setminus (E \cup E'), \\ 0 & \text{for } x \in E \cup E' \end{cases}$$

and

$$\psi: Y \longrightarrow \mathbb{R}$$

by

$$\psi(y) := \begin{cases} \int_X f^y \, d\nu & \text{for } y \in Y \setminus (F \cup F'), \\ 0 & \text{for } y \in F \cup F'. \end{cases}$$

Since

$$\phi(x) = \int_Y g_x \, d\nu \quad \text{for all } x \in X \setminus (E \cup E'),$$

it follows from part (i) of Theorem 7.20 for g that $\phi \in \mathcal{L}^1(\mu)$. The same argument, using part (ii) of Theorem 7.20 for g, shows that $\psi \in \mathcal{L}^1(\nu)$. Moreover, the three integrals in (7.20) for f agree with the corresponding integrals for g because

$$\mu(E \cup E') = \nu(F \cup F') = (\mu \otimes \nu)(Q) = 0.$$

Hence equation (7.20) for f follows from part (iii) of Theorem 7.20 for g. This proves Theorem 7.23. □

Example 7.24. Assume $(X, \mathcal{A}, \mu)$ is not complete. Then there exist a $E \in 2^X \setminus \mathcal{A}$ and $N \in \mathcal{A}$ such that $E \subset N$ and $\mu(N) = 0$. In this case $E \times Y$ is a null set in the completion $(X \times Y, (\mathcal{A} \otimes \mathcal{B})^*, (\mu \otimes \nu)^*)$. Therefore,

$$f := \chi_{E \times Y} \in \mathcal{L}^1((\mu \otimes \nu)^*).$$

However, the function $f^y = \chi_E$ is *not* measurable for *every* $y \in Y$. This shows that the hypothesis that $(X, \mathcal{A}, \mu)$ and $(Y, \mathcal{B}, \nu)$ are complete cannot be removed in Theorem 7.23.

Exercise 7.25. Continue the notation of Theorem 7.23 and suppose that the map $f: X \times Y \to [0, \infty]$ is $(\mathcal{A} \otimes \mathcal{B})^*$-measurable. Prove that f_x is $\mathcal{B}$-measurable for μ-almost all $x \in X$, that f^y is $\mathcal{A}$-measurable for ν-almost all $y \in Y$, and that (7.11) continues to hold. *Hint.* The proof of Theorem 7.23 carries over verbatim to nonnegative measurable functions.

We close this section with two remarks about the construction of product measures in the non σ-finite case, where the story is considerably more subtle. These remarks are not used elsewhere in this book and can safely be ignored.

Remark 7.26. Let $(X, \mathcal{A}, \mu)$ and $(Y, \mathcal{B}, \nu)$ be two arbitrary measure spaces. In [4, Chapter 251] Fremlin defines the function

$$\theta: 2^{X \times Y} \longrightarrow [0, \infty]$$

by

$$\theta(W) := \inf \left\{ \sum_{n=1}^{\infty} \mu(A_n) \cdot \mu(B_n) \;\middle|\; \begin{array}{l} A_n \in \mathcal{A},\ B_n \in \mathcal{B} \text{ for } n \in \mathbb{N} \\ \text{and } W \subset \bigcup_{n=1}^{\infty} (A_n \times B_n) \end{array} \right\} \tag{7.21}$$

for $W \subset X \times Y$ and proves that it is an outer measure. He shows that the σ-algebra $\mathcal{C} \subset 2^{X \times Y}$ of θ-measurable sets contains the product σ-algebra $\mathcal{A} \otimes \mathcal{B}$ and calls the measure

$$\lambda_1 := \theta|_{\mathcal{C}}: \mathcal{C} \longrightarrow [0, \infty]$$

the *primitive product measure.* By Carathéodory's theorem (Theorem 2.4), the measure space $(X \times Y, \mathcal{C}, \lambda_1)$ is complete. By definition,

$$\lambda_1(A \times B) = \mu(A) \cdot \nu(B)$$

for all $A \in \mathcal{A}$ and all $B \in \mathcal{B}$. Fremlin then defines the *complete locally determined (CLD) product measure*

$$\lambda_0: \mathcal{C} \longrightarrow [0, \infty]$$

by

$$\lambda_0(W) := \sup\left\{\lambda_1(W \cap (E \times F)) \;\middle|\; \begin{array}{l} E \in \mathcal{A},\ F \in \mathcal{B}, \\ \mu(E) < \infty,\ \nu(F) < \infty \end{array}\right\}. \tag{7.22}$$

He shows that $(X \times Y, \mathcal{C}, \lambda_0)$ is a complete measure space, that $\lambda_0 \le \lambda_1$, and

$$\lambda_1(W) < \infty \implies \lambda_0(W) = \lambda_1(W) \quad \text{for all } W \in \mathcal{C}.$$

(See [4, Theorem 251I].) One can also prove that a measure $\lambda\colon \mathcal{C} \to [0, \infty]$ satisfies $\lambda(E \times F) = \mu(E) \cdot \nu(F)$ for all $E \in \mathcal{A}$ and $F \in \mathcal{B}$ with $\mu(E) \cdot \nu(F) < \infty$ if and only if $\lambda_0 \le \lambda \le \lambda_1$. With these definitions Fubini's Theorem holds for λ_0 whenever the factor $(Y, \mathcal{B}, \nu)$ (over which the integral is performed first) is σ-finite and the factor $(X, \mathcal{A}, \mu)$ (over which the integral is performed second) is either *strictly localizable* (i.e., there is a partition $X = \bigcup_{i \in I} X_i$ into measurable sets with $\mu(X_i) < \infty$ such that a set $A \subset X$ is $\mathcal{A}$-measurable if and only if $A \cap X_i \in \mathcal{A}$ for all $i \in I$ and, moreover, $\mu(A) = \sum_{i \in I} \mu(A \cap X_i)$ for all $A \in \mathcal{A}$), or is complete and *locally determined* (i.e., it is semi-finite and a set $A \subset X$ is $\mathcal{A}$-measurable if and only if $A \cap E \in \mathcal{A}$ for all $E \in \mathcal{A}$ with $\mu(E) < \infty$). See Fremlin [4, Theorem 252B] for details.

If the measure spaces $(X, \mathcal{A}, \mu)$ and $(Y, \mathcal{B}, \nu)$ are both σ-finite, then the measures λ_0 and λ_1 agree and are equal to the completion of the product measure $\mu \otimes \nu$ on $\mathcal{A} \otimes \mathcal{B}$ (see [4, Proposition 251K]).

Remark 7.27. For topological spaces yet another approach to the product measure is based on the Riesz representation theorem (Theorem 3.15). Let $(X, \mathcal{U}_X)$ and $(Y, \mathcal{U}_Y)$ be two locally compact Hausdorff spaces, denote by $\mathcal{B}_X$ and $\mathcal{B}_Y$ their Borel σ-algebras, and let $\mu_X\colon \mathcal{B}_X \to [0, \infty]$ and $\mu_Y\colon \mathcal{B}_Y \to [0, \infty]$ be Borel measures. Define

$$\Lambda\colon C_c(X \times Y) \longrightarrow \mathbb{R}$$

by

$$\begin{aligned} \Lambda(f) &:= \int_X \left(\int_Y f(x, y)\, d\nu(y)\right) d\mu(x) \\ &= \int_Y \left(\int_X f(x, y)\, d\mu(x)\right) d\nu(y) \end{aligned} \tag{7.23}$$

for $f \in C_c(X \times Y)$. That the two integrals agree for every continuous function with compact support follows from Fubini's Theorem (Theorem 7.20) for finite measure spaces. (To see this, observe that every compact set $K \subset X \times Y$ is contained in the product of the compact sets $K_X := \{x \in X \mid (\{x\} \times Y) \cap K \neq \emptyset\}$ and

$K_Y := \{y \in Y \mid (X \times \{y\}) \cap K \neq \emptyset\}$.) Since Λ is a positive linear functional, the Riesz representation theorem (Theorem 3.15) asserts that there exists a unique outer regular Borel measure $\mu_1 \colon \mathcal{B}_{X\times Y} \to [0,\infty]$ that is inner regular on open sets and a unique Radon measure $\mu_0 \colon \mathcal{B}_{X\times Y} \to [0,\infty]$ such that

$$\Lambda(f) = \int_{X\times Y} f\, d\mu_0 = \int_{X\times Y} f\, d\mu_1$$

for all $f \in C_c(X \times Y)$. It turns out that in this situation the Borel σ-algebra $\mathcal{B}_{X\times Y}$ is contained in the σ-algebra $\mathcal{C} \subset 2^{X\times Y}$ of Remark 7.26 and

$$\mu_0 = \lambda_0|_{\mathcal{B}_{X\times Y}}, \quad \mu_1 = \lambda_1|_{\mathcal{B}_{X\times Y}}.$$

Recall from Lemma 7.6 that the product σ-algebra $\mathcal{B}_X \otimes \mathcal{B}_Y$ agrees with the Borel σ-algebra $\mathcal{B}_{X\times Y}$ whenever one of the spaces X or Y is second countable. If they are both second countable, then so is the product space $(X \times Y, \mathcal{U}_{X\times Y})$ (Appendix B). In this case

$$\mu_0 = \mu_1 = \mu_X \otimes \mu_Y$$

is the product measure of Definition 7.10 and $\lambda_0 = \lambda_1 \colon \mathcal{C} \to [0,\infty]$ is its completion. (See Theorem 3.15 and Remark 7.26.)

7.4 Fubini and Lebesgue

For $n \in \mathbb{N}$ denote by $(\mathbb{R}^n, \mathcal{A}_n, m_n)$ the Lebesgue measure space on $\mathbb{R}^n$ and by $\mathcal{B}_n \subset \mathcal{A}_n$ the Borel σ-algebra on $\mathbb{R}^n$ with respect to the standard topology. For $k, \ell \in \mathbb{N}$ we identify $\mathbb{R}^{k+\ell}$ with $\mathbb{R}^k \times \mathbb{R}^\ell$ in the standard manner. Since $\mathbb{R}^n$ is second countable for all n, it follows from Lemma 7.6 and Theorem 2.1 that

$$\mathcal{B}_k \otimes \mathcal{B}_\ell = \mathcal{B}_{k+\ell}, \quad (m_k|_{\mathcal{B}_k}) \otimes (m_\ell|_{\mathcal{B}_\ell}) = m_{k+\ell}|_{\mathcal{B}_{k+\ell}}. \tag{7.24}$$

(See Exercise 7.16.) Thus Theorem 7.17 has the following consequence.

Theorem 7.28 (Fubini and Borel). *Let $k, \ell \in \mathbb{N}$ and $n := k+\ell$. Let $f \colon \mathbb{R}^n \to [0,\infty]$ be Borel measurable. Then*

$$f_x := f(x,\cdot) \colon \mathbb{R}^\ell \longrightarrow [0,\infty]$$

and

$$f^y := f(\cdot,y) \colon \mathbb{R}^k \longrightarrow [0,\infty]$$

are Borel measurable for all $x \in \mathbb{R}^k$ and all $y \in \mathbb{R}^\ell$. Moreover, the functions

$$\mathbb{R}^k \longrightarrow [0,\infty] \colon x \longmapsto \int_{\mathbb{R}^\ell} f(x,y)\, dm_\ell(y)$$

and

$$\mathbb{R}^\ell \longrightarrow [0,\infty] \colon y \longmapsto \int_{\mathbb{R}^k} f(x,y)\, dm_k(x)$$

are Borel measurable, and

$$\begin{aligned} \int_{\mathbb{R}^n} f\, dm_n &= \int_{\mathbb{R}^k} \left(\int_{\mathbb{R}^\ell} f(x,y)\, dm_\ell(y) \right) dm_k(x) \\ &= \int_{\mathbb{R}^\ell} \left(\int_{\mathbb{R}^k} f(x,y)\, dm_k(x) \right) dm_\ell(y). \end{aligned} \tag{7.25}$$

Proof. The assertion follows directly from (7.24) and Theorem 7.17. □

For Lebesgue measurable functions $f \colon \mathbb{R}^n \to [0,\infty]$ the analogous statement is considerably more subtle. In that case the function f_x, respectively f^y, need not be Lebesgue measurable for all x, respectively all y. However, they are Lebesgue measurable for almost all $x \in \mathbb{R}^k$, respectively almost all $y \in \mathbb{R}^\ell$, and the three integrals in (7.25) can still be defined and agree. The key result that one needs to prove this is that the Lebesgue measure on $\mathbb{R}^n = \mathbb{R}^k \times \mathbb{R}^\ell$ is the completion of the product of the Lebesgue measures on $\mathbb{R}^k$ and $\mathbb{R}^\ell$. Then the assertion follows from Exercise 7.25.

Theorem 7.29. *Let $k, \ell \in \mathbb{N}$, define $n := k + \ell$, and identify $\mathbb{R}^n$ with the product space $\mathbb{R}^k \times \mathbb{R}^\ell$ in the canonical way. Denote the completion of the product space $(\mathbb{R}^k \times \mathbb{R}^\ell, \mathcal{A}_k \otimes \mathcal{A}_\ell, m_k \otimes m_\ell)$ by $(\mathbb{R}^k \times \mathbb{R}^\ell, (\mathcal{A}_k \otimes \mathcal{A}_\ell)^*, (m_k \otimes m_\ell)^*)$. Then*

$$\mathcal{A}_n = (\mathcal{A}_k \otimes \mathcal{A}_\ell)^* \quad \text{and} \quad m_n = (m_k \otimes m_\ell)^*.$$

Proof. Define

$$\mathcal{C}_n := \{[a_1, b_1) \times \cdots \times [a_n, b_n) \mid a_i, b_i \in \mathbb{R} \text{ and } a_i < b_i \text{ for } i = 1, \dots, n\},$$

so that $\mathcal{C}_n \subset \mathcal{B}_n \subset \mathcal{A}_n \subset 2^{\mathbb{R}^n}$ for all n. We prove the assertion in three steps.

Step 1. *$\mathcal{B}_n \subset \mathcal{A}_k \otimes \mathcal{A}_\ell$ and $m_n(B) = (m_k \otimes m_\ell)(B)$ for all $B \in \mathcal{B}_n$.*

By Lemma 7.6, we have $\mathcal{B}_n = \mathcal{B}_k \otimes \mathcal{B}_\ell \subset \mathcal{A}_k \otimes \mathcal{A}_\ell$. It then follows from the uniqueness of a normalized translation-invariant Borel measure on $\mathbb{R}^n$ in Theorem 2.1 that $m_n|_{\mathcal{B}_n} = (m_k \otimes m_\ell)|_{\mathcal{B}_n}$. Here is a more direct proof.

First, assume $B = E = [a_1, b_1) \times \cdots \times [a_n, b_n) \in \mathcal{C}_n$. Define

$$E' := [a_1, b_1) \times \cdots \times [a_k, b_k), \quad E'' := [a_{k+1}, b_{k+1}) \times \cdots \times [a_n, b_n).$$

Thus $E' \in \mathcal{C}_k \subset \mathcal{A}_k$, $E'' \in \mathcal{C}_\ell \subset \mathcal{A}_\ell$, and so $E = E' \times E'' \in \mathcal{A}_k \otimes \mathcal{A}_\ell$. Moreover

$$m_n(E) = \prod_{i=1}^{n} (b_i - a_i) = m_k(E') \cdot m_\ell(E'') = (m_k \otimes m_\ell)(E).$$

Second, assume $B = U \subset \mathbb{R}^n$ is open. Then there is a sequence of pairwise disjoint sets $E_i \in \mathcal{C}_n$ such that $U = \bigcup_{i=1}^{\infty} E_i$. Hence $U \in \mathcal{A}_k \otimes \mathcal{A}_\ell$ and

$$(m_k \otimes m_\ell)(U) = \sum_{i=1}^{\infty} (m_k \otimes m_\ell)(E_i) = \sum_{i=1}^{\infty} m_n(E_i) = m_n(U).$$

Thus every open set is an element of $\mathcal{A}_k \otimes \mathcal{A}_\ell$ and so $\mathcal{B}_n \subset \mathcal{A}_k \otimes \mathcal{A}_\ell$. Third, assume $B = K \subset \mathbb{R}^n$ is compact. Then there is an open set $U \subset \mathbb{R}^n$ such that $K \subset U$ and $m_n(U) < \infty$. Hence the set $V := U \setminus K$ is open. This implies that $K = U \setminus V \in \mathcal{A}_k \otimes \mathcal{A}_\ell$ and

$$\begin{aligned}(m_k \otimes m_\ell)(K) &= (m_k \otimes m_\ell)(U) - (m_k \otimes m_\ell)(V) \\ &= m_n(U) - m_n(V) \\ &= m_n(K).\end{aligned}$$

Now let $B \subset \mathbb{R}^n$ be any Borel set. Then $B \in \mathcal{A}_k \otimes \mathcal{A}_\ell$, as we have seen above. Moreover, it follows from Theorem 2.13 that

$$m_n(B) = \inf_{\substack{U \supset B \\ U \text{ is open}}} m_n(U) = \inf_{\substack{U \supset B \\ U \text{ is open}}} (m_k \otimes m_\ell)(U) \geq (m_k \otimes m_\ell)(B)$$

and

$$m_n(B) = \inf_{\substack{K \subset B \\ K \text{ is compact}}} m_n(K) = \inf_{\substack{K \subset B \\ K \text{ is compact}}} (m_k \otimes m_\ell)(K) \leq (m_k \otimes m_\ell)(B).$$

Hence, $m_n(B) = (m_k \otimes m_\ell)(B)$, and this proves Step 1.

Step 2. $\mathcal{A}_k \otimes \mathcal{A}_\ell \subset \mathcal{A}_n$.

We claim that

$$E \in \mathcal{A}_k \implies E \times \mathbb{R}^\ell \in \mathcal{A}_n. \tag{7.26}$$

To see this, fix a set $E \in \mathcal{A}_k$. Then there exist Borel sets $A, B \in \mathcal{B}_k$ such that $A \subset E \subset B$ and $m_k(B \setminus A) = 0$. Let $\pi: \mathbb{R}^n \times \mathbb{R}^k$ denote the projection onto the first k coordinates. This map is continuous and hence Borel measurable by Theorem 1.20. Thus $A \times \mathbb{R}^\ell = \pi^{-1}(A)$ and $B \times \mathbb{R}^\ell = \pi^{-1}(B)$ are Borel sets in $\mathbb{R}^n$. Moreover, by Step 1,

$$\begin{aligned} m_n((B \times \mathbb{R}^\ell) \setminus (A \times \mathbb{R}^\ell)) &= m_n((B \setminus A) \times \mathbb{R}^\ell) \\ &= (m_k \otimes m_\ell)((B \setminus A) \times \mathbb{R}^\ell) \\ &= m_k(B \setminus A) \cdot m_\ell(\mathbb{R}^\ell) \\ &= 0. \end{aligned}$$

Since $A \times \mathbb{R}^\ell \subset E \times \mathbb{R}^\ell \subset B \times \mathbb{R}^\ell$, it follows that $E \times \mathbb{R}^\ell \in \mathcal{A}_n$, as claimed in (7.26). A similar argument shows that

$$F \in \mathcal{A}_\ell \implies \mathbb{R}^k \times F \in \mathcal{A}_n.$$

Hence $E \times F = (E \times \mathbb{R}^\ell) \cap (\mathbb{R}^k \times F) \in \mathcal{A}_n$ for all $E \in \mathcal{A}_k$ and all $F \in \mathcal{A}_\ell$. Thus $\mathcal{A}_k \otimes \mathcal{A}_\ell \subset \mathcal{A}_n$ and this proves Step 2.

Step 3. *$(\mathcal{A}_k \otimes \mathcal{A}_\ell)^* = \mathcal{A}_n$ and $(m_k \otimes m_\ell)^* = m_n$.*

Let $A \in \mathcal{A}_n$. Then there are Borel sets $B_0, B_1 \in \mathcal{B}_n$ such that $B_0 \subset A \subset B_1$ and $m_n(B_1 \setminus B_0) = 0$. By Step 1, $B_0, B_1 \in \mathcal{A}_k \otimes \mathcal{A}_\ell$ and $(m_k \otimes m_\ell)(B_1 \setminus B_0) = 0$. Hence $A \in (\mathcal{A}_k \otimes \mathcal{A}_\ell)^*$ and

$$(m_k \otimes m_\ell)^*(A) = (m_k \otimes m_\ell)(B_0) = m_n(B_0) = m_n(A).$$

Thus we have proved that

$$\mathcal{A}_n \subset (\mathcal{A}_k \otimes \mathcal{A}_\ell)^*, \quad (m_k \otimes m_\ell)^*|_{\mathcal{A}_n} = m_n.$$

Since $\mathcal{A}_k \otimes \mathcal{A}_\ell \subset \mathcal{A}_n$ by Step 2, it follows that

$$m_n|_{\mathcal{A}_k \otimes \mathcal{A}_\ell} = (m_k \otimes m_\ell)^*|_{\mathcal{A}_k \otimes \mathcal{A}_\ell} = m_k \otimes m_\ell.$$

Now let $A \in (\mathcal{A}_k \otimes \mathcal{A}_\ell)^*$. Then there are sets $A_0, A_1 \in \mathcal{A}_k \otimes \mathcal{A}_\ell$ such that $A_0 \subset A \subset A_1$ and $(m_k \otimes m_\ell)(A_1 \setminus A_0) = 0$. Hence $A_0, A_1 \in \mathcal{A}_n$ by Step 2 and $m_n(A_1 \setminus A_0) = 0$. Since $(\mathbb{R}^n, \mathcal{A}_n, m_n)$ is complete, it follows that $A \setminus A_0 \in \mathcal{A}_n$ and so $A = A_0 \cup (A \setminus A_0) \in \mathcal{A}_n$. Hence $\mathcal{A}_n = (\mathcal{A}_k \otimes \mathcal{A}_\ell)^*$. This proves Step 3 and Theorem 7.29. □

The next result specializes Theorem 7.23 to the Lebesgue measure.

Theorem 7.30 (Fubini and Lebesgue). *Let $k, \ell \in \mathbb{N}$ and $n := k+\ell$. Let $f: \mathbb{R}^n \to \mathbb{R}$ be Lebesgue integrable and define*

$$f_x(y) := f^y(x) := f(x_1, \ldots, x_k, y_1, \ldots, y_\ell)$$

for $x = (x_1, \ldots, x_k) \in \mathbb{R}^k$ and $y = (y_1, \ldots, y_\ell) \in \mathbb{R}^\ell$. Then there are Lebesgue null sets $E \subset \mathbb{R}^k$ and $F \subset \mathbb{R}^\ell$ such that the following holds.

(i) *$f_x \in \mathcal{L}^1(\mathbb{R}^\ell)$ for every $x \in \mathbb{R}^k \setminus E$ and the map*

$$\phi: \mathbb{R}^k \longrightarrow \mathbb{R}$$

defined by

$$\phi(x) := \begin{cases} \int_{\mathbb{R}^\ell} f_x \, dm_\ell & \text{for } x \in \mathbb{R}^k \setminus E, \\ 0 & \text{for } x \in E, \end{cases} \tag{7.27}$$

is Lebesgue integrable.

(ii) *$f^y \in \mathcal{L}^1(\mathbb{R}^k)$ for every $y \in \mathbb{R}^\ell \setminus F$ and the map*

$$\psi: \mathbb{R}^\ell \longrightarrow \mathbb{R}$$

defined by

$$\psi(y) := \begin{cases} \int_{\mathbb{R}^k} f^y \, dm_k & \text{for } y \in \mathbb{R}^\ell \setminus F, \\ 0 & \text{for } y \in F, \end{cases} \tag{7.28}$$

is Lebesgue integrable.

(iii) *Let $\phi \in \mathcal{L}^1(\mu)$ and $\psi \in \mathcal{L}^1(\nu)$ be as in* (i) *and* (ii). *Then*

$$\int_{\mathbb{R}^k} \phi \, dm_k = \int_{\mathbb{R}^n} f \, dm_n = \int_{\mathbb{R}^\ell} \psi \, dm_\ell. \tag{7.29}$$

Proof. This follows directly from Theorem 7.23 and Theorem 7.29. □

7.5 Convolution

An application of Fubini's Theorem is the convolution product on the space of Lebesgue integrable functions on Euclidean space. Fix an integer $n \in \mathbb{N}$ and let $(\mathbb{R}^n, \mathcal{A}, m)$ be the Lebesgue measure space. The convolution of two Lebesgue integrable functions $f, g \in \mathcal{L}^1(\mathbb{R}^n)$ is defined by

$$(f * g)(x) := \int_{\mathbb{R}^n} f(x-y)g(y)\,dm(y) \quad \text{for almost all } x \in \mathbb{R}^n.$$

Here the function $\mathbb{R}^n \to \mathbb{R}\colon y \mapsto f(x-y)g(y)$ is Lebesgue integrable for almost every $x \in \mathbb{R}^n$ and the resulting almost everywhere defined function $f * g$ is again Lebesgue integrable. This is the content of Theorem 7.33. The convolution descends to a bilinear map

$$*\colon L^1(\mathbb{R}^n) \times L^1(\mathbb{R}^n) \longrightarrow L^1(\mathbb{R}^n).$$

This map is associative and endows $L^1(\mathbb{R}^n)$ with the structure of a Banach algebra. Throughout we use the notation $f \sim g$ for two Lebesgue measurable functions $f, g\colon \mathbb{R}^n \to \mathbb{R}$ to mean that they agree almost everywhere with respect to the Lebesgue measure.

Definition 7.31. Let $f, g\colon \mathbb{R}^n \to \mathbb{R}$ be Lebesgue measurable and define

$$E(f,g) := \left\{ x \in \mathbb{R}^n \,\middle|\, \begin{array}{l} \mathbb{R}^n \to \mathbb{R}\colon y \mapsto f(x-y)g(y) \\ \text{is not Lebesgue integrable} \end{array} \right\}. \tag{7.30}$$

The *convolution of f and g* is the function

$$f * g\colon \mathbb{R}^n \longrightarrow \mathbb{R}$$

defined by

$$(f * g)(x) := \begin{cases} \displaystyle\int_{\mathbb{R}^n} f(x-y)g(y)\,dm(y) & \text{for } x \in \mathbb{R}^n \setminus E(f,g), \\ 0 & \text{for } x \in E(f,g). \end{cases} \tag{7.31}$$

The next theorem shows that the convolution is *very robust* in that $f * g$ is always Borel measurable and depends only on the equivalence classes of f and g under equality almost everywhere.

Theorem 7.32. *Let $f, g, h, f', g' : \mathbb{R}^n \to \mathbb{R}$ be Lebesgue measurable. Then the following holds:*

(i) *the function $y \mapsto f(x-y)g(y)$ is Lebesgue measurable for all $x \in \mathbb{R}^n$;*

(ii) *if $f' \sim f$ and $g' \sim g$, then $E(f', g') = E(f, g)$ and $f' * g' = f * g$;*

(iii) *$E(f, g)$ is a Borel set and $f * g$ is Borel measurable;*

(iv) *$E(g, f) = E(f, g)$ and $g * f = f * g$;*

(v) *if $m(E(f, g)) = m(E(g, h)) = 0$, then*

$$E := E(|f|, |g| * |h|) = E(|f| * |g|, |h|)$$

and $f * (g * h) = (f * g) * h$ on $\mathbb{R}^n \setminus E$.

An example with $n = 1$, where $E(f, g * h) = E(f * g, h) = \emptyset$ and $E = \mathbb{R}$ in part (v) of Theorem 7.32, is discussed in Exercise 7.53 below.

Proof of Theorem 7.32. (i) For $x \in \mathbb{R}^n$ define

$$f_x : \mathbb{R}^n \longrightarrow \mathbb{R} \quad \text{and} \quad \phi_x : \mathbb{R}^n \longrightarrow \mathbb{R}^n$$

by

$$f_x(y) := f(x-y) \quad \text{and} \quad \phi_x(y) := x - y.$$

Then ϕ_x is a diffeomorphism and $|\det(d\phi_x)| \equiv 1$. Hence, Theorem 2.17 asserts that $f_x = f \circ \phi_x$ is Lebesgue measurable for all $x \in \mathbb{R}^n$.

(ii) By assumption,

$$A := \{y \in \mathbb{R}^n \mid f(y) \neq f'(y)\}$$

and

$$B := \{y \in \mathbb{R}^n \mid g(y) \neq g'(y)\}$$

are Lebesgue null sets. Hence so are the sets

$$C_x := \phi_x(A) \cup B = \{y \in \mathbb{R}^n \mid f(x-y) \neq f'(x-y) \text{ or } g(y) \neq g'(y)\}$$

for all $x \in \mathbb{R}^n$. Hence the functions $f_x g$ and $f'_x g'$ agree on the complement of a Lebesgue null set for every $x \in \mathbb{R}^n$. Hence they are either both integrable or both not integrable and when they are their integrals agree.

(iii) By (ii) and Theorem 1.55, it suffices to assume that f and g are Borel measurable. Now define

$$F, G: \mathbb{R}^{2n} \longrightarrow \mathbb{R}$$

and

$$\phi: \mathbb{R}^{2n} \longrightarrow \mathbb{R}^{2n}$$

by

$$F(x,y) := f(x-y)g(y),$$

$$G(x,y) := f(x)g(y),$$

$$\phi(x,y) := (x-y,y)$$

for $x, y \in \mathbb{R}^n$. Then G is Borel measurable and ϕ is a diffeomorphism. Hence ϕ preserves the Borel σ-algebra and this implies that

$$F = G \circ \phi$$

is Borel measurable. Hence the function

$$\mathbb{R}^n \longrightarrow [0,\infty]: x \longmapsto \int_{\mathbb{R}^n} |F(x,y)|\, dm(y)$$

is Borel measurable, by Fubini's Theorem (Theorem 7.28). Thus the set $E(f,g)$ where this function takes on the value ∞ is a Borel set. Moreover, the functions

$$F^{\pm} := \max\{\pm F, 0\}$$

are Borel measurable and so are the functions

$$\widetilde{F}^{\pm}: \mathbb{R}^{2n} \longrightarrow [0,\infty)$$

defined by

$$\widetilde{F}^{\pm}(x,y) := \begin{cases} F^{\pm}(x,y) & \text{if } x \in \mathbb{R}^n \setminus E(f,g), \\ 0 & \text{if } x \in E(f,g), \end{cases} \qquad \text{for } (x,y) \in \mathbb{R}^{2n}.$$

Since

$$(f * g)(x) = \int_{\mathbb{R}^n} \widetilde{F}^{+}(x,y)\, dm(y) - \int_{\mathbb{R}^n} \widetilde{F}^{-}(x,y)\, dm(y)$$

for all $x \in \mathbb{R}^n$, it follows from Theorem 7.28 that $f * g$ is Borel measurable.

(iv) Since $g_x f = (f_x g) \circ \phi_x$, it follows from Theorem 2.17 that

$$E(g, f) = \{x \in \mathbb{R}^n \mid g_x f \in \mathcal{L}^1(\mathbb{R}^n)\} = E(f, g)$$

and

$$\begin{aligned}(f * g)(x) &= \int_{\mathbb{R}^n} f_x g \, dm \\ &= \int_{\mathbb{R}^n} (f_x g) \circ \phi_x \, dm \\ &= \int_{\mathbb{R}^n} g_x f \, dm \\ &= (g * f)(x)\end{aligned}$$

for all $x \in \mathbb{R}^n \setminus E(f, g)$.

(v) By (ii) and Theorem 1.55, it suffices to assume that f, g, and h are Borel measurable. Let $x \in \mathbb{R}^n$ and define

$$F_x \colon \mathbb{R}^{2n} \longrightarrow \mathbb{R}$$

by

$$F_x(y, z) := f(z)g(x - y - z)h(y).$$

Thus F_x is the composition of the maps $\mathbb{R}^{2n} \to \mathbb{R}^{3n} \colon (y, z) \mapsto (z, x-y-z, y)$ and $\mathbb{R}^{3n} \to \mathbb{R}$, $(\xi, \eta, \zeta) \mapsto (f(\xi), g(\eta), h(\zeta))$. Since the first map is continuous and the second is Borel measurable, it follows that F_x is Borel measurable. We claim that

$$x \in E(|f|, |g| * |h|) \iff \int_{\mathbb{R}^{2n}} |F_x| = \infty \iff x \in E(|f| * |g|, |h|). \tag{7.32}$$

It follows from Theorem 7.28 that

$$\int_{\mathbb{R}^{2n}} |F_x| \, dm_{2n} = \int_{\mathbb{R}^n} \left(\int_{\mathbb{R}^n} |F_x(y, z)| \, dm(y) \right) dm(z).$$

This integral is finite if and only if $F_x \in \mathcal{L}^1(\mathbb{R}^{2n})$. Moreover,

$$\begin{aligned}\int_{\mathbb{R}^n} |F_x(y, z)| \, dm(y) &= |f(z)| \int_{\mathbb{R}^n} |g(x - y - z)||h(y)| \, dm(y) \\ &= |f(z)|(|g| * |h|)(x - z)\end{aligned}$$

for $z \in \mathbb{R}^n \setminus (x - E(g, h))$. Since $E(g, h)$ is a Lebesgue null set, it follows that

$$\|F_x\|_{L^1(\mathbb{R}^{2n})} = \int_{\mathbb{R}^n} |f(z)|(|g| * |h|)(x - z) \, dm(z).$$

The integral on the right is infinite if and only if $x \in E(|f|, |g| * |h|)$. This proves the first equivalence in (7.32). The proof of the second equivalence is analogous, with y and z interchanged.

Now let $x \in \mathbb{R}^n \setminus E$. Then $F_x \in \mathcal{L}^1(\mathbb{R}^{2n})$ and $x \in \mathbb{R}^n \setminus E(f, g * h)$. Moreover, for $z \in \mathbb{R}^n$, the function $\mathbb{R}^n \to \mathbb{R}: y \mapsto F_x(y, z)$ is integrable if and only $x - z \notin E(g, h)$, and in that case its integral is equal to

$$\begin{aligned}\int_{\mathbb{R}^n} F_x(y, z)\, dm(y) &= f(z) \int_{\mathbb{R}^n} g(x - y - z) h(y)\, dm(y) \\ &= f(z)(g * h)(x - z).\end{aligned}$$

Since $E(g, h)$ is a Lebesgue null set, Theorem 7.30 shows that

$$\begin{aligned}\int_{\mathbb{R}^{2n}} F_x\, dm_{2n} &= \int_{\mathbb{R}^n} f(z)(g * h)(x - z)\, dm(z) \\ &= (f * (g * h))(x)\end{aligned}$$

The last equality holds because $x \notin E(f, g * h)$. A similar argument with y and z interchanged shows that

$$\int_{\mathbb{R}^{2n}} F_x\, dm_{2n} = ((f * g) * h)(x) \quad \text{for all } x \in \mathbb{R}^n \setminus E.$$

This proves (v) and Theorem 7.32. □

Theorem 7.33. *Let $1 \le p, q, r \le \infty$ such that $1/p + 1/q = 1 + 1/r$ and let $f \in \mathcal{L}^p(\mathbb{R}^n)$ and $g \in \mathcal{L}^q(\mathbb{R}^n)$. Then $m(E(f, g)) = 0$ and*

$$\|f * g\|_r \le \|f\|_p \, \|g\|_q \,. \tag{7.33}$$

*Thus $f * g \in \mathcal{L}^r(\mathbb{R}^n)$. The estimate (7.33) is called **Young's inequality**.*

Proof. Define the function

$$h: \mathbb{R}^n \longrightarrow [0, \infty]$$

by

$$h(x) := \int_{\mathbb{R}^n} |f(x - y) g(y)|\, dm(y) \quad \text{for } x \in \mathbb{R}^n.$$

Then $|f * g| \le h$ and $E(f, g) = \{x \in \mathbb{R}^n \mid h(x) = \infty\}$. Hence it suffices to prove that $\|h\|_r \le \|f\|_p \, \|g\|_q$. For $r = \infty$ this follows from Hölder's inequality. So assume $r < \infty$. Then $1 \le p, q < \infty$. Define

$$\lambda := 1 - \frac{p}{r} = p - \frac{p}{q}, \quad q' := \frac{p}{\lambda}.$$

Then $0 \le \lambda < 1$ and $1/q + 1/q' = 1$. Also $\lambda = 0$ if and only if $q = 1$. If $\lambda > 0$, then Hölder's inequality in Theorem 4.1 shows that

$$h(x) = \int_{\mathbb{R}^n} |f_x|^\lambda |f_x|^{1-\lambda} |g| \, dm \le \| |f_x|^\lambda \|_{q'} \| |f_x|^{1-\lambda} |g| \|_q,$$

where $f_x(y) := f(x-y)$. Since $\lambda q' = p$, this implies

$$\begin{aligned} h(x)^q &\le \left(\int_{\mathbb{R}^n} |f_x|^{\lambda q'} \, dm \right)^{q/q'} \int_{\mathbb{R}^n} |f_x|^{(1-\lambda)q} |g|^q \, dm \\ &= \|f\|_p^{\lambda q} \int_{\mathbb{R}^n} |f(x-y)|^{(1-\lambda)q} |g(y)|^q \, dm(y) \end{aligned} \tag{7.34}$$

for all $x \in \mathbb{R}^n$. This continues to hold for $\lambda = 0$. Now it follows from Minkowski's inequality in Theorem 7.19 with the exponent $s := r/q \ge 1$ that

$$\begin{aligned} \|h\|_r^q &= \left(\int_{\mathbb{R}^n} h^r \, dm \right)^{q/r} \\ &= \left(\int_{\mathbb{R}^n} h^{qs} \, dm \right)^{1/s} \\ &\le \|f\|_p^{\lambda q} \left(\int_{\mathbb{R}^n} \left(\int_{\mathbb{R}^n} |f(x-y)|^{(1-\lambda)q} |g(y)|^q \, dm(y) \right)^s dm(x) \right)^{1/s} \\ &\le \|f\|_p^{\lambda q} \int_{\mathbb{R}^n} \left(\int_{\mathbb{R}^n} |f(x-y)|^{(1-\lambda)qs} |g(y)|^{qs} \, dm(x) \right)^{1/s} dm(y) \\ &= \|f\|_p^{\lambda q} \, \|f\|_p^{(1-\lambda)q} \, \|g\|_q^q . \end{aligned}$$

Here the last equality follows from the fact that $(1-\lambda)qs = (1-\lambda)r = p$. This proves Theorem 7.33. □

It follows from Theorem 7.33 and part (ii) of Theorem 7.32 that the convolution descends to a map

$$L^1(\mathbb{R}^n) \times L^1(\mathbb{R}^n) \longrightarrow L^1(\mathbb{R}^n) \colon (f, g) \longmapsto f * g. \tag{7.35}$$

This map is bilinear by Theorem 1.44, it is associative by part (v) of Theorem 7.32, and satisfies $\|f * g\|_1 \le \|f\|_1 \|g\|_1$ by Young's inequality in Theorem 7.33. Hence $L^1(\mathbb{R}^n)$ is a Banach algebra. By part (iv) of Theorem 7.32, $L^1(\mathbb{R}^n)$ is commutative, and, by Theorem 7.33 with $q = 1$ and $r = p$, it acts on $L^p(\mathbb{R}^n)$. (A *Banach algebra* is a Banach space $(\mathcal{X}, \|\cdot\|)$ equipped with an associative bilinear map $\mathcal{X} \times \mathcal{X} \to \mathcal{X} \colon (x, y) \mapsto xy$ that satisfies the inequality $\|xy\| \le \|x\| \, \|y\|$ for all $x, y \in \mathcal{X}$.)

Definition 7.34. Fix a constant $1 \leq p < \infty$. A Lebesgue measurable function $f: \mathbb{R}^n \to \mathbb{R}$ is called *locally p-integrable* if

$$\int_K |f|^p dm < \infty$$

for every compact set $K \subset \mathbb{R}^n$. It is called *locally integrable* if it is locally 1-integrable.

Theorem 7.33 carries over to locally integrable functions as follows. If $1/p + 1/q = 1 + 1/r$, f is locally p-integrable, and $g \in \mathcal{L}^q(\mathbb{R}^n)$ has compact support, then $E(f, g)$ is a Lebesgue null set and $f * g$ is locally r-integrable. To see this, let $K \subset \mathbb{R}^n$ be any compact set and choose a compactly supported smooth function β such that $\beta|_K \equiv 1$. Then $\beta f \in \mathcal{L}^p(\mathbb{R}^n)$ and $(\beta f) * g$ agrees with $f * g$ on the set $\{x \in \mathbb{R}^n \mid x - \mathrm{supp}(g) \subset K\}$. In the following theorem $C_0^\infty(\mathbb{R}^n)$ denotes the space of compactly supported smooth functions on $\mathbb{R}^n$.

Theorem 7.35. *Let $1 \leq p < \infty$ and $1 < q \leq \infty$ such that $1/p + 1/q = 1$.*

(i) *If $f: \mathbb{R}^n \to \mathbb{R}$ is locally p-integrable, then*

$$\lim_{\xi \to 0} \int_B |f(x + \xi) - f(x)|^p \, dm(x) = 0$$

for every bounded Lebesgue measurable subset $B \subset \mathbb{R}^n$. If $f \in \mathcal{L}^p(\mathbb{R}^n)$ this continues to hold for $B = \mathbb{R}^n$.

(ii) *If $f \in \mathcal{L}^p(\mathbb{R}^n)$ and $g \in \mathcal{L}^q(\mathbb{R}^n)$, then $f * g$ is uniformly continuous. If f is locally p-integrable and $g \in \mathcal{L}^q(\mathbb{R}^n)$ has compact support (or if $f \in \mathcal{L}^p(\mathbb{R}^n)$ has compact support and g is locally q-integrable), then $f * g$ is continuous.*

(iii) *If $f: \mathbb{R}^n \to \mathbb{R}$ is locally integrable and $g \in C_0^\infty(\mathbb{R}^n)$, then $f * g$ is smooth and $\partial^\alpha(f * g) = f * \partial^\alpha g$ for every multi-index α.*

(iv) *$C_0^\infty(\mathbb{R}^n)$ is dense in $L^p(\mathbb{R}^n)$ for $1 \leq p < \infty$.*

Proof. (i) Assume first that $f \in \mathcal{L}^p(\mathbb{R}^n)$ and fix a constant $\varepsilon > 0$. By Theorem 4.15, there is a function $g \in C_c(\mathbb{R}^n)$ such that $\|f - g\|_p < \varepsilon^{1/p}/3$. Since g is uniformly continuous, there is a $\delta > 0$ such that, for all $\xi \in \mathbb{R}^n$,

$$|\xi| < \delta \implies \sup_{x \in \mathbb{R}^n} |g(x + \xi) - g(x)| < \left(\frac{\varepsilon}{3^p m(\mathrm{supp}(g))}\right)^{1/p}.$$

Take $\xi \in \mathbb{R}^n$ such that $|\xi| < \delta$. Then

$$\begin{aligned}
&\left(\int_{\mathbb{R}^n} |f(x+\xi) - f(x)|^p \, dm(x)\right)^{1/p} \\
&\le 2\,\|f-g\|_p + \left(\int_{\mathbb{R}^n} |g(x+\xi) - g(x)|^p \, dm(x)\right)^{1/p} \\
&\le \frac{2\varepsilon^{1/p}}{3} + \Big(m(\mathrm{supp}(g)) \sup_{x\in\mathbb{R}^n} |g(x+\xi) - g(x)|^p\Big)^{1/p} \\
&< \varepsilon^{1/p}.
\end{aligned}$$

This proves (i) for $f \in \mathcal{L}^p(\mathbb{R}^n)$. To prove the result in general, choose a compact set $K \subset \mathbb{R}^n$ such that $B_1(x) \subset K$ for all $x \in B$ and multiply f by a smooth compactly supported cutoff function to obtain a function $f' \in \mathcal{L}^p(\mathbb{R}^n)$ that agrees with f on K. Then (i) holds for f' and hence also for f.

(ii) Assume first that $f \in \mathcal{L}^p(\mathbb{R}^n)$ and $g \in \mathcal{L}^q(\mathbb{R}^n)$ and fix a constant $\varepsilon > 0$. By part (i), there exists a $\delta > 0$ such that, for all $\xi \in \mathbb{R}^n$,

$$|\xi| < \delta \implies \int_{\mathbb{R}^n} |f(y+\xi) - f(y)|^p \, dm(y) < \left(\frac{\varepsilon}{\|g\|_q}\right)^p$$

Fix two elements $x, \xi \in \mathbb{R}^n$ such that $|\xi| < \delta$ and denote

$$f_x(y) := f(x-y).$$

Then, by Hölder's inequality in Theorem 4.1,

$$\begin{aligned}
|(f*g)(x+\xi) - (f*g)(x)| &= \left|\int_{\mathbb{R}^n} (f_{x+\xi} - f_x) g \, dm\right| \\
&\le \|f_{x+\xi} - f_x\|_p \|g\|_q \\
&= \left(\int_{\mathbb{R}^n} |f(y+\xi) - f(y)|^p \, dm(y)\right)^{1/p} \|g\|_q \\
&< \varepsilon.
\end{aligned}$$

This shows that $f * g$ is uniformly continuous. If f is locally p-integrable and $g \in \mathcal{L}^q(\mathbb{R}^n)$ has compact support, continuity follows by taking the integral over a suitable compact set. In the opposite case continuity follows by taking the L^q-norm of g over a suitable compact set.

(iii) Fix an index $i \in \{1, \dots, n\}$ and denote by $e_i \in \mathbb{R}^n$ the ith unit vector. Fix an element $x \in \mathbb{R}^n$ and choose a compact set $K \subset \mathbb{R}^n$ such that $B_1(y) \subset K$ whenever $x - y \in \mathrm{supp}(g)$. Let $\varepsilon > 0$. Since $\partial_i g$ is continuous, there is a constant $0 < \delta < 1$ such that

$$|\partial_i g(y + he_i) - \partial_i g(y)| < \frac{\varepsilon}{\int_K |f| \, dm}$$

for all $y \in \mathbb{R}^n$ and all $h \in \mathbb{R}$ with $|h| < \delta$. Hence the fundamental theorem of calculus asserts that

$$\sup_{y \in \mathbb{R}^n} \left| \frac{g(y + he_i - y) - g(y)}{h} - \partial_i g(y) \right| < \frac{\varepsilon}{\int_K |f| \, dm}$$

for all $h \in \mathbb{R}$ with $0 < |h| < \delta$. Take $h \in \mathbb{R}^n$ with $0 < |h| < \delta$. Then

$$\begin{aligned} &\left| \frac{(f * g)(x + he_i) - (f * g)(x)}{h} - (f * \partial_i g)(x) \right| \\ &= \left| \int_{\mathbb{R}^n} f(y) \Big(\frac{g(x + he_i - y) - g(x - y)}{h} - \partial_i g(x - y) \Big) \, dm(y) \right| \\ &\le \int_{\mathbb{R}^n} |f(y)| \left| \frac{g(x + he_i - y) - g(x - y)}{h} - \partial_i g(x - y) \, dm(y) \right| \\ &< \varepsilon. \end{aligned}$$

By part (ii), the function $\partial_i (f * g) = f * \partial_i g$ is continuous for $i = 1, \dots, n$. For higher derivatives the assertion follows by induction.

(iv) Let $f \in \mathcal{L}^p(\mathbb{R}^n)$ and choose a compactly supported smooth function $\rho \colon \mathbb{R}^n \to [0, \infty)$ such that

$$\mathrm{supp}(\rho) \subset B_1, \qquad \int_{\mathbb{R}^n} \rho \, dm = 1.$$

Define

$$\rho_\delta \colon \mathbb{R}^n \longrightarrow \mathbb{R}$$

by

$$\rho_\delta(x) := \frac{1}{\delta^n} \rho\Big(\frac{x}{\delta}\Big)$$

for $\delta > 0$ and $x \in \mathbb{R}^n$. Then

$$\mathrm{supp}(\rho_\delta) \subset B_\delta, \qquad \int_{\mathbb{R}^n} \rho_\delta \, dm = 1$$

by Theorem 2.17. By part (iii), the function

$$f_\delta := \rho_\delta * f : \mathbb{R}^n \longrightarrow \mathbb{R}$$

is smooth for all $\delta > 0$. Now fix a constant $\varepsilon > 0$. By part (i), there exists a constant $\delta > 0$ such that, for all $y \in \mathbb{R}^n$,

$$|y| < \delta \implies \int_{\mathbb{R}^n} |f(x-y) - f(x)|^p \, dm(x) < \varepsilon^p.$$

Hence, by Minkowski's inequality in Theorem 7.19,

$$\begin{aligned}
\|f_\delta - f\|_p &= \left(\int_{\mathbb{R}^n} \left| \int_{\mathbb{R}^n} (f(x-y) - f(x))\rho_\delta(y)\, dm(y) \right|^p dm(x) \right)^{1/p} \\
&\le \int_{\mathbb{R}^n} \left(\int_{\mathbb{R}^n} |f(x-y) - f(x)|^p \rho_\delta(y)^p \, dm(x) \right)^{1/p} dm(y) \\
&\le \sup_{|y|<\delta} \left(\int_{\mathbb{R}^n} |f(x-y) - f(x)|^p \, dm(x) \right)^{1/p} \\
&\le \varepsilon.
\end{aligned}$$

If f has compact support, then so does f_δ. If not, choose a function $g \in \mathcal{L}^p(\mathbb{R}^n)$ with compact support such that $\|f - g\|_p < \varepsilon/2$ and then a smooth function $h: \mathbb{R}^n \to \mathbb{R}$ with compact support such that $\|g - h\|_p < \varepsilon/2$. This proves (iv) and Theorem 7.35. □

The method explained in the proof of part (iv) of Theorem 7.35 is called the *mollifier technique.* The functions ρ_δ can be viewed as approximate Dirac delta functions that concentrate near the origin as δ tends to zero.

7.6 Marcinkiewicz interpolation

Another interesting application of Fubini's Theorem is Marcinkiewicz interpolation, which provides a criterion for a linear operator on $L^2(\mu)$ to induce a linear operator on $L^p(\mu)$ for $1 < p < 2$. Marcinkiewicz interpolation applies to all measure spaces, although it is used here only for the Lebesgue measure space on $\mathbb{R}^n$. In particular, Marcinkiewicz interpolation plays a central role in the proof of the Calderón–Zygmund inequality in Section 7.7.

Let $(X, \mathcal{A}, \mu)$ be a measure space. For a measurable function $f\colon X \to \mathbb{R}$ define the function

$$\kappa_f\colon [0,\infty) \longrightarrow [0,\infty]$$

by (6.1), i.e.,

$$\kappa_f(t) := \mu(A(t,f)), \quad A(t,f) := \{x \in X \mid |f(x)| > t\}, \quad \text{for } t \geq 0.$$

The function κ_f is nonincreasing and hence Borel measurable.

Lemma 7.36. *Let $1 \leq p < \infty$ and let $f, g\colon X \to \mathbb{R}$ be measurable. Then*

$$\kappa_{f+g}(t) \leq \kappa_f(t/2) + \kappa_g(t/2), \tag{7.36}$$

$$t^p \kappa_f(t) \leq \int_X |f|^p \, d\mu = p \int_0^\infty s^{p-1} \kappa_f(s) \, ds \tag{7.37}$$

for all $t \geq 0$.

Proof. Inequality (7.36) was established in the proof of Lemma 6.2. We prove (7.37) in four steps.

Step 1. *For all $t \geq 0$,*

$$t^p \kappa_f(t) \leq \int_X |f|^p \, d\mu.$$

Since $t^p \chi_{A(t,f)} \leq |f|^p$, it follows that

$$t^p \kappa_f(t) = \int_X t^p \chi_{A(t,f)} \, d\mu \leq \int_X |f|^p \, d\mu \quad \text{for all } t \geq 0.$$

This proves Step 1.

Step 2. *If $\kappa_f(t) = \infty$ for some $t > 0$, then*

$$\int_X |f|^p \, d\mu = \infty = \int_0^\infty t^{p-1} \kappa_f(t) \, dt.$$

By Step 1, we have

$$\int_X |f|^p \, d\mu = \infty.$$

Moreover, $t^{p-1}\kappa_f(t) = \infty$ for $t > 0$ sufficiently small, and hence

$$\int_0^\infty t^{p-1} \kappa_f(t) \, dt = \infty.$$

This proves Step 2.

Step 3. *Assume* $(X, \mathcal{A}, \mu)$ *is* σ*-finite and* $\kappa_f(t) < \infty$ *for all* $t > 0$. *Then equation* (7.37) *holds.*

Let $\mathcal{B} \subset 2^{[0,\infty)}$ be the Borel σ-algebra and denote by $m: \mathcal{B} \to [0,\infty]$ the restriction of the Lebesgue measure to $\mathcal{B}$. Let $(X \times [0,\infty), \mathcal{A} \otimes \mathcal{B}, \mu \otimes m)$ be the product measure space of Definition 7.10. We prove that

$$Q(f) := \{(x,t) \in X \times [0,\infty) \mid 0 \le t < |f(x)|\} \in \mathcal{A} \otimes \mathcal{B}.$$

To see this, assume first that f is an $\mathcal{A}$-measurable step-function. Then there exist finitely many pairwise disjoint measurable sets $A_1, \dots, A_\ell \in \mathcal{A}$ and positive real numbers $\alpha_1, \dots, \alpha_\ell$ such that

$$|f| = \sum_{i=1}^{\ell} \alpha_i \chi_{A_i}.$$

In this case $Q(f) = \bigcup_{i=1}^{\ell} A_i \times [0, \alpha_i) \in \mathcal{A} \otimes \mathcal{B}$. Now consider the general case. Then Theorem 1.26 asserts that there is a sequence of $\mathcal{A}$-measurable step-functions $f_i: X \to [0,\infty)$ such that $0 \le f_1 \le f_2 \le \cdots$ and f_i converges pointwise to $|f|$. Then $Q(f_i) \in \mathcal{A} \otimes \mathcal{B}$ for all i, and so $Q(f) = \bigcup_{i=1}^{\infty} Q(f_i) \in \mathcal{A} \otimes \mathcal{B}$.

Now define

$$h: X \times [0,\infty) \longrightarrow [0,\infty)$$

by

$$h(x,t) := pt^{p-1}.$$

This function is $\mathcal{A} \otimes \mathcal{B}$-measurable, and so is $h\chi_{Q(f)}$. Hence, by Fubini's Theorem (Theorem 7.17),

$$\begin{aligned}
\int_X |f|^p \, d\mu &= \int_X \left(\int_0^{|f(x)|} pt^{p-1} \, dt \right) d\mu(x) \\
&= \int_X \left(\int_0^{\infty} (h\chi_{Q(f)})(x,t) \, dm(t) \right) d\mu(x) \\
&= \int_0^{\infty} \left(\int_X (h\chi_{Q(f)})(x,t) \, d\mu(x) \right) dm(t) \\
&= \int_0^{\infty} pt^{p-1} \mu(A(t,f)) \, dt.
\end{aligned}$$

This proves Step 3.

Step 4. *Assume $\kappa_f(t) < \infty$ for all $t > 0$. Then (7.37) holds.*

Define

$$X_0 := \{x \in X \mid f(x) \neq 0\},$$

$$\mathcal{A}_0 := \{A \in \mathcal{A} \mid A \subset X_0\},$$

and

$$\mu_0 := \mu|_{\mathcal{A}_0}.$$

Then the measure space $(X_0, \mathcal{A}_0, \mu_0)$ is σ-finite because $X_n := A(1/n, f)$ is a sequence of $\mathcal{A}_n$-measurable sets such that $\mu_0(X_n) = \kappa_f(1/n) < \infty$ for all n and $X_0 = \bigcup_{n=1}^{\infty} X_n$. Moreover, $f_0 := f|_{X_0} \colon X_0 \to \mathbb{R}$ is $\mathcal{A}_0$-measurable and $\kappa_f = \kappa_{f_0}$. Hence it follows from Step 3 that

$$\int_X |f|^p \, d\mu = \int_{X_0} |f_0|^p \, d\mu_0 = \int_0^\infty t^{p-1} \kappa_{f_0}(t) \, dt = \int_0^\infty t^{p-1} \kappa_f(t) \, dt.$$

This proves Step 4 and Lemma 7.36. □

Fix real numbers $1 \leq p \leq q$. Then the inequality

$$\|f\|_p \leq \|f\|_1^{\frac{q-p}{p(q-1)}} \, \|f\|_q^{\frac{q(p-1)}{p(q-1)}} \tag{7.38}$$

in Exercise 4.44 shows that

$$L^1(\mu) \cap L^q(\mu) \subset L^p(\mu).$$

Since the intersection $L^1(\mu) \cap L^q(\mu)$ contains (the equivalences classes of) all characteristic functions of measurable sets with finite measure, it is dense in $L^p(\mu)$ by Lemma 4.12. The following theorem was proved in 1939 by Józef Marcinkiewicz (a Ph.D. student of Antoni Zygmund). To formulate the result it will be convenient to slightly abuse notation and use the same letter f to denote an element of $\mathcal{L}^p(\mu)$ and its equivalence class in $L^p(\mu)$.

Theorem 7.37 (Marcinkiewicz). *Let $q > 1$ and let $T \colon L^q(\mu) \to L^q(\mu)$ be a linear operator. Suppose that there exist constants $c_1 > 0$ and $c_q > 0$ such that*

$$\|Tf\|_{1,\infty} \leq c_1 \|f\|_1, \quad \|Tf\|_q \leq c_q \|f\|_q \tag{7.39}$$

for all $f \in L^1(\mu) \cap L^q(\mu)$. Fix a constant $1 < p < q$. Then

$$\|Tf\|_p \leq c_p \|f\|_p, \quad c_p := 2\Big(\frac{p(q-1)}{(q-p)(p-1)}\Big)^{1/p} c_1^{\frac{q-p}{p(q-1)}} c_q^{\frac{q(p-1)}{p(q-1)}}, \tag{7.40}$$

for all $f \in L^1(\mu) \cap L^q(\mu)$. Thus the restriction of T to $L^1(\mu) \cap L^q(\mu)$ extends (uniquely) to a bounded linear operator from $L^p(\mu)$ to itself for $1 < p < q$.

Proof. Let $c > 0$ and let $f \in \mathcal{L}^1(\mu) \cap \mathcal{L}^2(\mu)$. For $t \geq 0$ define

$$f_t(x) := \begin{cases} f(x) & \text{if } |f(x)| > ct, \\ 0 & \text{if } |f(x)| \leq ct, \end{cases}$$

and

$$g_t(x) := \begin{cases} 0 & \text{if } |f(x)| > ct, \\ f(x) & \text{if } |f(x)| \leq ct. \end{cases}$$

Then

$$A(s, f_t) = \begin{cases} A(s, f) & \text{if } s > ct, \\ A(ct, f) & \text{if } s \leq ct, \end{cases}$$

$$A(s, g_t) = \begin{cases} \emptyset & \text{if } s \geq ct, \\ A(s, f) \setminus A(ct, f) & \text{if } s < ct, \end{cases}$$

and

$$\kappa_{f_t}(s) = \begin{cases} \kappa_f(s) & \text{if } s > ct, \\ \kappa_f(ct) & \text{if } s \leq ct, \end{cases}$$

$$\kappa_{g_t}(s) = \begin{cases} 0 & \text{if } s \geq ct, \\ \kappa_f(s) - \kappa_f(ct) & \text{if } s < ct. \end{cases}$$

By Lemma 7.36 and Fubini's Theorem (Theorem 7.28), this implies that

$$\begin{aligned} \int_0^\infty t^{p-2} \|f_t\|_1 \, dt &= \int_0^\infty t^{p-2} \left(\int_0^\infty \kappa_{f_t}(s) \, ds \right) dt \\ &= \int_0^\infty t^{p-2} \left(ct\kappa_f(ct) + \int_{ct}^\infty \kappa_f(s) \, ds \right) dt \\ &= c^{1-p} \int_0^\infty t^{p-1} \kappa_f(t) \, dt + \int_0^\infty \int_0^{s/c} t^{p-2} \, dt \, \kappa_f(s) \, ds \\ &= c^{1-p} \int_0^\infty t^{p-1} \kappa_f(t) \, dt + \int_0^\infty \frac{(s/c)^{p-1}}{p-1} \kappa_f(s) \, ds \\ &= \frac{c^{1-p} p}{p-1} \int_0^\infty t^{p-1} \kappa_f(t) \, dt \\ &= \frac{c^{1-p}}{p-1} \int_X |f|^p \, d\mu, \end{aligned}$$

and

$$\begin{aligned}
&\int_0^\infty t^{p-q-1} \|g_t\|_q^q \, dt \\
&= \int_0^\infty t^{p-q-1} \left(\int_0^\infty q s^{q-1} \kappa_{g_t}(s) \, ds \right) dt \\
&= \int_0^\infty t^{p-q-1} \left(\int_0^{ct} q s^{q-1} (\kappa_f(s) - \kappa_f(ct)) \, ds \right) dt \\
&= q \int_0^\infty \int_{s/c}^\infty t^{p-q-1} \, dt \, s^{q-1} \kappa_f(s) \, ds - c^q \int_0^\infty t^{p-1} \kappa_f(ct) \, dt \\
&= q \int_0^\infty \frac{s^{p-1} c^{q-p}}{q-p} \kappa_f(s) \, ds - c^{q-p} \int_0^\infty t^{p-1} \kappa_f(t) \, dt \\
&= \frac{c^{q-p} p}{q-p} \int_0^\infty t^{p-1} \kappa_f(t) \, dt \\
&= \frac{c^{q-p}}{q-p} \int_X |f|^p \, d\mu.
\end{aligned}$$

Moreover, $f = f_t + g_t$ for all $t \geq 0$. Hence, by Lemma 7.36 and (7.39),

$$\begin{aligned}
\kappa_{Tf}(t) &\leq \kappa_{Tf_t}(t/2) + \kappa_{Tg_t}(t/2) \\
&\leq \frac{2}{t} \|Tf_t\|_{1,\infty} + \frac{2^q}{t^q} \|Tg_t\|_q^q \\
&\leq \frac{2c_1}{t} \|f_t\|_1 + \frac{(2c_q)^q}{t^q} \|g_t\|_q^q .
\end{aligned}$$

Hence, by Lemma 7.36,

$$\begin{aligned}
\int_X |Tf|^p \, d\mu &= p \int_0^\infty t^{p-1} \kappa_{Tf}(t) \, dt \\
&\leq p 2c_1 \int_0^\infty t^{p-2} \|f_t\|_1 \, dt + p(2c_q)^q \int_0^\infty t^{p-q-1} \|g_t\|_q^q \, dt \\
&= \left(\frac{p 2 c_1 c^{1-p}}{p-1} + \frac{p(2c_q)^q c^{q-p}}{q-p} \right) \int_X |f|^p \, d\mu \\
&= \frac{p(q-1) 2^p c_1^{(q-p)/(q-1)} c_q^{(qp-q)/(q-1)}}{(q-p)(p-1)} \int_X |f|^p \, d\mu.
\end{aligned}$$

Here the last equality follows with the choice of $c := (2c_1)^{1/(q-1)}/(2c_q)^{q/(q-1)}$. This proves Theorem 7.37. $\square$

7.7 The Calderón–Zygmund inequality

The convolution product discussed in Section 7.5 has many applications, notably in the theory of partial differential equations. One such application is the Calderón–Zygmund inequality, which plays a central role in the regularity theory for elliptic equations. Its proof requires many results from measure theory, including Fubini's Theorem, convolution, Marcinkiewicz interpolation, Lebesgue's' differentiation theorem, and the dual space of L^p. Denote the standard *Laplace operator* on $\mathbb{R}^n$ by

$$\Delta := \sum_{i=1}^{n} \frac{\partial^2}{\partial x_i^2} \tag{7.41}$$

and, for $i = 1, \dots, n$, denote the partial derivative with respect to the ith coordinate by $\partial_i = \partial/\partial x_i$. Denote the open ball of radius $r > 0$ centered at the origin by $B_r := \{x \in \mathbb{R}^n \mid |x| < r\}$. Call a function $u\colon \mathbb{R}^n \to \mathbb{R}$ *smooth* if all its partial derivatives exist and are continuous. Denote by $C_0^\infty(\mathbb{R}^n)$ the space of compactly supported smooth functions on $\mathbb{R}^n$.

Definition 7.38. Fix an integer $n \geq 2$. The *fundamental solution* of Laplace's equation is the function $K\colon \mathbb{R}^n \setminus \{0\} \to \mathbb{R}$ defined by

$$K(x) := \begin{cases} (2\pi)^{-1}\log(|x|) & \text{if } n = 2, \\ (2-n)^{-1}\omega_n^{-1}|x|^{2-n} & \text{if } n > 2. \end{cases} \tag{7.42}$$

Here ω_n denotes the area of the unit sphere $S^{n-1} \subset \mathbb{R}^n$ or, equivalently,

$$\omega_n/n := m(B_1)$$

denotes the Lebesgue measure of the unit ball in $\mathbb{R}^n$. The first and second partial derivatives

$$K_i := \partial_i K \quad \text{and} \quad K_{ij} := \partial_i \partial_j K$$

of the fundamental solution are given by

$$K_i(x) = \frac{x_i}{\omega_n |x|^n}, \quad K_{ij}(x) = \frac{-n x_i x_j}{\omega_n |x|^{n+2}}, \quad K_{ii}(x) = \frac{|x|^2 - n x_i^2}{\omega_n |x|^{n+2}} \tag{7.43}$$

for $1 \leq i, j \leq n$ with $i \neq j$. Extend the functions K, K_i, K_{ij} to all of $\mathbb{R}^n$ by setting

$$K(0) := K_i(0) := K_{ij}(0) := 0 \quad \text{for all } i, j.$$

Exercise 7.39. Prove that $\Delta K = 0$. Prove that K and K_i are locally integrable, while K_{ij} is not Lebesgue integrable over any neighborhood of the origin.

Hint. Use Fubini's Theorem in polar coordinates (Exercise 7.47).

Exercise 7.40. Prove that

$$m(B_1) = \omega_n/n$$

and that

$$\omega_n = \frac{2\pi^{n/2}}{\Gamma(n/2)} = \begin{cases} \dfrac{2\pi^{n/2}}{(n/2-1)!} & \text{if } n \text{ is even,} \\ \dfrac{2^{(n+1)/2}\pi^{(n-1)/2}}{1\cdot 3\cdots(n-2)} & \text{if } n \text{ is odd.} \end{cases} \tag{7.44}$$

Hint. Use Fubini's Theorem to prove that $\int_{\mathbb{R}^n} e^{-|x|^2}\,dm(x) = \pi^{n/2}$. Use polar coordinates to express the integral in terms of ω_n (Exercise 7.47).

Theorem 7.41. *Fix an integer $n \geq 2$ and let $f \in C_0^\infty(\mathbb{R}^n)$. Then*

$$f = K * \Delta f. \tag{7.45}$$

Moreover, the function

$$u\colon \mathbb{R}^n \longrightarrow \mathbb{R},$$

defined by

$$u(x) := (K * f)(x) = \int_{\mathbb{R}^n} K(x-y)f(y)\,dm(y) \tag{7.46}$$

for $x \in \mathbb{R}^n$, is smooth and satisfies

$$\Delta u = f, \quad \partial_i u = K_i * f \quad \text{for } i = 1,\dots,n. \tag{7.47}$$

The equations (7.45) *and* (7.47) *are called* ***Poisson's identities****.*

Proof. The proof relies on *Green's formula*

$$\int_\Omega (u\Delta v - v\Delta u)\,dm = \int_{\partial\Omega}\left(u\frac{\partial v}{\partial \nu} - v\frac{\partial u}{\partial \nu}\right)d\sigma \tag{7.48}$$

for a bounded open set $\Omega \subset \mathbb{R}^n$ with smooth boundary $\partial\Omega$ and two smooth functions $u, v\colon \mathbb{R}^n \to \mathbb{R}$. The term

$$\frac{\partial u}{\partial \nu}(x) := \sum_{i=1}^n \nu_i(x)\frac{\partial u}{\partial x_i}(x)$$

where $x \in \partial\Omega$ denotes the outward normal derivative and $\nu\colon \partial\Omega \to S^{n-1}$ denotes the outward pointing unit normal vector field on the boundary. The integral over the boundary is understood with respect to the Borel measure σ induced by the geometry of the ambient Euclidean space. We do not give a precise definition because the boundary integral will only be needed here when the boundary component is a sphere (see Exercise 7.47 below). Formula (7.48) can be viewed as a higher-dimensional analogue of the fundamental theorem of calculus.

Now let $f \in C_0^\infty(\mathbb{R}^n)$ and choose $r > 0$ so large that $\mathrm{supp}(f) \subset B_r$. Fix an element $\xi \in \mathrm{supp}(f)$ and a constant $\varepsilon > 0$ such that $\overline{B}_\varepsilon(\xi) \subset B_r$. Choose

$$\Omega := B_r \setminus \overline{B}_\varepsilon(\xi), \quad u(x) := K_\xi(x) := K(\xi - x), \quad v := f.$$

Then $\partial\Omega = \partial B_r \cup \partial B_\varepsilon(\xi)$ and the functions v and $\partial v/\partial\nu$ vanish on ∂B_r. Moreover, $\Delta K_\xi \equiv 0$. Hence Green's formula (7.48) asserts that

$$\int_{\mathbb{R}^n \setminus B_\varepsilon(\xi)} K_\xi \Delta f \, dm = \int_{\partial B_\varepsilon(\xi)} \left(f \frac{\partial K_\xi}{\partial\nu} - K_\xi \frac{\partial f}{\partial\nu} \right) d\sigma. \tag{7.49}$$

Here the reversal of sign arises from the fact that the outward unit normal vector on $\partial B_\varepsilon(\xi)$ is inward pointing with respect to Ω. Moreover,

$$\nu(x) = |x - \xi|^{-1}(x - \xi) \quad \text{for } x \in \partial B_\varepsilon(\xi),$$

so

$$\partial K_\xi/\partial\nu(x) = \omega_n^{-1}\varepsilon^{1-n}$$

by (7.43). Also, by (7.42),

$$K_\xi(x) = \begin{cases} 2\pi^{-1}\log(\varepsilon) & \text{if } n = 2, \\ (2-n)^{-1}\omega_n^{-1}\varepsilon^{2-n} & \text{if } n > 2, \end{cases} =: \psi(\varepsilon) \quad \text{for } x \in \partial B_\varepsilon(\xi).$$

Hence it follows from (7.49) that

$$\int_{\mathbb{R}^n \setminus B_\varepsilon(\xi)} K_\xi \Delta f \, dm = \frac{1}{\omega_n \varepsilon^{n-1}} \int_{\partial B_\varepsilon(\xi)} u \, d\sigma - \psi(\varepsilon) \int_{B_\varepsilon(\xi)} \Delta f \, dm. \tag{7.50}$$

The last summand is obtained from (7.48) with $u = 1$, $v = f$, $\Omega = B_\varepsilon(\xi)$. Now take the limit $\varepsilon \to 0$. Then the first term on the right in (7.50) converges to $f(\xi)$ and the second term converges to zero. This proves (7.45). It follows from Theorem 7.35 and equation (7.45) that

$$\Delta u = \Delta(K * f) = K * \Delta f = f.$$

To prove the second equation in (7.47), fix an index $i \in \{1, \dots, n\}$ and a point $\xi \in \mathbb{R}^n$. Then the divergence theorem on

$$\Omega := B_r \setminus \bar{B}_\varepsilon(\xi)$$

asserts that

$$\begin{aligned}
&\int_{\mathbb{R}^n \setminus B_\varepsilon(\xi)} (K_i(\xi - x) f(x) - K(\xi - x)\partial_i f(x))\, dm(x) \\
&= -\int_{\mathbb{R}^n \setminus B_\varepsilon(\xi)} ((\partial_i K_\xi) f + K_\xi \partial_i f)\, dm \\
&= -\int_{\mathbb{R}^n \setminus B_\varepsilon(\xi)} \partial_i (K_\xi f)\, dm \\
&= \int_{\partial B_\varepsilon(\xi)} \nu_i K_\xi f \, d\sigma \\
&= \psi(\varepsilon) \int_{\partial B_\varepsilon(\xi)} \frac{x_i - \xi_i}{\varepsilon} f(x)\, d\sigma(x).
\end{aligned}$$

The last term converges to zero as ε tends to zero. Hence

$$(K_i * f)(\xi) = (K * \partial_i f)(\xi) = \partial_i (K * f)(\xi)$$

by Theorem 7.35. This proves Theorem 7.41. □

Remark 7.42. Theorem 7.41 extends to compactly supported C^1-functions $f: \mathbb{R}^n \to \mathbb{R}$ and asserts that $K * f$ is C^2. However, this does not hold for continuous functions with compact support. A counterexample is $u(x) = |x|^3$, which is not C^2 and satisfies $f := \Delta u = 3(n+1)|x|$. It then follows that $K * \beta f$ (for any $\beta \in C_0^\infty(\mathbb{R}^n)$ equal to 1 near the origin) cannot be C^2.

Theorem 7.43 (Calderón–Zygmund). *Fix both an integer $n \geq 2$ and a number $1 < p < \infty$. Then there exists a constant $c = c(n, p) > 0$ such that*

$$\sum_{i,j=1}^{n} \left\| \partial_i \partial_j u \right\|_p \leq c \left\| \Delta u \right\|_p \tag{7.51}$$

for all $u \in C_0^\infty(\mathbb{R}^n)$.

Proof. See page 304. The proof is based on the exposition in Gilbarg and Trudinger [5]. □

The Calderón–Zygmund inequality is a beautiful and deep result in the theory of partial differential equations. It extends to all functions $u = K * f$ with $f \in C_0^\infty(\mathbb{R}^n)$ and thus can be viewed as a result about the convolution operator $f \mapsto K * f$. Theorem 7.35 shows that a derivative of a convolution is equal to the convolution with the derivative. This extends to the case where the derivative only exists in the weak sense and is locally integrable. For the function K this is spelled out in equation (7.47) in Theorem 7.41. Thus the convolution of an L^p function with a function whose derivatives are integrable has derivatives in L^p. The same holds for second derivatives. (The precise formulation of this observation requires the theory of *Sobolev spaces.*) The remarkable fact is that the second derivatives of the fundamental solution K of Laplace's equation are not locally integrable and, nevertheless, the Calderón–Zygmund inequality still asserts that the second derivatives of its convolution $u = K * f$ with a p-integrable function f are p-integrable. Despite this subtlety, the proof is elementary in the case $p = 2$. Denote by

$$\nabla u := (\partial_1 u, \ldots, \partial_n u) \colon \mathbb{R}^n \longrightarrow \mathbb{R}^n$$

the gradient of a smooth function $u \colon \mathbb{R}^n \to \mathbb{R}$.

Lemma 7.44. *Fix an integer $n \geq 2$ and let $f \in C_0^\infty(\mathbb{R}^n)$. Then*

$$\|\nabla(K_j * f)\|_2 \leq \|f\|_2 \quad \textit{for } j = 1, \ldots, n. \tag{7.52}$$

Proof. Define $u := K_j * f$. This function is smooth by Theorem 7.35, but it need not have compact support. By the divergence theorem,

$$\int_{B_r} |\nabla u|^2 \, dm + \int_{B_r} u \Delta u \, dm = \int_{B_r} \sum_{i=1}^{n} \partial_i (u \partial_i u) \, dm = \int_{\partial B_r} u \frac{\partial u}{\partial \nu} \, d\sigma \tag{7.53}$$

for all $r > 0$. By Poisson's identities (7.45) and (7.47), we have

$$\Delta u = \Delta(K_j * f) = \Delta \partial_j (K * f) = \partial_j (K * \Delta f) = \partial_j f.$$

Since f has compact support, it follows from (7.43) that there is a constant $c > 0$ such that $|u(x)| + |\partial u / \partial \nu(x)| \leq c|x|^{1-n}$ for $|x|$ sufficiently large. Hence the integral on the right in (7.53) tends to zero as r tends to infinity. Thus

$$\begin{aligned}
\|\nabla u\|_2^2 &= \int_{\mathbb{R}^n} |\nabla u|^2 \, dm \\
&= -\int_{\mathbb{R}^n} u \partial_j f \, dm \\
&= \int_{\mathbb{R}^n} (\partial_j u) f \, dm \leq \|\nabla u\|_2 \, \|f\|_2 \, .
\end{aligned}$$

This proves Lemma 7.44. □

By Theorem 7.35, the space $C_0^\infty(\mathbb{R}^n)$ is dense in $L^2(\mathbb{R}^n)$. Thus Lemma 7.44 shows that the linear operator $f \mapsto \partial_k(K_j * f)$ extends uniquely to a bounded linear operator from $L^2(\mathbb{R}^n)$ to $L^2(\mathbb{R}^n)$. The heart of the proof of the Calderón–Zygmund inequality is the following delicate argument which shows that this operator also extends to a continuous linear operator from the Banach space $L^1(\mathbb{R}^n)$ to the topological vector space $L^{1,\infty}(\mathbb{R}^n)$ of weakly integrable functions introduced in Section 6.1. This argument occupies the next six pages. Recall the definition

$$\|f\|_{1,\infty} := \sup_{t>0} t\kappa_f(t),$$

where

$$\kappa_f(t) := m(A(t,f)), \quad A(t,f) := \{x \in \mathbb{R}^n \mid |f(x)| > t\}.$$

(See equation (6.1).)

Lemma 7.45. *Fix an integer $n \geq 2$. Then there is a constant $c = c(n) > 0$ such that*

$$\|\partial_k(K_j * f)\|_{1,\infty} \leq c\|f\|_1 \tag{7.54}$$

for all $f \in C_0^\infty(\mathbb{R}^n)$ and all indices $j, k = 1, \dots, n$.

Proof. Fix two integers $j, k \in \{1, \dots, n\}$ and let $T\colon L^2(\mathbb{R}^n) \to L^2(\mathbb{R}^n)$ be the unique bounded linear operator that satisfies

$$Tf = \partial_k(K_j * f) \tag{7.55}$$

for $f \in C_0^\infty(\mathbb{R}^n)$. This operator is well defined by Lemma 7.44. We prove in three steps that there is a constant $c = c(n) > 0$ such that $\|Tf\|_{1,\infty} \leq c\,\|f\|_1$ for all $f \in L^1(\mathbb{R}^n) \cap L^2(\mathbb{R}^n)$. Throughout we abuse notation and use the same letter f to denote a function in $\mathcal{L}^2(\mathbb{R}^n)$ and its equivalence class in $L^2(\mathbb{R}^n)$.

Step 1. *There is a constant $c = c(n) \geq 1$ such that the following holds. If $B \subset \mathbb{R}^n$ is a countable union of closed cubes $Q_i \subset \mathbb{R}^n$ with pairwise disjoint interiors and if $h \in \mathcal{L}^2(\mathbb{R}^n) \cap \mathcal{L}^1(\mathbb{R}^n)$ satisfies*

$$h|_{\mathbb{R}^n \setminus B} \equiv 0, \qquad \int_{Q_i} h\,dm = 0 \quad \text{for all } i \in \mathbb{N}, \tag{7.56}$$

then

$$\kappa_{Th}(t) \leq c\Big(m(B) + \frac{1}{t}\,\|h\|_1\Big) \tag{7.57}$$

for all $t > 0$.

For $i \in \mathbb{N}$ define

$$h_i : \mathbb{R}^n \longrightarrow \mathbb{R}$$

by

$$h_i(x) := \begin{cases} h(x) & \text{if } x \in Q_i, \\ 0 & \text{if } x \notin Q_i. \end{cases}$$

Denote by $q_i \in Q_i$ the center of the cube Q_i and by $2r_i > 0$ its side length. Then $|x - q_i| \le \sqrt{n} r_i$ for all $x \in Q_i$. Fix an element $x \in \mathbb{R}^n \setminus Q_i$. Then K_j is smooth on $x - Q_i$ and so Theorem 7.35 asserts that

$$\begin{aligned} (Th_i)(x) &= (\partial_k K_j * h_i)(x) \\ &= \int_{Q_i} \big(\partial_k K_j(x-y) - \partial_k K_j(x-q_i)\big) h_i(y)\, dm(y). \end{aligned} \tag{7.58}$$

This identity is more delicate than it looks at first glance. To see this, note that equation (7.55) only holds for compactly supported smooth functions, but is not meaningful for all L^2 functions f because $K_j * f$ may not be differentiable. The function h_i is not smooth, so care must be taken. Since $x \notin Q_i = \operatorname{supp}(h_i)$ one can approximate h_i in $L^2(\mathbb{R}^n)$ by a sequence of compactly supported smooth functions that vanish near x (by using the mollifier method in the proof of Theorem 7.35). For the approximating sequence part (iii) of Theorem 7.35 asserts that the partial derivative with respect to the kth variable of the convolution with K_j is equal to the convolution with $\partial_k K_j$ near x. Now the first equality in (7.58) follows by taking the limit. The second equality follows from (7.56). It follows from (7.58) that

$$\begin{aligned} |(Th_i)(x)| &\le \int_{Q_i} |\partial_k K_j(x-y) - \partial_k K_j(x-q_i)||h_i(y)|\, dm(y) \\ &\le \sup_{y \in Q_i} |\partial_k K_j(x-y) - \partial_k K_j(x-q_i)|\, \|h_i\|_1 \\ &\le \sqrt{n} r_i \sup_{y \in Q_i} |\nabla \partial_k K_j(x-y)|\, \|h_i\|_1 \\ &\le c_1 r_i \sup_{y \in Q_i} \frac{1}{|x-y|^{n+1}} \|h_i\|_1 \\ &= \frac{c_1 r_i}{d(x, Q_i)^{n+1}} \|h_i\|_1 . \end{aligned}$$

Here $d(x, Q_i) := \inf_{y \in Q_i} |x - y|$ and

$$c_1 = c_1(n) := \max_{j,k} \sup_{y \in \mathbb{R}^n \setminus \{0\}} |y|^{n+1} |\nabla \partial_k K_j(y)| \leq \frac{n(n+3)}{\omega_n}.$$

Here the last inequality follows by differentiating equation (7.43).

Now define

$$P_i := \{x \in \mathbb{R}^n \mid |x - q_i| < 2\sqrt{n} r_i\} \supset Q_i.$$

Then

$$d(x, Q_i) \geq |x - q_i| - \sqrt{n} r_i \quad \text{for all } x \in \mathbb{R}^n \setminus P_i.$$

Hence

$$\begin{aligned}
\int_{\mathbb{R}^n \setminus P_i} |Th_i|\, dm &\leq c_1 r_i \int_{\mathbb{R}^n \setminus P_i} \frac{1}{(|x - q_i| - \sqrt{n} r_i)^{n+1}}\, dm(x)\, \|h_i\|_1 \\
&= c_1 r_i \int_{|y| > 2\sqrt{n} r_i} \frac{1}{(|y| - \sqrt{n} r_i)^{n+1}}\, dm(y)\, \|h_i\|_1 \\
&= c_1 r_i \int_{2\sqrt{n} r_i}^{\infty} \frac{\omega_n s^{n-1}\, ds}{(s - \sqrt{n} r_i)^{n+1}}\, \|h_i\|_1 \\
&= c_1 \omega_n r_i \int_{\sqrt{n} r_i}^{\infty} \frac{(s + \sqrt{n} r_i)^{n-1}\, ds}{s^{n+1}}\, \|h_i\|_1 \\
&\leq c_1 2^{n-1} \omega_n r_i \int_{\sqrt{n} r_i}^{\infty} \frac{ds}{s^2}\, \|h_i\|_1 \\
&= c_2 \|h_i\|_1 .
\end{aligned}$$

Here $c_2 = c_2(n) := c_1(n) 2^{n-1} \omega_n \sqrt{n} \leq 2^{n-1} n^{3/2} (n + 3)$. The third step in the above computation follows from Fubini's Theorem in polar coordinates (Exercise 7.47). Thus we have proved that

$$\int_{\mathbb{R}^n \setminus P_i} |Th_i|\, dm \leq c_2 \|h_i\|_1 \quad \text{for all } i \in \mathbb{N}. \tag{7.59}$$

Recall that Th and Th_i are only equivalence classes in $L^2(\mathbb{R}^n)$. Choose square integrable functions on $\mathbb{R}^n$ representing these equivalence classes and denote them by the same letters $Th, Th_i \in \mathcal{L}^2(\mathbb{R}^n)$. We prove that there is a Lebesgue null set $E \subset \mathbb{R}^n$ such that

$$|Th(x)| \leq \sum_{i=1}^{\infty} |Th_i(x)| \quad \text{for all } x \in \mathbb{R}^n \setminus E. \tag{7.60}$$

To see this, note that the sequence $\sum_{i=1}^{\ell} h_i$ converges to h in $L^2(\mathbb{R}^n)$ as ℓ tends to infinity. So the sequence $\sum_{i=1}^{\ell} Th_i$ converges to Th in $L^2(\mathbb{R}^n)$ by Lemma 7.44. By Corollary 4.10, a subsequence converges almost everywhere. Hence there exist a Lebesgue null set $E \subset \mathbb{R}^n$ and a sequence of integers $0 < \ell_1 < \ell_2 < \ell_3 < \cdots$ such that the sequence $\sum_{i=1}^{\ell_\nu} Th_i(x)$ converges to $Th(x)$ as ν tends to infinity for all $x \in \mathbb{R}^n \setminus E$. Since $|\sum_{i=1}^{\ell_\nu} Th_i(x)| \le \sum_{i=1}^{\infty} |Th_i(x)|$ for all $x \in \mathbb{R}^n$, this proves (7.60).

Now put

$$A := \bigcup_{i=1}^{\infty} P_i.$$

Then it follows from (7.59), (7.60), and Theorem 1.38 that

$$\begin{aligned}
\int_{\mathbb{R}^n \setminus A} |Th|\, dm &\le \int_{\mathbb{R}^n \setminus A} \sum_{i=1}^{\infty} |Th_i|\, dm \\
&= \sum_{i=1}^{\infty} \int_{\mathbb{R}^n \setminus A} |Th_i|\, dm \\
&\le \sum_{i=1}^{\infty} \int_{\mathbb{R}^n \setminus P_i} |Th_i|\, dm \\
&\le c_2 \sum_{i=1}^{\infty} \|h_i\|_1 \\
&= c_2 \|h\|_1 .
\end{aligned}$$

Moreover,

$$m(A) \le \sum_{i=1}^{\infty} m(P_i) = c_3 \sum_{i=1}^{\infty} m(Q_i) = c_3 m(B),$$

where

$$c_3 = c_3(n) := \frac{m(B_{2\sqrt{n}})}{m([-1,1]^n)} = m(B_{\sqrt{n}}) = \omega_n n^{n/2-1}.$$

Hence

$$\begin{aligned}
t\kappa_{Th}(t) &\le tm(A) + tm(\{x \in \mathbb{R}^n \setminus A \mid |Th(x)| > t\}) \\
&\le tm(A) + \int_{\mathbb{R}^n \setminus A} |Th|\, dm \\
&\le c_3 tm(B) + c_2 \|h\|_1 \\
&\le c_4 (tm(B) + \|h\|_1)
\end{aligned}$$

for all $t > 0$, where

$$c_4 = c_4(n) := \max\{c_2(n), c_3(n)\} \le \max\{2^{n-1}n^{3/2}(n+3), \omega_n n^{n/2-1}\}.$$

This proves Step 1.

Step 2 (Calderón–Zygmund decomposition). *Let $f \in \mathcal{L}^2(\mathbb{R}^n) \cap \mathcal{L}^1(\mathbb{R}^n)$ and $t > 0$. Then there exists a countable collection of closed cubes $Q_i \subset \mathbb{R}^n$ with pairwise disjoint interiors such that*

$$m(Q_i) < \frac{1}{t}\int_{Q_i} |f| \le 2^n m(Q_i) \quad \textit{for all } i \in \mathbb{N} \tag{7.61}$$

and

$$|f(x)| \le t \quad \textit{for almost all } x \in \mathbb{R}^n \setminus B, \tag{7.62}$$

where $B := \bigcup_{i=1}^{\infty} Q_i$.

For $\xi \in \mathbb{Z}^n$ and $\ell \in \mathbb{Z}$ define

$$Q(\xi,\ell) := \{x \in \mathbb{R}^n \mid 2^{-\ell}\xi_i \le x_i \le 2^{-\ell}(\xi_i + 1)\}.$$

Let

$$\mathcal{Q} := \{Q(\xi,\ell) \mid \xi \in \mathbb{Z}^n,\ \ell \in \mathbb{Z}\}$$

and define the subset $\mathcal{Q}_0 \subset \mathcal{Q}$ by

$$\mathcal{Q}_0 := \left\{ Q \in \mathcal{Q} \;\middle|\; \begin{array}{l} tm(Q) < \int_Q |f|\,dm \text{ and, for all } Q' \in \mathcal{Q}, \\ Q \subsetneq Q' \implies \int_{Q'} |f|\,dm \le tm(Q') \end{array} \right\}.$$

Then every decreasing sequence of cubes in $\mathcal{Q}$ contains at most one element of $\mathcal{Q}_0$. Hence every element of $\mathcal{Q}_0$ satisfies (7.61) and any two cubes in $\mathcal{Q}_0$ have disjoint interiors. Define $B := \bigcup_{Q \in \mathcal{Q}_0} Q$. We prove that

$$x \in \mathbb{R}^n \setminus B, \quad x \in Q \in \mathcal{Q} \implies \frac{1}{m(Q)}\int_Q |f|\,dm \le t. \tag{7.63}$$

Suppose, by contradiction, that there exists an element $x \in \mathbb{R}^n \setminus B$ and a cube $Q \in \mathcal{Q}$ such that $x \in Q$ and $tm(Q) < \int_Q |f|\,dm$. Then, since $\|f\|_1 < \infty$, there is a maximal cube $Q \in \mathcal{Q}$ such that $x \in Q$ and $tm(Q) < \int_Q |f|\,dm$. Such a maximal cube would be an element of $\mathcal{Q}_0$ and hence $x \in B$, a contradiction. This proves (7.63). Now Theorem 6.14 asserts that there exists a Lebesgue null set $E \subset \mathbb{R}^n \setminus B$ such that every element of $\mathbb{R}^n \setminus (B \cup E)$ is a Lebesgue point of f. By (7.63), every point $x \in \mathbb{R}^n \setminus (B \cup E)$ is the intersection point of a decreasing sequence of cubes over which $|f|$ has mean value at most t. Hence it follows from Theorem 6.16 that $|f(x)| \le t$ for all $x \in \mathbb{R}^n \setminus (B \cup E)$. This proves Step 2.

Step 3. *Let $c = c(n) \geq 1$ be the constant in Step 1. Then*

$$\|Tf\|_{1,\infty} \leq (2^{n+1} + 6c)\,\|f\|_1 \tag{7.64}$$

for all $f \in L^2(\mathbb{R}^n) \cap L^1(\mathbb{R}^n)$.

Fix a function $f \in \mathcal{L}^2(\mathbb{R}^n) \cap \mathcal{L}^1(\mathbb{R}^n)$ and a constant $t > 0$. Let the Q_i be as in Step 2 and define

$$B := \bigcup_i Q_i.$$

Then $m(Q_i) < \frac{1}{t}\int_{Q_i}|f|\,dm$ for all i by Step 2, and hence

$$m(B) = \sum_i m(Q_i) \leq \frac{1}{t}\sum_i \int_{Q_i}|f|\,dm \leq \frac{1}{t}\,\|f\|_1\,.$$

Define

$$g, h\colon \mathbb{R}^n \longrightarrow \mathbb{R}$$

by

$$g := f\chi_{\mathbb{R}^n\setminus B} + \sum_i \frac{\int_{Q_i} f\,dm}{m(Q_i)}\chi_{Q_i}, \qquad h := f - g.$$

Then

$$\|g\|_1 \leq \|f\|_1\,, \qquad \|h\|_1 \leq 2\,\|f\|_1\,.$$

Further, h vanishes on $\mathbb{R}^n \setminus B$ and $\int_{Q_i} h\,dm = 0$ for all i. Hence it follows from Step 1 that

$$\kappa_{Th}(t) \leq c\Big(m(B) + \frac{1}{t}\,\|h\|_1\Big) \leq \frac{3c}{t}\,\|f\|_1\,. \tag{7.65}$$

Moreover, it follows from Step 2 that $|g(x)| \leq t$ for almost every $x \in \mathbb{R}^n \setminus B$ and $|g(x)| \leq 2^n t$ for every $x \in \mathrm{int}(Q_i)$. Thus $|g| \leq 2^n t$ almost everywhere. Hence it follows from Lemma 7.36 that

$$\kappa_{Tg}(t) \leq \frac{1}{t^2}\int_{\mathbb{R}^n}|g|^2\,dm \leq \frac{2^n}{t}\int_{\mathbb{R}^n}|g|\,dm \leq \frac{2^n}{t}\,\|f\|_1\,. \tag{7.66}$$

Now combine (7.65) and (7.66) with the inequality (7.36) in Lemma 7.36 to obtain the estimate

$$\kappa_{Tf}(2t) \leq \kappa_{Tg}(t) + \kappa_{Th}(t) \leq \frac{2^{n+1} + 6c}{2t}\,\|f\|_1\,.$$

Here the splitting $f = g + h$ depends on t, but the constant c does not. Multiply the inequality by $2t$ and take the supremum over all t to obtain (7.64). This proves Step 3 and Lemma 7.45. □

Theorem 7.46 (Calderón–Zygmund). *Fix both an integer $n \geq 2$ and a number $1 < p < \infty$. Then there exists a constant $c = c(n, p) > 0$ such that*

$$\|\partial_i (K_j * f)\|_p \leq c\|f\|_p \tag{7.67}$$

for all $f \in C_0^\infty(\mathbb{R}^n)$ and all $i, j = 1, \dots, n$.

Proof. For $p = 2$ this estimate was established in Lemma 7.44 with $c = 1$. Second, suppose $1 < p < 2$ and let $c_1(n)$ be the constant of Lemma 7.45. For $i, j = 1, \dots, n$ denote by

$$T_{ij} \colon L^2(\mathbb{R}^n) \longrightarrow L^2(\mathbb{R}^n)$$

the unique bounded linear operator that satisfies

$$T_{ij} f = \partial_i (K_j * f) \quad \text{for } f \in C_0^\infty(\mathbb{R}^n).$$

Then $\|T_{ij} f\|_{1,\infty} \leq c_1(n) \|f\|_1$ for all $f \in C_0^\infty(\mathbb{R}^n)$ and all i, j by Lemma 7.45. Since $C_0^\infty(\mathbb{R}^n)$ is dense in $L^2(\mathbb{R}^n) \cap L^1(\mathbb{R}^n)$ by Theorem 7.35, it follows that $\|T_{ij} f\|_{1,\infty} \leq c_1(n) \|f\|_1$ for all $f \in L^2(\mathbb{R}^n) \cap L^1(\mathbb{R}^n)$. Hence Theorem 7.37 (with $q = 2$) asserts that (7.67) holds with

$$c = c(n, p) := 2\Big(\frac{p}{(2-p)(p-1)}\Big)^{1/p} c_1(n)^{2/p-1}.$$

Third, suppose $2 < p < \infty$ and choose $1 < q < 2$ such that $1/p + 1/q = 1$. Then it follows from Theorem 7.41, integration by parts, Hölder's inequality, and from what we have just proved that, for all $f, g \in C_0^\infty(\mathbb{R}^n)$,

$$\begin{aligned}
\int_{\mathbb{R}^n} (\partial_i (K_j * f)) g \, dm &= \int_{\mathbb{R}^n} (\partial_i \partial_j f) g \, dm \\
&= \int_{\mathbb{R}^n} f (\partial_i \partial_j g) \, dm \\
&= \int_{\mathbb{R}^n} f (\partial_i (K_j * g)) \, dm \\
&\leq \|f\|_p \, \|\partial_i (K_j * g)\|_q \\
&\leq c(n, q) \, \|f\|_p \, \|g\|_q \, .
\end{aligned}$$

Since $C_0^\infty(\mathbb{R}^n)$ is dense in $L^q(\mathbb{R}^n)$ by Theorem 7.35, and the Lebesgue measure is semi-finite, it follows from Lemma 4.34 that $\|\partial_i (K_j * f)\|_p \leq c(n, q) \, \|f\|_p$ for all $f \in C_0^\infty(\mathbb{R}^n)$. This proves Theorem 7.46. □

Proof of Theorem 7.43. Fix both an integer $n \geq 2$ and a number $1 < p < \infty$. Let $c = c(n, p)$ be the constant of Theorem 7.46 and let $u \in C_0^\infty(\mathbb{R}^n)$. Then

$$\partial_j u = \partial_j (K * \Delta u) = K_j * \Delta u$$

by Theorem 7.41. Hence it follow from Theorem 7.46 with $f = \Delta u$ that

$$\begin{aligned} \|\partial_i \partial_j u|_p &= \|\partial_i (K_j * \Delta u)|_p \\ &\leq c(n, p) \, \|\Delta u\|_p \end{aligned}$$

for $i, j = 1, \ldots, n$. This proves Theorem 7.43. □

7.8 Exercises

Exercise 7.47 (Lebesgue measure on the sphere). For $n \in \mathbb{N}$, let $(\mathbb{R}^n, \mathcal{A}_n, m_n)$ the Lebesgue measure space, denote the open unit ball by $B^n := \{x \in \mathbb{R}^n \mid |n| < 1\}$, and the unit sphere by

$$S^{n-1} := \partial B^n = \{x \in \mathbb{R}^n \mid |x| = 1\}.$$

For $A \subset S^{n-1}$ define

$$A^{\pm} := \{x \in B^{n-1} \mid (x, \pm\sqrt{1 - |x|^2}) \in A\}.$$

Prove that the collection

$$\mathcal{A}_S := \{A \subset S^{n-1} \mid A^+, A^- \in \mathcal{A}_{n-1}\}$$

is a σ-algebra and that the map

$$\sigma \colon \mathcal{A}_S \longrightarrow [0, \infty]$$

defined by

$$\sigma(A) := \int_{A^+} \frac{1}{\sqrt{1 - |x|^2}} \, dm_{n-1}(x) + \int_{A^-} \frac{1}{\sqrt{1 - |x|^2}} \, dm_{n-1}(x)$$

for $A \in \mathcal{A}_S$ is a measure. Prove Fubini's Theorem in polar coordinates as stated below. Use it to prove that

$$\omega_n := \sigma(S^{n-1}) < \infty.$$

FUBINI'S THEOREM FOR POLAR COORDINATES. *Let $f : \mathbb{R}^n \to \mathbb{R}$ be Lebesgue integrable. For $r \geq 0$ and $x \in S^{n-1}$ define*

$$f^r(x) := f_x(r) := f(rx).$$

Then there is a set $E \in \mathcal{A}_S$ such that

$$\sigma(E) = 0$$

and

$$f_x \in \mathcal{L}^1([0,\infty)) \quad \text{for all } x \in S^{n-1} \setminus E,$$

and there is a Lebesgue null set $F \subset [0,\infty)$ such that

$$f^r \in \mathcal{L}^1(\sigma) \quad \text{for all } r \in [0,\infty) \setminus F.$$

Define

$$g : S^{n-1} \longrightarrow \mathbb{R} \quad \text{and} \quad h : [0,\infty) \longrightarrow \mathbb{R}$$

by

$$g(x) := \begin{cases} \displaystyle\int_{[0,\infty)} r^{n-1} f_x(r)\, dm_1(r) & \text{for } x \in S^{n-1} \setminus E, \\ 0 & \text{for } x \in E, \end{cases} \tag{7.68a}$$

$$h(r) := \begin{cases} \displaystyle r^{n-1} \int_{S^{n-1}} f^r \, d\sigma & \text{for } r \in [0,\infty) \setminus F, \\ 0 & \text{for } r \in F. \end{cases} \tag{7.68b}$$

Then

$$g \in \mathcal{L}^1(\sigma), \quad h \in \mathcal{L}^1([0,\infty)),$$

and

$$\int_{\mathbb{R}^n} f\, dm_n = \int_{S^{n-1}} g\, d\sigma = \int_{[0,\infty)} h\, dm_1. \tag{7.69}$$

Hint. Define the diffeomorphism

$$\phi : B^{n-1} \times (0,\infty) \longrightarrow \{x \in \mathbb{R}^n \mid x_n > 0\}$$

by

$$\phi(x,r) := (rx, r(1-|x|^2)^{1/2}).$$

Prove that

$$\det(d\phi(x,r)) = (1-|x|^2)^{-1/2} r^{n-1} \quad \text{for } x \in B^{n-1} \text{ and } r > 0.$$

Use Theorem 2.17 and Fubini's Theorem (Theorem 7.30).

Exercise 7.48 (divergence theorem). Let $f\colon \mathbb{R}^n \to \mathbb{R}$ be a smooth function. Prove that

$$\int_{B^n} \partial_i f \, dm_n = \int_{S^{n-1}} x_i f(x) \, d\sigma(x). \tag{7.70}$$

Hint. Assume first that $i = n$. Use the fundamental theorem of calculus and Fubini's Theorem (Theorem 7.30) for $\mathbb{R}^n = \mathbb{R}^{n-1} \times \mathbb{R}$.

Exercise 7.49. Prove that

$$\int_0^\infty \frac{\sin(x)}{x} \, dx := \lim_{\varepsilon \to 0} \int_\varepsilon^{1/\varepsilon} \frac{\sin(x)}{x} \, dx = \frac{\pi}{2}.$$

Hint. Use the identity

$$\int_0^\infty e^{-rt} \, dt = 1/r \quad \text{for } r > 0$$

and Fubini's Theorem.

Exercise 7.50. Define the function

$$f\colon \mathbb{R}^2 \longrightarrow \mathbb{R}$$

by

$$f(x, y) := \frac{\operatorname{sign}(xy)}{x^2 + y^2}, \quad \operatorname{sign}(z) := \begin{cases} 1 & \text{if } z > 0, \\ 0 & \text{if } z = 0, \\ -1 & \text{if } z < 0, \end{cases} \quad \text{for } (x, y) \neq 0,$$

and by

$$f(0, 0) := 0.$$

Prove that $f_x, f^y\colon \mathbb{R} \to \mathbb{R}$ are Lebesgue integrable for all $x, y \in \mathbb{R}$. Prove that the functions $\mathbb{R} \to \mathbb{R}\colon x \mapsto \int_{\mathbb{R}} f_x \, dm_1$ and $\mathbb{R} \to \mathbb{R}\colon y \mapsto \int_{\mathbb{R}} f^y \, dm_1$ are Lebesgue integrable and

$$\int_{\mathbb{R}} \left(\int_{\mathbb{R}} f(x, y) \, dm_1(x) \right) dm_1(y) = \int_{\mathbb{R}} \left(\int_{\mathbb{R}} f(x, y) \, dm_1(y) \right) dm_1(x).$$

Prove that f is not Lebesgue integrable.

Exercise 7.51. Let $(X, \mathcal{A}, \mu)$ and $(Y, \mathcal{B}, \nu)$ be two σ-finite measure spaces and let $f \in \mathcal{L}^1(\mu)$ and $g \in \mathcal{L}^1(\nu)$. Define

$$h: X \times Y \longrightarrow \mathbb{R}$$

by

$$h(x, y) := f(x)g(y), \quad \text{for } x \in X \text{ and } y \in Y.$$

Prove that $h \in \mathcal{L}^1(\mu \otimes \nu)$ and

$$\int_{X \times Y} h \, d(\mu \otimes \nu) = \int_X f \, d\mu \int_Y g \, d\nu.$$

Exercise 7.52. Let $(X, \mathcal{A}, \mu)$ and $(Y, \mathcal{B}, \nu)$ be two σ-finite measure spaces and let $\lambda: \mathcal{A} \otimes \mathcal{B} \to \mathbb{R}$ be any measure such that $\lambda(A \times B) = \mu(A)\nu(B)$ for all $A \in \mathcal{A}$ and all $B \in \mathcal{B}$. Prove that $\lambda = \mu \otimes \nu$.

Exercise 7.53. Define

$$\phi: \mathbb{R} \longrightarrow \mathbb{R}$$

by

$$\phi(x) := \begin{cases} 1 - \cos(x) & \text{for } 0 \leq x \leq 2\pi, \\ 0 & \text{otherwise.} \end{cases}$$

Define the functions

$$f, g, h: \mathbb{R} \longrightarrow \mathbb{R}$$

by

$$f(x) := 1, \quad g(x) := \phi'(x), \quad h(x) := \int_{-\infty}^{x} \phi(t) \, dt$$

for $x \in \mathbb{R}$. Prove that $(f * g) * h = 0$ and $f * (g * h) > 0$. Thus the convolution need not be associative on nonintegrable functions. Compare this with part (v) of Theorem 7.32. Prove that

$$E(|f| * |g|, |h|) = E(|f|, |g| * |h|) = \mathbb{R}$$

while

$$E(f * g, h) = E(f, g * h) = \emptyset.$$

Exercise 7.54. Let $(\mathbb{R}, \mathcal{A}, m)$ be the Lebesgue measure space, let $\mathcal{B} \subset \mathcal{A}$ be the Borel σ-algebra, and denote by $\mathcal{M}$ the Banach space of all signed Borel measures $\mu: \mathcal{B} \to [0, \infty)$ with the norm $\|\mu\| := |\mu|(\mathbb{R})$. (See Exercise 5.34.) The *convolution of two signed measures* $\mu, \nu \in \mathcal{M}$ is the map

$$\mu * \nu: \mathcal{B} \longrightarrow \mathbb{R}$$

defined by

$$(\mu * \nu)(B) := (\mu \otimes \nu)(\{(x, y) \in \mathbb{R}^2 \mid x + y \in B\}) \tag{7.71}$$

for $B \in \mathcal{B}$, where

$$(\mu \otimes \nu) := \mu^+ \otimes \nu^+ + \mu^- \otimes \nu^- - \mu^+ \otimes \nu^- - \mu^- \otimes \nu^+.$$

(See Definition 5.13 and Theorem 5.20.)

Prove the following.

(i) If $\mu, \nu \in \mathcal{M}$, then $\mu * \nu \in \mathcal{M}$ and

$$\|\mu * \nu\| \leq \|\mu\| \, \|\nu\| \, .$$

(ii) There exists a unique element $\delta \in \mathcal{M}$ such that

$$\delta * \mu = \mu$$

for all $\mu \in \mathcal{M}$.

(iii) The convolution product on $\mathcal{M}$ is commutative, associative, and distributive. Thus $\mathcal{M}$ is a commutative Banach algebra with unit.

(iv) If $f \in \mathcal{L}^1(\mathbb{R})$ and

$$\mu_f : \mathcal{B} \longrightarrow \mathbb{R}$$

is defined by

$$\mu_f(B) := \int_B f \, dm \quad \text{for } B \in \mathcal{B},$$

then $\mu_f \in \mathcal{M}$ and $\|\mu_f\| = \|f\|_1$.

(v) If $f, g \in \mathcal{L}^1(\mathbb{R})$, then

$$\mu_f * \mu_g = \mu_{f*g}.$$

(vi) Let $\lambda, \mu, \nu \in \mathcal{M}$. Then $\lambda = \mu * \nu$ if and only if

$$\int_{\mathbb{R}} f \, d\lambda = \int_{\mathbb{R}^2} f(x + y) d(\mu \otimes \nu)(x, y)$$

for all bounded Borel measurable functions $f : \mathbb{R} \to \mathbb{R}$.

(vii) If $\mu, \nu \in \mathcal{M}$ and $B \in \mathcal{B}$, then

$$(\mu * \nu)(B) = \int_{\mathbb{R}} \mu(B - t) \, d\nu(t).$$

Chapter 8

The Haar measure

The purpose of this last chapter is to prove the existence and uniqueness of a normalized invariant Radon measure on any compact Hausdorff group. In the case of a locally compact Hausdorff group the theorem asserts the existence of a left-invariant Radon measure that is unique up to a scaling factor. An example is the Lebesgue measure on $\mathbb{R}^n$. A useful exposition is the paper by Gert K. Pedersen [16], which also discusses the original references.

8.1 Topological groups

Let G be a group, in multiplicative notation, with the group operation

$$\mathrm{G} \times \mathrm{G} \longrightarrow \mathrm{G}\colon (x, y) \longmapsto xy \tag{8.1}$$

the unit $\mathbb{1} \in \mathrm{G}$, and the inverse map

$$\mathrm{G} \longrightarrow \mathrm{G}\colon x \longmapsto x^{-1}. \tag{8.2}$$

A *topological group* is a pair $(\mathrm{G}, \mathcal{U})$ consisting of a group G and a topology

$$\mathcal{U} \subset 2^{\mathrm{G}}$$

such that the group multiplication (8.1) and the inverse map (8.2) are continuous. Here the continuity of the group multiplication (8.1) is understood with respect to the product topology on $\mathrm{G} \times \mathrm{G}$ (see Appendix B). A *locally compact Hausdorff group* is a topological group $(\mathrm{G}, \mathcal{U})$ such that the topology is locally compact and Hausdorff (see page 95).

Example 8.1. Let G be any group and define $\mathcal{U} := \{\emptyset, \mathrm{G}\}$. Then $(\mathrm{G}, \mathcal{U})$ is a compact topological group, but is not Hausdorff unless $\mathrm{G} = \{\mathbb{1}\}$.

Example 8.2 (discrete groups). Let G be any group. Then the pair $(\mathrm{G}, \mathcal{U})$ with the discrete topology $\mathcal{U} := 2^{\mathrm{G}}$ is a locally compact Hausdorff group, called a *discrete group*. Examples of discrete groups (where the discrete topology appears naturally) are the additive group $\mathbb{Z}^n$, the multiplicative group $\mathrm{SL}(n, \mathbb{Z})$ of integer

$n \times n$-matrices with determinant one, the mapping class group of isotopy classes of diffeomorphisms of any manifold, and every finite group.

Example 8.3 (Lie groups). Let $G \subset \mathrm{GL}(n, \mathbb{C})$ be any subgroup of the general linear group of invertible complex $n \times n$-matrices that is closed as a subset of $\mathrm{GL}(n, \mathbb{C})$ with respect to the relative topology, i.e., $\mathrm{GL}(n, \mathbb{C}) \setminus G$ is an open set in $\mathbb{C}^{n \times n}$. Let $\mathcal{U} \subset 2^G$ be the relative topology on G, i.e., $U \subset G$ is open if and only if there is an open subset $V \subset \mathbb{C}^{n \times n}$ such that $U = G \cap V$. Then $(G, \mathcal{U})$ is a locally compact Hausdorff group. In fact, it is a basic result in the theory of Lie groups that every closed subgroup of $\mathrm{GL}(n, \mathbb{C})$ is a smooth submanifold of $\mathbb{C}^{n \times n}$ and hence is a *Lie group*. Specific examples of Lie groups are the general linear groups $\mathrm{GL}(n, \mathbb{R})$ and $\mathrm{GL}(n, \mathbb{C})$, the special linear groups $\mathrm{SL}(n, \mathbb{R})$ and $\mathrm{SL}(n, \mathbb{C})$ of real and complex $n \times n$-matrices with determinant one, the orthogonal group $\mathrm{O}(n)$ of matrices $x \in \mathbb{R}^{n \times n}$ such that $x^T x = \mathbb{1}$, the special orthogonal group $\mathrm{SO}(n) := \mathrm{O}(n) \cap \mathrm{SL}(n, \mathbb{R})$, the unitary group $\mathrm{U}(n)$ of matrices $x \in \mathbb{C}^{n \times n}$ such that $x^* x = \mathbb{1}$, the special unitary group $\mathrm{SU}(n) := \mathrm{U}(n) \cap \mathrm{SL}(n, \mathbb{C})$, the group $\mathrm{Sp}(1)$ of the unit quaternions, the unit circle $S^1 = \mathrm{U}(1)$ in the complex plane, the torus $\mathbb{T}^n := S^1 \times \cdots \times S^1$ (n times), or, for any multi-linear form $\tau \colon (\mathbb{C}^n)^k \to \mathbb{C}$, the group of all matrices $x \in \mathrm{GL}(n, \mathbb{C})$ that preserve τ. The additive groups $\mathbb{R}^n$ and $\mathbb{C}^n$ are also Lie groups. Lie groups form an important class of locally compact Hausdorff groups and play a central role in differential geometry.

Example 8.4. If $(V, \|\cdot\|)$ is a normed vector space (Example 1.11), then the additive group V is a Hausdorff topological group. It is locally compact if and only if V is finite-dimensional.

Example 8.5. The rational numbers $\mathbb{Q}$ with the additive structure form a Hausdorff topological group with the relative topology as a subset of $\mathbb{R}$. It is totally disconnected (every connected component is a single point), but does not have the discrete topology. It is not locally compact.

Example 8.6 (p-adic integers). Fix a prime number $p \in \mathbb{N}$ and denote by

$$\mathbb{N}_0 := \mathbb{N} \cup \{0\}$$

the set of nonnegative integers. For $x, y \in \mathbb{Z}$ define

$$d_p(x, y) := |x - y|_p := \inf\{p^{-k} \mid k \in \mathbb{N}_0,\ x - y \in p^k \mathbb{Z}\}. \tag{8.3}$$

Then the function

$$d_p\colon \mathbb{Z} \times \mathbb{Z} \longrightarrow [0, 1]$$

is a distance function and so $(\mathbb{Z}, d_p)$ is a metric space. It is not complete. Its completion is denoted by $\mathbb{Z}_p$ and called the *ring of p-adic integers.* Here is another description of the p-adic integers. Consider the sequence of projections

$$\cdots \stackrel{\pi_{k+1}}{\longrightarrow} \mathbb{Z}/p^k\mathbb{Z} \stackrel{\pi_k}{\longrightarrow} \mathbb{Z}/p^{k-1}\mathbb{Z} \stackrel{\pi_{k-1}}{\longrightarrow} \cdots \stackrel{\pi_3}{\longrightarrow} \mathbb{Z}/p^2\mathbb{Z} \stackrel{\pi_2}{\longrightarrow} \mathbb{Z}/p\mathbb{Z} \stackrel{\pi_1}{\longrightarrow} \{1\}.$$

The *inverse limit* of this sequence of maps is the set of sequences

$$\mathbb{Z}_p := \{x = (x_k)_{k\in\mathbb{N}_0} \mid x_k \in \mathbb{Z}/p^k\mathbb{Z},\ \pi_k(x_k) = x_{k-1} \text{ for all } k \in \mathbb{N}\}.$$

This set is a commutative ring with unit. Addition and multiplication are defined term by term, i.e.,

$$x + y := (x_k + y_k)_{x\in\mathbb{N}_0}, \quad xy := (x_k y_k)_{x\in\mathbb{N}_0}$$

for $x = (x_k)_{k\in\mathbb{N}_0} \in \mathbb{Z}_p$ and $y = (y_k)_{k\in\mathbb{N}_0} \in \mathbb{Z}_p$. The ring of p-adic integers is a compact metric space with

$$d_p(x, y) := \inf\{p^{-k} \mid k \in \mathbb{N}_0,\ x_k = y_k\}. \tag{8.4}$$

The inclusion of $\mathbb{Z}$ into the p-adic integers is given by

$$\iota_p\colon \mathbb{Z} \longrightarrow \mathbb{Z}_p, \quad \iota_p(x) := (x \bmod p^k)_{k\in\mathbb{N}}.$$

This is an isometric embedding with respect to the distance functions (8.3) and (8.4). The additive p-adic integers form an uncountable compact Hausdorff group (with the topology of a Cantor set) that is not a Lie group.

Example 8.7 (p-adic Rationals). Fix a prime number $p \in \mathbb{N}$. Write a nonzero rational number $x \in \mathbb{Q}$ in the form $x = p^k a/b$ where $k \in \mathbb{Z}$ and the numbers $a \in \mathbb{Z}$ and $b \in \mathbb{N}$ are relatively prime to p, and define $|x|_p := p^{-k}$. For $x = 0$, define $|0|_p := 0$. Define the function

$$d_p\colon \mathbb{Q} \times \mathbb{Q} \longrightarrow [0, \infty)$$

by

$$\begin{aligned} d_p(x, y) &:= |x - y|_p \\ &:= \inf\left\{p^{-k} \,\middle|\, k \in \mathbb{Z},\ x - y = p^k \frac{a}{b},\ a \in \mathbb{Z},\ b \in \mathbb{N} \setminus p\mathbb{N}\right\}. \end{aligned} \tag{8.5}$$

Then $(\mathbb{Q}, d_p)$ is a metric space. The completion of $\mathbb{Q}$ with respect to d_p is denoted by $\mathbb{Q}_p$ and is called the *field of p-adic rational numbers.* It can also be described as the quotient field of the ring of p-adic integers in Example 8.6. The multiplicative group of nonzero p-adic rationals is a locally compact Hausdorff group that is not a Lie group. One can also consider groups of matrices whose entries are p-adic rationals. Such groups play an important role in number theory.

Example 8.8 (infinite products). Let I be any index set and, for $i \in I$, let G_i be a compact Hausdorff group. Then the product

$$\mathrm{G} := \prod_{i \in I} \mathrm{G}_i$$

is a compact Hausdorff group. Its elements are maps $I \to \bigsqcup_{i \in I} \mathrm{G}_i : i \mapsto x_i$ such that $x_i \in \mathrm{G}_i$ for all $i \in I$. Write such a map as $x = (x_i)_{i \in I}$. The *product topology* on G is defined as the smallest topology such that the obvious projections $\pi_i : \mathrm{G} \to \mathrm{G}_i$ are continuous. Thus the (infinite) products $U = \prod_{i \in I} U_i$ of open sets $U_i \subset \mathrm{G}_i$, such that $U_i = \mathrm{G}_i$ for all, but finitely many i, form a basis for the topology of G. (See Appendix B for $\#I = 2$.) The product topology is obviously Hausdorff and *Tychonoff's theorem* asserts that it is compact (see Munkres [14]). An uncountable product of nontrivial groups G_i is not first countable.

Example 8.9. Let $(\mathcal{X}, \|\cdot\|)$ be a Banach algebra with a unit $\mathbb{1}$ and the product $\mathcal{X} \times \mathcal{X} \to \mathcal{X} : (x, y) \mapsto xy$. (See page 282.) Then the group of invertible elements $\mathcal{G} := \{x \in \mathcal{X} \mid \text{there exists } y \in \mathcal{X} \text{ such that } xy = yx = \mathbb{1}\}$ is a Hausdorff topological group. Examples include the group of nonzero quaternions, the general linear group of a finite-dimensional vector space, the group of bijective bounded linear operators on a Banach space, and the multiplicative group of nowhere vanishing real-valued continuous functions on a compact topological space. In general $\mathcal{G}$ is not locally compact.

8.2 Haar measures

Throughout let G be a locally compact Hausdorff group, in multiplicative notation, and denote by $\mathcal{B} \subset 2^{\mathrm{G}}$ its Borel σ-algebra. We begin our discussion with a technical lemma about continuous functions on G.

Lemma 8.10. *Let $f \in C_c(\mathrm{G})$ and fix a constant $\varepsilon > 0$. Then there exists an open neighborhood U of $\mathbb{1}$ such that, for all $x, y \in \mathrm{G}$,*

$$x^{-1}y \in U \implies |f(x) - f(y)| < \varepsilon. \tag{8.6}$$

Proof. Choose an open neighborhood $U_0 \subset \mathrm{G}$ with compact closure and define

$$K := \{ab^{-1} \mid a \in \mathrm{supp}(f),\ b \in \bar{U}_0\}.$$

This set is compact because the maps (8.1) and (8.2) are continuous. Also,

$$x \notin K,\ x^{-1}y \in U_0 \implies f(x) = f(y) = 0 \tag{8.7}$$

for all $x, y \in \mathrm{G}$. (If $y \in \mathrm{supp}(f)$ and $x^{-1}y \in U_0$, then $x = y(x^{-1}y)^{-1} \in K$.) Since f is continuous there exists, for each $x \in K$, an open neighborhood $V(x) \subset \mathrm{G}$ of x such that

$$y \in V(x) \implies |f(x) - f(y)| < \frac{\varepsilon}{2}. \tag{8.8}$$

Since the map $\mathrm{G} \to \mathrm{G}\colon y \mapsto x^{-1}y$, is a homeomorphism, the set $x^{-1}V(x)$ is an open neighborhood of $\mathbb{1}$ for every $x \in K$. Since the map (8.1) is continuous it follows from the definition of the product topology in Appendix B that, for every $x \in K$, there exists an open neighborhood $U(x) \subset \mathrm{G}$ of $\mathbb{1}$ such that the product neighborhood $U(x) \times U(x)$ of the pair $(\mathbb{1}, \mathbb{1})$ is contained in the pre-image of $x^{-1}V(x)$ under the multiplication map (8.1). In other words,

$$a, b \in U(x) \implies xab \in V(x). \tag{8.9}$$

Since the map $\mathrm{G} \to \mathrm{G}\colon y \mapsto xy$, is a homeomorphism, the set $xU(x)$ is an open neighborhood of x for every $x \in K$. Since K is compact, there exist finitely many elements $x_1, \dots, x_\ell \in K$ such that $K \subset \bigcup_{i=1}^{\ell} x_i U(x_i)$. Define

$$U := U_0 \cap \bigcap_{i=1}^{\ell} U(x_i)$$

and let $x, y \in \mathrm{G}$ such that $x^{-1}y \in U$. If $x \notin K$, then $f(x) = f(y) = 0$ by (8.7). Hence assume $x \in K$. Then there exists an index $i \in \{1, \dots, \ell\}$ such that $x \in x_i U(x_i)$ and therefore $x_i^{-1}x \in U(x_i)$. Hence $x = x_i(x_i^{-1}x)\mathbb{1} \in V(x_i)$ and $y = x_i(x_i^{-1}x)(x^{-1}y) \in V(x_i)$ by (8.9). Now, it follows from (8.8) that

$$|f(x) - f(y)| \le |f(x) - f(x_i)| + |f(x_i) - f(y)| < \varepsilon.$$

This proves Lemma 8.10. □

For $x \in \mathrm{G}$ define the homeomorphisms

$$L_x, R_x : \mathrm{G} \longrightarrow \mathrm{G}$$

by

$$L_x(a) := xa, \quad R_x(a) := ax \quad \text{for } x \in \mathrm{G}. \tag{8.10}$$

They satisfy

$$L_x \circ L_y = L_{xy}, \quad R_x \circ R_y = R_{yx}, \quad L_x \circ R_y = R_y \circ L_x. \tag{8.11}$$

For $A \subset \mathrm{G}$ and $x \in \mathrm{G}$ define

$$xA := \{xa \mid a \in A\}, \quad Ax := \{ax \mid a \in A\}, \quad A^{-1} := \{a^{-1} \mid a \in A\}.$$

Thus $xA = L_x(A)$ and $Ax = R_x(A)$.

Definition 8.11. A measure $\mu : \mathcal{B} \to [0, \infty]$ is called *left-invariant* if

$$\mu(xB) = \mu(B) \quad \text{for all } B \in \mathcal{B} \text{ and } x \in \mathrm{G}.$$

It is called *right-invariant* if

$$\mu(Bx) = \mu(B) \quad \text{for all } B \in \mathcal{B} \text{ and } x \in \mathrm{G}.$$

It is called *invariant* if it is both left- and right-invariant. A *left Haar measure on* G is a left-invariant Radon measure that does not vanish identically. A *right Haar measure on* G is a right-invariant Radon measure that does not vanish identically. An *invariant Haar measure on* G is an invariant Radon measure that does not vanish identically.

Theorem 8.12 (Haar). *Let* G *be a locally compact Hausdorff group. Then the following holds.*

(i) G *admits a left Haar measure μ, unique up to a positive factor. Every such measure satisfies $\mu(U) > 0$ for every nonempty open set $U \subset \mathrm{G}$.*

(ii) G *admits a right Haar measure μ, unique up to a positive factor. Every such measure satisfies $\mu(U) > 0$ for every nonempty open set $U \subset \mathrm{G}$.*

(iii) *Assume* G *is compact. Then* G *admits a unique invariant Haar measure μ such that $\mu(\mathrm{G}) = 1$. This measure satisfies $\mu(B^{-1}) = \mu(B)$ for all $B \in \mathcal{B}$ and $\mu(U) > 0$ for every nonempty open set $U \subset \mathrm{G}$.*

Proof. See page 327. □

Examples of Haar measures are the restriction to the Borel σ-algebra of the Lebesgue measure on $\mathbb{R}^n$ (where the group structure is additive), the restriction to the Borel σ-algebra of the measure σ on $S^1 = \mathrm{U}(1)$ or on $S^3 = \mathrm{Sp}(1)$ in Exercise 7.47, the measure dt/t on the multiplicative group of the positive real numbers, and the counting measure on any discrete group. The proof of Theorem 8.12 rests on the Riesz representation theorem (Theorem 3.15) and the following result about positive linear functionals.

Definition 8.13. Let G be a locally compact Hausdorff group. A linear functional $\Lambda\colon C_c(\mathrm{G}) \to \mathbb{R}$ is called *left-invariant* if

$$\Lambda(f \circ L_x) = \Lambda(f) \tag{8.12}$$

for all $f \in C_c(\mathrm{G})$ and all $x \in \mathrm{G}$. It is called *right-invariant*

$$\Lambda(f \circ R_x) = \Lambda(f) \tag{8.13}$$

for all $f \in C_c(\mathrm{G})$ and all $x \in \mathrm{G}$. It is called *invariant* if it is both left- and right-invariant. It is called a *left Haar integral* if it is left-invariant, positive, and nonzero. It is called a *right Haar integral* if it is right-invariant, positive, and nonzero.

Theorem 8.14 (Haar). *Every locally compact Hausdorff group* G *admits a left Haar integral, unique up to a positive factor. Moreover, if* $\Lambda\colon C_c(\mathrm{G}) \to \mathbb{R}$ *is a left Haar integral and* $f \in C_c(\mathrm{G})$ *is a nonnegative function that does not vanish identically, then* $\Lambda(f) > 0$.

Proof. See page 318. □

The proof of Theorem 8.14 given below follows the notes of Pedersen [16], which are based on a proof by Weil. Our exposition benefits from elegant simplifications due to Urs Lang [11]. In preparation for the proof it is convenient to introduce some notation. Let

$$C_c^+(\mathrm{G}) := \{f \in C_c(\mathrm{G}) \mid f \geq 0,\ f \not\equiv 0\} \tag{8.14}$$

be the space of nonnegative continuous functions with compact support that do not vanish identically. A function $\Lambda\colon C_c^+(\mathrm{G}) \to [0,\infty)$ is called

- *additive* if $\Lambda(f+g) = \Lambda(f) + \Lambda(g)$ for all $f, g \in C_c^+(\mathrm{G})$,
- *subadditive* if $\Lambda(f+g) \leq \Lambda(f) + \Lambda(g)$ for all $f, g \in C_c^+(\mathrm{G})$,
- *homogeneous* if $\Lambda(cg) = c\Lambda(f)$ for all $f \in C_c^+(\mathrm{G})$ and all $c > 0$,
- *monotone* if $f \leq g$ implies $\Lambda(f) \leq \Lambda(g)$ for all $f, g \in C_c^+(\mathrm{G})$,
- *left-invariant* if $\Lambda(f \circ L_x) = \Lambda(f)$ for all $f \in C_c^+(\mathrm{G})$ and all $x \in \mathrm{G}$.

Every additive functional $\Lambda\colon C_c^+(G) \to [0,\infty)$ is necessarily homogeneous and monotone. Moreover, every positive linear functional on $C_c(G)$ restricts to an additive functional $\Lambda\colon C_c^+(G) \to [0,\infty)$ and, conversely, every additive functional $\Lambda\colon C_c^+(G) \to [0,\infty)$ extends uniquely to a positive linear functional on $C_c(G)$. This is the content of the next lemma.

Lemma 8.15. *Let $\Lambda\colon C_c^+(G) \to [0,\infty)$ be an additive functional. Then there is a unique positive linear functional on $C_c(G)$ whose restriction to $C_c^+(G)$ agrees with Λ. If Λ is left-invariant, then so is its linear extension to $C_c(G)$.*

Proof. We prove that Λ is monotone. Let $f, g \in C_c^+(G)$ such that $f \le g$. If $f \neq g$, then $g - f \in C_c^+(G)$ and hence

$$\Lambda(f) \le \Lambda(f) + \Lambda(g-f) = \Lambda(g)$$

by additivity. If $f = g$ there is nothing to prove.

We prove that Λ is homogeneous. Let $f \in C_c^+(G)$. Then $\Lambda(nf) = n\Lambda(f)$ for all $n \in \mathbb{N}$ by additivity and induction. If $c = m/n$ is a positive rational number, then $\Lambda(f) = n\Lambda(f/n)$ and hence $\Lambda(cf) = m\Lambda(f/n) = c\Lambda(f)$. If $c > 0$ is irrational, then it follows from monotonicity that

$$a\Lambda(f) = \Lambda(af) \le \Lambda(cf) \le \Lambda(bf) = b\Lambda(f)$$

for all $a, b \in \mathbb{Q}$ with $0 < a < c < b$, and this implies $\Lambda(cf) = c\Lambda(f)$.

Now define

$$\Lambda(0) := 0$$

and, for $f \in C_c(G)$,

$$\Lambda(f) := \Lambda(f^+) - \Lambda(f^-).$$

If $f, g \in C_c(G)$, then $f^+ + g^+ + (f+g)^- = f^- + g^- + (f+g)^+$, hence

$$\Lambda(f^+) + \Lambda(g^+) + \Lambda((f+g)^-) = \Lambda(f^-) + \Lambda(g^-) + \Lambda((f+g)^+)$$

by additivity, and hence $\Lambda(f) + \Lambda(g) = \Lambda(f+g)$. Moreover, $(-f)^+ = f^-$ and $(-f)^- = f^+$, and so

$$\Lambda(-f) = \Lambda(f^-) - \Lambda(f^+) = -\Lambda(f).$$

Hence it follows from homogeneity that $\Lambda(cf) = c\Lambda(f)$ for all $f \in C_c(G)$ and all $c \in \mathbb{R}$. This shows that the extended functional is linear.

If the original functional $\Lambda\colon C_c^+(G) \to [0,\infty)$ is left-invariant, then so is the extended linear functional on $C_c(G)$ because $(f \circ L_x)^\pm = f^\pm \circ L_x$ for all $f \in C_c(G)$ and all $x \in G$. This proves Lemma 8.15. □

Consider the space

$$\mathcal{L} := \left\{ \Lambda \colon C_c^+(G) \to (0,\infty) \;\middle|\; \begin{array}{l} \Lambda \text{ is subadditive, monotone,} \\ \text{homogeneous, and left-invariant} \end{array} \right\}. \tag{8.15}$$

The strategy of the proof of Theorem 8.14 is to construct certain functionals $\Lambda_g \in \mathcal{L}$ associated to functions $g \in C_c^+(G)$ supported near the identity element and to construct the required positive linear functional $\Lambda \colon C_c(G) \to \mathbb{R}$ as a suitable limit where the functions g converge to the Dirac δ-function at the identity. The precise definition of the Λ_g involves the following construction.

Denote by $\mathcal{P}$ the set of all Borel measures $\mu \colon \mathcal{B} \to [0,\infty)$ of the form

$$\mu := \sum_{i=1}^{k} \alpha_i \delta_{x_i} \tag{8.16}$$

where $k \in \mathbb{N}$, $\alpha_1, \dots, \alpha_k$ are positive real numbers, $x_1, \dots, x_k \in G$, and δ_{x_i} is the Dirac measure at x_i (see Example 1.31). The norm of a measure $\mu \in \mathcal{P}$ of the form (8.16) is defined by

$$\|\mu\| := \mu(G) = \sum_{i=1}^{k} \alpha_i > 0. \tag{8.17}$$

If $\nu := \sum_{j=1}^{\ell} \beta_j \delta_{y_j} \in \mathcal{P}$ is any other such measure, define the convolution product of μ and ν by

$$\mu * \nu := \sum_{i=1}^{k} \sum_{j=1}^{\ell} \alpha_i \beta_j \delta_{x_i y_j}.$$

This product is not commutative in general. It satisfies $\|\mu * \nu\| = \|\mu\| \, \|\nu\|$. Associated to a measure $\mu \in \mathcal{P}$ of the form (8.16) are two linear operators

$$L_\mu, R_\mu \colon C_c(G) \longrightarrow C_c(G)$$

defined by

$$(L_\mu f)(a) := \sum_{i=1}^{k} \alpha_i f(x_i a), \quad (R_\mu f)(a) := \sum_{i=1}^{k} \alpha_i f(a x_i) \tag{8.18}$$

for $f \in C_c(G)$ and $a \in G$. The next two lemmas establish some basic properties of the operators L_μ and R_μ. Denote by

$$\|f\|_\infty := \sup_{x \in G} |f(x)|$$

the supremum norm of a compactly supported function $f \colon G \to \mathbb{R}$.

Lemma 8.16. *Let $\mu, \nu \in \mathcal{P}$, $f \in C_c(G)$, and $x \in G$. Then*

$$L_\mu \circ L_\nu = L_{\nu * \mu} \qquad f \circ L_x = L_{\delta_x} f, \qquad \|L_\mu f\|_\infty \le \|\mu\| \, \|f\|_\infty,$$
$$R_\mu \circ R_\nu = R_{\mu * \nu}, \qquad f \circ R_x = R_{\delta_x} f, \qquad \|R_\mu f\|_\infty \le \|\mu\| \, \|f\|_\infty,$$
$$L_\mu \circ R_\nu = R_\nu \circ L_\mu.$$

Proof. The assertions follow directly from the definitions. □

Lemma 8.17. *Let $f, g \in C_c^+(G)$. Then there exists a $\mu \in \mathcal{P}$ such that*

$$f \le L_\mu g.$$

Proof. Fix an element $y \in G$ such that $g(y) > 0$. For $x \in G$ define

$$U_x := \left\{ a \in G \,\middle|\, f(a) < \frac{f(x) + 1}{g(y)} g(yx^{-1}a) \right\}$$

This set is an open neighborhood of x. Since f has compact support, there exist finitely many points $x_1, \dots, x_k \in G$ such that $\mathrm{supp}(f) \subset U_{x_1} \cup \cdots \cup U_{x_k}$. Define

$$\mu := \sum_{i=1}^{k} \frac{f(x_i) + 1}{g(y)} \delta_{yx_i^{-1}}.$$

Then

$$f(a) \le \sum_{i=1}^{k} \frac{f(x_i) + 1}{g(y)} g(yx_i^{-1}a) = (L_\mu g)(a)$$

for all $a \in \mathrm{supp}(f)$ and hence $f \le L_\mu g$. This proves Lemma 8.17. □

Proof of Theorem 8.14. The proof has five steps. Step 1 is the main construction of a subadditive functional

$$M_g \colon C_c^+(G) \longrightarrow (0, \infty)$$

associated to a function $g \in C_c^+(G)$. Step 2 shows that M_g is asymptotically linear as g *concentrates* near the unit $\mathbb{1}$. The heart of the convergence proof is Step 3 and is due to Cartan. Step 4 proves uniqueness and Step 5 proves existence.

Step 1. *For $f \in C_c^+(G)$ define*

$$M_g(f) := M(f;g) := \inf\{\|\mu\| \mid \mu \in \mathcal{P},\ f \le L_\mu g\}. \tag{8.19}$$

Then the following holds.

(i) *$M(f;g) > 0$ for all $f, g \in C_c^+(G)$.*

(ii) *For every $g \in C_c^+(G)$, the functional $M_g : C_c^+(G) \to (0,\infty)$ is subadditive, homogeneous, monotone, and left-invariant, and hence is an element of $\mathcal{L}$.*

(iii) *Let $\Lambda \in \mathcal{L}$. Then*

$$\Lambda(f) \le M(f;g)\Lambda(g) \quad \textit{for all } f, g \in C_c^+(G). \tag{8.20}$$

In particular, $M(f;h) \le M(f;g)M(g;h)$ for all $f, g, h \in C_c^+(G)$.

(iv) *$M(f;f) = 1$ for all $f \in C_c^+(G)$.*

We prove part (ii). Monotonicity follows directly from the definition. Homogeneity follows from the identities $L_{c\mu}g = cL_\mu g$ and $\|c\mu\| = c\,\|\mu\|$. To prove left-invariance, observe that

$$(L_\mu g) \circ L_x = L_{\mu * \delta_x} g, \quad \|\mu * \delta_x\| = \|\mu\|$$

for all $\mu \in \mathcal{P}$ by Lemma 8.16. Since $f \le L_\mu g$ if and only if $f \circ L_x \le (L_\mu g) \circ L_x$, this proves left-invariance. To prove subadditivity, fix a constant $\varepsilon > 0$ and choose $\mu, \mu' \in \mathcal{P}$ such that

$$f \le L_\mu g, \quad f' \le L_{\mu'} g, \quad \|\mu\| < M(f;g) + \frac{\varepsilon}{2}, \quad \|\mu'\| < M(f';g) + \frac{\varepsilon}{2}.$$

Then $f + f' \le L_\mu g + L_{\mu'} g = L_{\mu+\mu'} g$ and hence

$$M(f+f';g) \le \|\mu + \mu'\| = \|\mu\| + \|\mu'\| < M(f;g) + M(f';g) + \varepsilon.$$

Thus

$$M(f+f';g) < M(f;g) + M(f';g) + \varepsilon \quad \text{for all } \varepsilon > 0.$$

This proves subadditivity and part (ii).

We prove part (iii). Fix a functional $\Lambda \in \mathcal{L}$. We prove first that

$$\Lambda(L_\mu f) \leq \|\mu\| \, \Lambda(f) \tag{8.21}$$

for all $f \in C_c^+(\mathrm{G})$ and all $\mu \in \mathcal{P}$. To see this write $\mu = \sum_{i=1}^k \alpha_i \delta_{x_i}$. Then $L_\mu f = \sum_{i=1}^k \alpha_i (f \circ L_{x_i})$, and hence

$$\begin{aligned} \Lambda(L_\mu f) &\leq \sum_{i=1}^k \Lambda(\alpha_i (f \circ L_{x_i})) \\ &= \sum_{i=1}^k \alpha_i \Lambda(f \circ L_{x_i}) \\ &= \sum_{i=1}^k \alpha_i \Lambda(f) \\ &= \|\mu\| \, \Lambda(f). \end{aligned}$$

Here the first step follows from subadditivity, the second step follows from homogeneity, the third step follows from left-invariance, and the last step follows from the definition of $\|\mu\|$. This proves (8.21). Now let $f, g \in C_c^+(\mathrm{G})$. By Lemma 8.17, there is a $\mu \in \mathcal{P}$ such that $f \leq L_\mu g$. Since Λ is monotone this implies $\Lambda(f) \leq \Lambda(L_\mu g) \leq \|\mu\| \, \Lambda(g)$ by (8.21). Now take the infimum over all $\mu \in \mathcal{P}$ such that $f \leq L_\mu g$ to obtain $\Lambda(f) \leq M(f; g)\Lambda(g)$.

We prove parts (i) and (iv). Since the map $C_c^+(\mathrm{G}) \to (0, \infty)\colon f \mapsto \|f\|_\infty$, is an element of $\mathcal{L}$, it follows from part (iii) that

$$\|f\|_\infty \leq M(f; g) \, \|g\|_\infty \tag{8.22}$$

and hence $M(f; g) > 0$ for all $f, g \in C_c^+(\mathrm{G})$. Next observe that $M(f; f) \geq 1$ by (8.22) and $M(f; f) \leq 1$ because $f = L_{\delta_{\mathbb{1}}} f$. This proves Step 1.

Step 2. *Let $f, f' \in C_c^+(\mathrm{G})$ and let $\varepsilon > 0$. Then there is an open neighborhood $U \subset \mathrm{G}$ of $\mathbb{1}$ such that every $g \in C_c^+(\mathrm{G})$ with* $\mathrm{supp}(g) \subset U$ *satisfies*

$$M_g(f) + M_g(f') < (1 + \varepsilon) M_g(f + f'). \tag{8.23}$$

By Urysohn's lemma (Theorem A.1), there is a function $\rho \in C_c(\mathrm{G})$ such that $\rho(x) = 1$ for all $x \in \mathrm{supp}(f) \cup \mathrm{supp}(f')$. Choose a constant $0 < \delta \leq 1/2$ such that

$$2\delta + 2\delta \|f + f'\|_\infty M(\rho; f + f') < \varepsilon. \tag{8.24}$$

Define

$$h := f + f' + \delta\|f + f'\|_\infty \rho.$$

Then f/h and f'/h extend to continuous functions on G with compact support by setting them equal to zero on $\mathrm{G} \setminus \mathrm{supp}(\rho)$. By Lemma 8.10, there exists an open neighborhood $U \subset \mathrm{G}$ of $\mathbb{1}$ such that

$$x^{-1}y \in U \implies \left|\frac{f(x)}{h(x)} - \frac{f(y)}{h(y)}\right| + \left|\frac{f'(x)}{h(x)} - \frac{f'(y)}{h(y)}\right| < \delta$$

for all $x, y \in \mathrm{G}$. Let $g \in C_c^+(\mathrm{G})$ with $\mathrm{supp}(g) \subset U$. If $\mu = \sum_{i=1}^{\ell} \alpha_i \delta_{x_i} \in \mathcal{P}$ such that $h \leq L_\mu g$, then for all $a \in \mathrm{supp}(f)$,

$$f(a) \leq \frac{L_\mu g(a)}{h(a)} f(a) = \sum_{i=1}^{\ell} \alpha_i \frac{f(a)}{h(a)} g(x_i a) \leq \sum_{i=1}^{\ell} \alpha_i \Big(\frac{f(x_i^{-1})}{h(x_i^{-1})} + \delta\Big) g(x_i a).$$

Thus $f \leq \mathcal{L}_\nu g$, where

$$\nu := \sum_{i=1}^{\ell} \alpha_i \Big(\frac{f(x_i^{-1})}{h(x_i^{-1})} + \delta\Big) \delta_{x_i}.$$

This implies

$$M_g(f) \leq \sum_{i=1}^{\ell} \alpha_i \Big(\frac{f(x_i^{-1})}{h(x_i^{-1})} + \delta\Big).$$

The same inequality holds for f'. Since $f + f' \leq h$, we find

$$M_g(f) + M_g(f') \leq \sum_{i=1}^{\ell} \alpha_i \Big(\frac{f(x_i^{-1}) + f'(x_i^{-1})}{h(x_i^{-1})} + 2\delta\Big) \leq \|\mu\| \,(1 + 2\delta).$$

Now take the infimum over all $\mu \in \mathcal{P}$ such that $h \leq L_\mu g$ to obtain

$$\begin{aligned} M_g(f) + M_g(f') &\leq (1 + 2\delta) M_g(h) \\ &\leq (1 + 2\delta)(M_g(f + f') + \delta\|f + f'\|_\infty M_g(\rho)) \\ &\leq (1 + 2\delta + 2\delta\|f + f'\|_\infty M(\rho; f + f')) M_g(f + f') \\ &\leq (1 + \varepsilon) M_g(f + f'). \end{aligned}$$

Here we have used the inequalities $1 + 2\delta \leq 2$ and (8.24). This proves Step 2.

Step 3. *Let $f \in C_c^+(G)$ and let $\varepsilon > 0$. Then there is an open neighborhood $U \subset G$ of $\mathbb{1}$ with the following significance. For every $g \in C_c^+(G)$ such that*

$$\operatorname{supp}(g) \subset U, \quad g(x) = g(x^{-1}) \quad \text{for all } x \in G, \tag{8.25}$$

there exists an open neighborhood $W \subset G$ of $\mathbb{1}$ such that every $h \in C_c^+(G)$ with $\operatorname{supp}(h) \subset W$ satisfies the inequality

$$M(f; g) M_h(g) \le (1 + \varepsilon) M_h(f). \tag{8.26}$$

This inequality continues to hold with M_h replaced by any left-invariant positive linear functional $\Lambda\colon C_c(G) \to \mathbb{R}$.

By Urysohn's lemma (Theorem A.1) there is a function $\rho \in C_c^+(G)$ such that $\rho(x) = 1$ for all $x \in K := \operatorname{supp}(f)$. Choose ε_0 and ε_1 such that

$$0 < \varepsilon_0 < 1, \quad \frac{1 + \varepsilon_0}{1 - \varepsilon_0} \le 1 + \varepsilon, \quad \varepsilon_1 := \frac{\varepsilon_0}{2M(\rho; f)}. \tag{8.27}$$

By Lemma 8.10, there exists an open neighborhood $U \subset G$ of $\mathbb{1}$ such that

$$x^{-1}y \in U \implies |f(x) - f(y)| < \varepsilon_1 \tag{8.28}$$

for all $x, y \in G$. We prove that the assertion of Step 3 holds with this neighborhood U. Fix a function $g \in C_c^+(G)$ that satisfies (8.25). Put

$$\varepsilon_2 := \frac{\varepsilon_1}{2M(f; g)}. \tag{8.29}$$

Use Lemma 8.10 to find an open neighborhood $V \subset G$ of $\mathbb{1}$ such that

$$xy^{-1} \in V \implies |g(x) - g(y)| < \varepsilon_2 \tag{8.30}$$

for all $x, y \in G$. Then the sets xV for $x \in K$ form an open cover of K. Hence there exist finitely many points $x_1, \dots, x_\ell \in K$ such that $K \subset \bigcup_{i=1}^{\ell} x_i V$. By Theorem A.4, there exist functions $\rho_1, \dots, \rho_\ell \in C_c^+(G)$ such that

$$0 \le \rho_i \le 1, \quad \operatorname{supp}(\rho_i) \subset x_i V, \quad \sum_{i=1}^{\ell} \rho_i|_K \equiv 1. \tag{8.31}$$

It follows from Step 2 by induction that there exists an open neighborhood $W \subset G$ of $\mathbb{1}$ such that every $h \in C_c^+(G)$ with $\operatorname{supp}(h) \subset W$ satisfies

$$\sum_{i=1}^{\ell} M_h(\rho_i f) < (1 + \varepsilon_0)\, M_h(f). \tag{8.32}$$

We prove that every $h \in C_c^+(G)$ with $\operatorname{supp}(h) \subset W$ satisfies (8.26).

For $x \in \mathrm{G}$ define the function $F_x \in C_c(\mathrm{G})$ by

$$F_x(y) := f(y)g(y^{-1}x) \quad \text{for } y \in \mathrm{G}. \tag{8.33}$$

It follows from (8.25) and (8.28) that $f(x)g(y^{-1}x) - f(y)g(y^{-1}x) \le \varepsilon_1 g(y^{-1}x)$ for all $x, y \in \mathrm{G}$. Since $g(y^{-1}x) = g(x^{-1}y) = (g \circ L_{x^{-1}})(y)$ by (8.25), this implies $f(x)g \circ L_{x^{-1}} \le F_x + \varepsilon_1 g \circ L_{x^{-1}}$. Hence, for all $x \in \mathrm{G}$ and all $h \in C_c^+(\mathrm{G})$,

$$f(x)M_h(g) \le M_h(F_x) + \varepsilon_1 M_h(g) \tag{8.34}$$

Now fix a function $h \in C_c^+(\mathrm{G})$ with $\mathrm{supp}(h) \subset W$. By (8.30) and (8.31),

$$\rho_i(y)F_x(y) = \rho_i(y)f(y)g(y^{-1}x) \le \rho_i(y)f(y)(g(x_i^{-1}x) + \varepsilon_2)$$

for all $x, y \in \mathrm{G}$ and all $i = 1, \dots, \ell$. Since $F_x = \sum_i \rho_i F_x$, this implies

$$\begin{aligned} M_h(F_x) &\le \sum_i M_h(\rho_i F_x) \\ &\le \sum_i M_h(\rho_i f)(g(x_i^{-1}x) + \varepsilon_2) \\ &\le \sum_i M_h(\rho_i f)g(x_i^{-1}x) + 2\varepsilon_2 M_h(f). \end{aligned} \tag{8.35}$$

Here the last step uses (8.32). By (8.34) and (8.35),

$$\begin{aligned} f(x)M_h(g) &\le \sum_i M_h(\rho_i f)g(x_i^{-1}x) + 2\varepsilon_2 M_h(f) + \varepsilon_1 M_h(g) \\ &\le \sum_i M_h(\rho_i f)g(x_i^{-1}x) + 2\varepsilon_1 M_h(g). \end{aligned}$$

Here the second step uses (8.29) and the inequality

$$M_h(f) \le M(f; g)M_h(g).$$

Thus

$$(f - 2\varepsilon_1)^+ M_h(g) \le L_\mu g,$$

where

$$\mu := \sum_i M_h(\rho_i f)\delta_{x_i^{-1}}.$$

This implies

$$M_g((f - 2\varepsilon_1)^+)M_h(g) \le \sum_i M_h(\rho_i f) \le (1 + \varepsilon_0)M_h(f).$$

Here the second step uses (8.32). Since $f \le (f - 2\varepsilon_1)^+ + 2\varepsilon_1\rho$, we have

$$\begin{aligned} M_g(f)M_h(g) &\le M_g((f-2\varepsilon_1)^+)M_h(g) + 2\varepsilon_1 M_g(\rho)M_h(g) \\ &\le (1+\varepsilon_0)M_h(f) + 2\varepsilon_1 M(\rho; f)M_g(f)M_h(g) \\ &= (1+\varepsilon_0)M_h(f) + \varepsilon_0 M_g(f)M_h(g). \end{aligned}$$

Hence

$$M_g(f)M_h(g) \le \frac{1+\varepsilon_0}{1-\varepsilon_0}M_h(f) \le (1+\varepsilon)M_h(f)$$

and this proves Step 3 for M_h. This inequality and its proof carry over to any left-invariant positive linear functional $\Lambda\colon C_c(\mathrm{G}) \to \mathbb{R}$.

Step 4. *We prove uniqueness.*

Let $\Lambda, \Lambda'\colon C_c(\mathbb{R}) \to \mathbb{R}$ be two left-invariant positive linear functionals that do not vanish identically. Then there exists a function $f \in C_c^+(\mathrm{G})$ such that $\Lambda(f) > 0$ by Lemma 8.15. Hence

$$\Lambda(g) \ge M(f; g)^{-1}\Lambda(f) > 0$$

for all $g \in C_c^+(\mathrm{G})$ by (8.20). The same argument shows that $\Lambda'(g) > 0$ for all $g \in C_c^+(\mathrm{G})$.

Now fix two functions $f, f_0 \in C_c^+(\mathrm{G})$ and a constant $\varepsilon > 0$. Choose an open neighborhood $U \subset \mathrm{G}$ of $\mathbb{1}$ that satisfies the requirements of Step 3 for both f and f_0 and this constant ε. By Urysohn's lemma (Theorem A.1), there exists a function $g \in C_c^+(\mathrm{G})$ such that

$$g(\mathbb{1}) > 0, \quad \operatorname{supp}(g) \subset \{x \in \mathrm{G} \mid x \in U \text{ and } x^{-1} \in U\}.$$

Replacing g by the function $x \mapsto g(x) + g(x^{-1})$, if necessary, we may assume that g satisfies (8.25). Hence it follows from Step 1 and Step 3 that

$$\Lambda(f) \le M(f; g)\Lambda(g) \le (1+\varepsilon)\Lambda(f)$$

and

$$(1+\varepsilon)\Lambda(f_0) \ge M(f_0; g)\Lambda(g) \ge \Lambda(f_0).$$

Take the quotient of these inequalities to obtain

$$(1+\varepsilon)^{-1}\frac{\Lambda(f)}{\Lambda(f_0)} \le \frac{M(f;g)}{M(f_0;g)} \le (1+\varepsilon)\frac{\Lambda(f)}{\Lambda(f_0)}.$$

The same inequality holds with Λ replaced by Λ'. Hence

$$(1+\varepsilon)^{-2}\frac{\Lambda(f)}{\Lambda(f_0)} \le \frac{\Lambda'(f)}{\Lambda'(f_0)} \le (1+\varepsilon)^2\frac{\Lambda(f)}{\Lambda(f_0)}.$$

Since this holds for all $\varepsilon > 0$, it follows that

$$\Lambda'(f) = c\Lambda(f), \quad c := \frac{\Lambda'(f_0)}{\Lambda(f_0)}.$$

Since this holds for all $f \in C_c^+(\mathrm{G})$, Λ' and $c\Lambda$ agree on $C_c^+(\mathrm{G})$. Hence $\Lambda' = c\Lambda$ by Lemma 8.15. This proves Step 4.

Step 5. *We prove existence.*

The proof follows the elegant exposition [11] by Urs Lang. Fix a reference function $f_0 \in C_c^+(\mathrm{G})$ and, for $g \in C_c^+(\mathrm{G})$, define

$$\Lambda_g : C_c^+(\mathrm{G}) \longrightarrow (0, \infty)$$

by

$$\Lambda_g(f) := \frac{M(f; g)}{M(f_0; g)} \quad \text{for } f \in C_c^+(\mathrm{G}). \tag{8.36}$$

Then $\Lambda_g \in \mathcal{L}$ for all $g \in C_c^+(\mathrm{G})$ by Step 1. It follows also from Step 1 that $M(f_0; g) \le M(f_0; f)M(f; g)$ and $M(f; g) \le M(f; f_0)M(f_0; g)$ and hence

$$M(f_0; f)^{-1} \le \Lambda_g(f) \le M(f; f_0) \tag{8.37}$$

for all $f, g \in C_c^+(\mathrm{G})$. Fix a function $f \in C_c^+(\mathrm{G})$ and a number $\varepsilon > 0$. Define

$$\mathcal{L}_\varepsilon(f) := \left\{ \Lambda \in \mathcal{L} \;\middle|\; \begin{array}{l} \Lambda(f_0) = 1 \text{ and there exists a neighborhood} \\ W \subset \mathrm{G} \text{ of } \mathbb{1} \text{ such that for all } h \in C_c^+(\mathrm{G}) \\ \mathrm{supp}(h) \subset W \implies \Lambda(f) \le (1+\varepsilon)\Lambda_h(f) \end{array} \right\}.$$

We prove that $\mathcal{L}_\varepsilon(f) \ne \emptyset$. To see this let $U \subset \mathrm{G}$ be the open neighborhood of $\mathbb{1}$ constructed in Step 3 for f and ε. Choose a function $g \in C_c^+(\mathrm{G})$ that satisfies (8.25) and choose an open neighborhood $W \subset \mathrm{G}$ of $\mathbb{1}$ associated to g that satisfies the requirements of Step 3. Let $h \in C_c^+(\mathrm{G})$ with $\mathrm{supp}(h) \subset W$. Then $M(f; g)M(g; h) \le (1+\varepsilon)M(f; h)$ and $M(f_0; g)M(g; h) \ge M(f_0; h)$ by Step 3 and Step 1. Take the quotient of these inequalities to obtain $\Lambda_g(f) \le (1+\varepsilon)\Lambda_h(f)$. Since $\Lambda_g(f_0) = 1$, it follows that $\Lambda_g \in \mathcal{L}_\varepsilon(f)$. This shows that $\mathcal{L}_\varepsilon(f) \ne \emptyset$, as claimed. Next we observe that

$$\Lambda(f) \le M(f; f_0)\Lambda(f_0) = M(f; f_0)$$

for all $\Lambda \in \mathcal{L}_\varepsilon(f)$ by Step 1. Hence the supremum

$$\Lambda_\varepsilon(f) := \sup\{\Lambda(f) \mid \Lambda \in \mathcal{L}_\varepsilon(f)\} \tag{8.38}$$

is a real number, bounded above by $M(f;f_0)$. Since $\mathcal{L}_\varepsilon(f)$ contains an element of the form Λ_g for some $g \in C_c^+(\mathrm{G})$, it follows from (8.37) that

$$M(f_0;f)^{-1} \le \Lambda_\varepsilon(f) \le M(f;f_0) \tag{8.39}$$

for all $f \in C_c^+(\mathrm{G})$ and all $\varepsilon > 0$. Moreover, the function $\varepsilon \mapsto \Lambda_\varepsilon(f)$ is nondecreasing by definition. Hence the limit

$$\Lambda_0(f) := \lim_{\varepsilon\to 0} \Lambda_\varepsilon(f) = \inf_{\varepsilon>0} \Lambda_\varepsilon(f) \tag{8.40}$$

exists and is a positive real number for every $f \in C_c^+(\mathrm{G})$.

We prove that the functional $\Lambda_0 \colon C_c^+(\mathrm{G}) \to (0,\infty)$ is left-invariant. To see this, fix a function $f \in C_c^+(\mathrm{G})$ and an element $x \in \mathrm{G}$. Then

$$\mathcal{L}_\varepsilon(f) = \mathcal{L}_\varepsilon(f \circ L_x)$$

for all $\varepsilon > 0$. Namely, if $W \subset \mathrm{G}$ is an open neighborhood of $\mathbb{1}$ such that $\Lambda(f) \le (1+\varepsilon)\Lambda_h(f)$ for all $h \in C_c^+(\mathrm{G})$ with $\mathrm{supp}(h) \subset W$, then the same inequality holds with f replaced by $f \circ L_x$ because both Λ and Λ_h are left-invariant. Hence $\Lambda_\varepsilon(f) = \Lambda_\varepsilon(f \circ L_x)$ for all $\varepsilon > 0$, and so $\Lambda_0(f) = \Lambda_0(f \circ L_x)$.

We prove that the functional $\Lambda_0 \colon C_c^+(\mathrm{G}) \to (0,\infty)$ is additive. To see this, fix two functions $f, f' \in C_c^+(\mathrm{G})$. We show that

$$(1+\varepsilon)^{-1}\Lambda_\varepsilon(f+f') \le \Lambda_\varepsilon(f) + \Lambda_\varepsilon(f') \le (1+\varepsilon)\Lambda_\varepsilon(f+f') \tag{8.41}$$

for all $\varepsilon > 0$. To establish the first inequality in (8.41), choose any functional $\Lambda \in \mathcal{L}_\varepsilon(f+f')$. Then there exists an open neighborhood $W \subset \mathrm{G}$ of $\mathbb{1}$ such that $\Lambda(f+f') \le (1+\varepsilon)\Lambda_h(f+f')$ for all $h \in C_c^+(\mathrm{G})$ with $\mathrm{supp}(h) \subset W$. Moreover, we have seen above that $h \in C_c^+(\mathrm{G})$ can be chosen such that $\mathrm{supp}(h) \subset W$ and also $\Lambda_h \in \mathcal{L}_\varepsilon(f) \cap \mathcal{L}_\varepsilon(f')$. Any such h satisfies

$$(1+\varepsilon)^{-1}\Lambda(f+f') \le \Lambda_h(f+f') \le \Lambda_h(f) + \lambda_h(f') \le \Lambda_\varepsilon(f) + \Lambda_\varepsilon(f').$$

Take the supremum over all $\Lambda \in \mathcal{L}_\varepsilon(f+f')$ to obtain the first inequality in (8.41). To prove the second inequality in (8.41), fix a constant $\alpha > 1$ and choose two functionals $\Lambda \in \mathcal{L}_\varepsilon(f)$ and $\Lambda' \in \mathcal{L}_\varepsilon(f')$. Then there exists an open neighborhood $W \subset \mathrm{G}$ of $\mathbb{1}$ such that $\Lambda(f) \le (1+\varepsilon)\Lambda_h(f)$ and $\Lambda'(f') \le (1+\varepsilon)\Lambda_h(f')$ for all $h \in C_c^+(\mathrm{G})$ with $\mathrm{supp}(h) \subset W$. By Step 2, the function $h \in C_c^+(\mathrm{G})$ can be chosen such that $\mathrm{supp}(h) \subset W$ and also $\Lambda_h(f) + \Lambda_h(f') \le \alpha\Lambda_h(f+f')$ and $\Lambda_h \in \mathcal{L}_\varepsilon(f+f')$. Any such h satisfies

$$\begin{aligned}(1+\varepsilon)^{-1}(\Lambda(f)+\Lambda'(f')) &\le \Lambda_h(f) + \Lambda_h(f') \\ &\le \alpha\Lambda_h(f+f') \\ &\le \alpha\Lambda_\varepsilon(f+f').\end{aligned}$$

Take the supremum over all pairs of functionals $\Lambda \in \mathcal{L}_\varepsilon(f)$ and $\Lambda' \in \mathcal{L}_\varepsilon(f')$ to obtain $(1-\varepsilon)^{-1}(\Lambda_\varepsilon(f) + \Lambda_\varepsilon(f')) \le \alpha\Lambda_\varepsilon(f + f')$ for all $\alpha > 1$. This proves the second inequality in (8.41). Take the limit $\varepsilon \to 0$ in (8.41) to obtain that Λ_0 is additive. Moreover, it follows directly from the definition that $\Lambda_0(f_0) = 1$. Hence by Lemma 8.15, Λ_0 extends to a nonzero left-invariant positive linear functional on $C_c(\mathrm{G})$. This proves Step 5 and Theorem 8.14. □

If one is prepared to use some abstract concepts from general topology, then the existence proof in Theorem 8.14 is essentially complete after Step 2. This approach is taken in Pedersen [16]. In this language the space

$$\mathcal{G} := \{g \in C_c^+(\mathrm{G}) \mid 0 \le g \le 1,\ g(\mathbb{1}) = 1\}$$

is a directed set equipped with a map $g \mapsto \Lambda_g$ that takes values in the space

$$\mathcal{L} := \left\{ \Lambda\colon C_c^+(\mathrm{G}) \to \mathbb{R} \;\middle|\; \begin{array}{l} M(f_0; f)^{-1} \le \Lambda(f) \le M(f_0; f) \\ \text{for all } f \in C_c^+(\mathrm{G}) \end{array} \right\}.$$

The map $\mathcal{G} \to \mathcal{L}\colon g \mapsto \Lambda_g$ is a net. A net can be thought of as an uncountable analogue of a sequence and a subnet as an analogue of a subsequence. The existence of a universal subnet is guaranteed by the general theory and its convergence for each f by the fact that the target space is compact. Instead Step 3 in the proof of Theorem 8.14 implies that the original net $g \mapsto \Lambda_g$ converges and so there is no need to choose a universal subnet. That this can be proved with a refinement of the uniqueness argument (Λ in Step 3) is pointed out in Pedersen [16]. That paper also contains two further uniqueness proofs. One is based on Fubini's Theorem and the other on the Radon–Nikodým theorem. Another existence proof for compact second countable Hausdorff groups is due to Pontryagin. It uses the Arzelà–Ascoli theorem to establish the existence of a sequence $\mu_i \in \mathcal{P}$ with $\|\mu_i\| = 1$ such that $L_{\mu_i} f$ converges to a constant function whose value is then taken to be $\Lambda(f)$.

Proof of Theorem 8.12. Existence and uniqueness in (i) follow directly from Theorem 8.14 and the Riesz representation theorem (Theorem 3.15). That nonempty open sets have positive measure follows from Urysohn's lemma (Theorem A.1). To prove (ii), consider the map

$$\phi\colon \mathrm{G} \longrightarrow \mathrm{G}$$

defined by

$$\phi(x) := x^{-1} \quad \text{for } x \in \mathrm{G}.$$

Since ϕ is a homeomorphism, it preserves the Borel σ-algebra $\mathcal{B}$. Since

$$\phi \circ R_x = L_{x^{-1}} \circ \phi,$$

a measure $\mu\colon \mathcal{B} \to [0, \infty]$ is a left Haar measure if and only if the measure

$$\nu\colon \mathcal{B} \longrightarrow [0, \infty]$$

defined by

$$\nu(B) := \mu(\phi(B)) = \mu(B^{-1})$$

is a right Haar measure. Hence assertion (ii) follows from (i).

We prove (iii). Assume G is compact and let $\mu\colon \mathcal{B} \to [0, 1]$ be the unique left Haar measure such that $\mu(\mathrm{G}) = 1$. For $x \in \mathrm{G}$ define

$$\mu_x\colon \mathcal{B} \longrightarrow [0, 1]$$

by

$$\mu_x(B) := \mu(R_x(B)) \quad \text{for } B \in \mathcal{B}.$$

Since R_x commutes with L_y for all y by (8.11), μ_x is a left Haar measure. Since $\mu_x(\mathrm{G}) = \mu(R_x(\mathrm{G})) = \mu(\mathrm{G}) = 1$, it follows that $\mu_x = \mu$ for all $x \in \mathrm{G}$. Hence μ is right-invariant. Therefore, the map

$$\mathcal{B} \longrightarrow [0, 1]\colon B \longmapsto \nu(B) := \mu(\phi(B)) = \mu(B^{-1})$$

is a left Haar measure and, since $\nu(\mathrm{G}) = 1$, it agrees with μ. This proves Theorem 8.12. □

In the noncompact case the left and right Haar measures need not agree. The above argument then shows that the measure μ_x differs from μ by a positive factor. Thus there exists a unique map $\rho\colon \mathrm{G} \to (0, \infty)$ such that

$$\mu(R_x(B)) = \rho(x)\mu(B) \tag{8.42}$$

for all $x \in \mathrm{G}$ and all $B \in \mathcal{B}$. The map $\rho\colon \mathrm{G} \to (0, \infty)$ in (8.42) is a continuous group homomorphism, called the *modular character*. It is independent of the choice of μ. A locally compact Hausdorff group is called *unimodular* if its modular character is trivial or, equivalently, if its left and right Haar measures agree. Thus every compact Hausdorff group is unimodular.

Exercise 8.18. Prove that ρ is a continuous homomorphism.

Exercise 8.19. Prove that the group of all real 2×2-matrices of the form

$$\begin{pmatrix} a & b \\ 0 & 1 \end{pmatrix}, \qquad a, b \in \mathbb{R}, a > 0,$$

is not unimodular. Prove that the additive group $\mathbb{R}^n$ is unimodular. Prove that every discrete group is unimodular.

Exercise 8.20. Let $\nu\colon \mathcal{B} \to [0, \infty]$ be a right Haar measure. Show that the modular character is characterized by the condition $\nu(L_{x^{-1}}(B)) = \rho(x)\nu(B)$ for all $x \in \mathrm{G}$ and all $B \in \mathcal{B}$.

Haar measures are extremely useful tools in geometry, especially when the group in question is compact. For example, if a compact Hausdorff group G acts on a topological space X continuously via

$$\mathrm{G} \times X \longrightarrow X\colon (g, x) \longmapsto g_* x \tag{8.43}$$

one can use the Haar measure to produce G-invariant continuous functions by *averaging*. Namely, if $f\colon X \to \mathbb{R}$ is any continuous function, and μ is the Haar measure on G with $\mu(\mathrm{G}) = 1$, then the function $F\colon X \to \mathbb{R}$ defined by

$$F(x) := \int_{\mathrm{G}} f(a_* x)\, d\mu(a) \tag{8.44}$$

for $x \in X$ is G-invariant in that

$$F(g_* x) = F(x)$$

for all $x \in X$ and all $g \in \mathrm{G}$.

Exercise 8.21. Give a precise definition of what it means for a topological group to act continuously on a topological space.

Exercise 8.22. Show that the map F in (8.44) is continuous and G-invariant.

Exercise 8.23. Let $\rho\colon \mathrm{G} \to \mathrm{GL}(V)$ be a homomorphism from a compact Hausdorff group to the general linear group of automorphisms of a finite-dimensional vector space. (Such a homomorphism is called a *representation* of G.) Prove that V admits a G-invariant inner product. This observation does not extend to noncompact groups. Show that the standard representation of $\mathrm{SL}(2, \mathbb{R})$ on $\mathbb{R}^2$ does not admit an invariant inner product.

Exercise 8.24. Show that the Haar measure on a discrete group is a multiple of the counting measure. Deduce that for a finite group the formula (8.44) defines $F(x)$ as the average (with multiplicities) of the values of f over the group orbit of x.

Exercise 8.25. Let G be a locally compact Hausdorff group and let μ be a left Haar measure on G. Define the convolution product on $L^1(\mu)$. Show that $L^1(\mu)$ is a Banach algebra. (See page 282.) Find conditions under which $f * g = g * f$. Show that the convolution is not commutative in general. *Hint.* See Section 7.5 for $G = \mathbb{R}^n$. See also Step 3 in the proof of Theorem 8.14.

Appendices

Appendix A

Urysohn's lemma

Theorem A.1 (Urysohn's lemma). *Let X be a locally compact Hausdorff space and let $K \subset X$ be a compact set and $U \subset X$ be an open set such that*

$$K \subset U.$$

Then there exists a compactly supported continuous function

$$f: X \longrightarrow [0, 1]$$

such that

$$f|_K \equiv 1, \quad \operatorname{supp}(f) = \overline{\{x \in X \mid f(x) \neq 0\}} \subset U. \tag{A.1}$$

Proof. See page 334. □

Lemma A.2. *Let X be a topological space and let $K \subset X$ be compact. Then the following holds*:

(i) *every closed subset of K is compact*;

(ii) *if X is Hausdorff, then for every $y \in X \setminus K$, there exist open sets $U, V \subset X$ such that $K \subset U$, $y \in V$, and $U \cap V = \emptyset$*;

(iii) *if X is Hausdorff, then K is closed.*

Proof. (i) Let $F \subset K$ be closed and let $\{U_i\}_{i \in I}$ be an open cover of F. Then the sets $\{U_i\}_{i \in I}$ together with $V := X \setminus F$ form an open cover of K. Therefore, there exist finitely many indices $i_1, \dots, i_n$ such that the sets $U_{i_1}, \dots, U_{i_n}, V$ cover K. Hence $F \subset U_{i_1} \cup \dots \cup U_{i_n}$. This shows that every open cover of F has a finite subcover, and so F is compact.

(ii) Assume X is Hausdorff and let $y \in X \setminus K$. Define

$$\mathcal{U} := \{U \subset X \mid U \text{ is open and } y \notin \bar{U}\}.$$

Since X is Hausdorff, the collection $\mathcal{U}$ is an open cover of K. Since K is compact, there exists finitely many set $U_1, \dots, U_n \in \mathcal{U}$ such that

$$K \subset U := U_1 \cup \cdots \cup U_n.$$

Since $y \notin \bar{U}_i$ for all i, it follows that $y \in V := X \setminus \bar{U}$ and $U \cap V = \emptyset$. Hence the sets U and V satisfy the requirements of (ii).

(iii) Assume X is Hausdorff. Then it follows from part (ii) that, for every $y \in X \setminus K$, there exists an open set $V \subset X$ such that $y \in V$ and $V \cap K = \emptyset$. Hence $X \setminus K$ is the union of all open sets in X that are disjoint from K. Thus $X \setminus K$ is open and so K is closed. This proves Lemma A.2. □

Lemma A.3. *Let X be a locally compact Hausdorff space and let K, U be subsets of X such that K is compact, U is open, and $K \subset U$. Then there exists an open set $V \subset X$ such that $\bar{V}$ is compact and*

$$K \subset V \subset \bar{V} \subset U. \tag{A.2}$$

Proof. We first prove the assertion in the case where $K = \{x\}$ consist of a single element. Choose a compact neighborhood $B \subset X$ of x. Then $F := B \setminus U$ is a closed subset of B and hence is compact by part (i) of Lemma A.2. Since $x \notin F$, it follows from part (ii) of Lemma A.2 that there exist open sets $W, W' \subset X$ such that $x \in W$, $F \subset W'$ and $W \cap W' = \emptyset$. Hence $V := \mathrm{int}(B) \cap W$ is an open neighborhood of x, its closure is a closed subset of B and hence compact, and

$$\bar{V} \subset B \cap \bar{W} \subset B \setminus W' \subset B \setminus F \subset U.$$

This proves the lemma in the case $\#K = 1$.

Now consider the general case. By the first part of the proof, the open sets whose closures are compact and contained in U form an open cover of K. Since K is compact, there exist finitely any open sets $V_1, \dots, V_n$ such that $\bar{V}_i$ is a compact subset of U for all i and $K \subset \bigcup_{i=1}^n V_i$. Hence $V := \bigcup_{i=1}^n V_i$ is an open set containing K and its closure $\bar{V} = \bigcup_{i=1}^n \bar{V}_i$ is a compact subset of U. This proves Lemma A.3. □

Proof of Theorem A.1. The proof has three steps. The first step requires the Axiom of Dependent Choice.[1]

[1] See http://en.wikipedia.org/wiki/Axiom_of_dependent_choice.

Step 1. *There exists a family of open sets $V_r \subset X$ with compact closure, parametrized by $r \in \mathbb{Q} \cap [0,1]$, such that*

$$K \subset V_1 \subset \overline{V}_1 \subset V_0 \subset \overline{V}_0 \subset U \tag{A.3}$$

and

$$s > r \implies \overline{V}_s \subset V_r \tag{A.4}$$

for all $r, s \in \mathbb{Q} \cap [0,1]$.

The existence of open sets V_0 and V_1 with compact closure satisfying (A.3) follows from Lemma A.3. Now choose a bijective map $\mathbb{N}_0 \to \mathbb{Q} \cap [0,1]: i \mapsto q_i$ such that $q_0 = 0$ and $q_1 = 1$. Suppose by induction that the open sets $V_i = V_{q_i}$ have been constructed for $i = 0, 1, \dots, n$ such that (A.4) holds for $r, s \in \{q_0, q_1, \dots, q_n\}$. Choose $k, \ell \in \{0, 1, \dots, n\}$ such that

$$q_k := \max\{q_i \mid 0 \le i \le n,\ q_i < q_{n+1}\},$$

$$q_\ell := \min\{q_i \mid 0 \le i \le n,\ q_i > q_{n+1}\}.$$

Then $\overline{V}_\ell \subset V_k$. Hence it follows from Lemma A.3 that there exists an open set $V_{n+1} = V_{q_{n+1}} \subset X$ with compact closure such that $\overline{V}_\ell \subset V_{n+1} \subset \overline{V}_{n+1} \subset V_k$. This completes the induction argument and Step 1 then follows from the Axiom of Dependent Choice. (Denote by $\mathcal{V}$ the collection of all open sets $V \subset X$ such that $K \subset V \subset \overline{V} \subset U$. Denote by $\mathcal{V}$ the set of all finite sequences $\mathrm{v} = (V_0, \dots, V_n)$ in $\mathcal{V}$ that satisfy (A.3) and $q_i < q_j \implies \overline{V}_j \subset V_i$ for all i, j. Define a relation on $\mathcal{V}$ by $\mathrm{v} = (V_1, \dots, V_n) \prec \mathrm{v}' = (V_1', \dots, V_{n'}')$ if $n < n'$ and $V_i = V_i'$ for $i = 0, \dots, n$. Then $\mathcal{V}$ is nonempty and for every $\mathrm{v} \in \mathcal{V}$ there is a $\mathrm{v}' \in \mathcal{V}$ such that $\mathrm{v} \prec \mathrm{v}'$. Hence, by the Axiom of Dependent Choice, there exists a sequence $\mathrm{v}_j = (V_{j,0}, \dots, V_{j,n_j}) \in \mathcal{V}$ such that $\mathrm{v}_j \prec \mathrm{v}_{j+1}$ for all $j \in \mathbb{N}$. Define the map $\mathbb{Q} \cap [0,1] \to \mathcal{V}: q \mapsto V_q$ by $V_{q_i} := V_{j,i}$ for $i, j \in \mathbb{N}$ with $n_j \ge i$. This map is well defined and satisfies (A.3) and (A.4) by the definition of $\mathcal{V}$ and $\prec$.)

Step 2. *Let $V_r \subset X$ be as in Step 1 for $r \in \mathbb{Q} \cap [0,1]$. Then*

$$\begin{aligned} f(x) &:= \sup\{r \in \mathbb{Q} \cap [0,1] \mid x \in V_r\} \\ &= \inf\{s \in \mathbb{Q} \cap [0,1] \mid x \notin \overline{V}_s\} \end{aligned} \tag{A.5}$$

for all $x \in X$. (Here the supremum of the empty set is zero and the infimum over the empty set is one.)

We prove equality in (A.5). Fix a point $x \in X$ and define

$$a := \sup\{r \in \mathbb{Q} \cap [0,1] \mid x \in V_r\},$$

$$b := \inf\{s \in \mathbb{Q} \cap [0,1] \mid x \notin \overline{V}_s\}.$$

We prove that $a \le b$. If $b = 1$ this follows directly from the definitions. Hence assume $b < 1$ and choose an element $s \in \mathbb{Q} \cap [0, 1]$ such that

$$x \notin \overline{V}_s.$$

If $r \in \mathbb{Q} \cap [0, 1]$ is such that $x \in V_r$, then $V_r \setminus \overline{V}_s \neq \emptyset$, hence $\overline{V}_s \subset V_r$, and hence $r \le s$. Thus we have proved that

$$x \in V_r \implies r \le s$$

for all $r \in \mathbb{Q} \cap [0, 1]$. Take the supremum over all $r \in \mathbb{Q} \cap [0, 1]$ with $x \in V_r$ to obtain $a \le s$. Then take the infimum over all $s \in \mathbb{Q} \cap [0, 1]$ with $x \notin \overline{V}_s$ to obtain $a \le b$. Now suppose, by contradiction, that $a < b$. Choose rational numbers $r, s \in \mathbb{Q} \cap [0, 1]$ such that $a < r < s < b$. Since $a < r$, it follows that $x \notin V_r$, since $s < b$, it follows that $x \in \overline{V}_s$, and since $r < s$, it follows from Step 1 that $\overline{V}_s \subset V_r$. This is a contradiction and shows that our assumption that $a < b$ was wrong. Thus $a = b$ and this proves Step 2.

Step 3. *The function $f\colon X \to [0, 1]$ in Step 2 is continuous and*

$$f(x) = \begin{cases} 0 & \text{for } x \in X \setminus V_0, \\ 1 & \text{for } x \in \overline{V}_1. \end{cases} \tag{A.6}$$

Thus f satisfies the requirements of Theorem A.1.

That f satisfies (A.6) follows directly from the definition of f in (A.5). We prove that f is continuous. To see this, fix a constant $c \in \mathbb{R}$. Then $f(x) < c$ if and only if there exists a rational number $s \in \mathbb{Q} \cap [0, 1]$ such that $s < c$ and $x \notin \overline{V}_s$. Likewise, $f(x) > c$ if and only if there exists a rational number $r \in \mathbb{Q} \cap [0, 1]$ such that $r > c$ and $x \in V_r$. Hence,

$$f^{-1}((c, \infty)) = \bigcup_{r \in \mathbb{Q} \cap (c,1]} V_r, \quad f^{-1}((-\infty, c)) = \bigcup_{s \in \mathbb{Q} \cap [0,c)} (X \setminus \overline{V}_s).$$

This implies that the pre-image under f of every open interval in $\mathbb{R}$ is an open subset of X. Hence also the pre-image under f of every union of open intervals is open in X and so f is continuous. This proves Step 3 and Theorem A.1. □

Theorem A.4 (partition of unity). *Let X be a locally compact Hausdorff space, let $U_1, \dots, U_n \subset X$ be open sets, and let $K \subset U_1 \cup \dots \cup U_n$ be a compact set. Then there exist continuous functions $f_1, \dots, f_n\colon X \to \mathbb{R}$ with compact support such that*

$$f_i \ge 0, \quad \sum_{i=1}^{n} f_i \le 1, \quad \operatorname{supp}(f_i) \subset U_i$$

for all i and $\sum_{i=1}^{n} f_i(x) = 1$ for all $x \in K$.

Proof. Define the set

$$\mathcal{V} := \left\{ V \subset X \;\middle|\; \begin{array}{l} V \text{ is open, } \bar{V} \text{ is compact, and there exists} \\ \text{an index } i \in \{1, \dots, n\} \text{ such that } \bar{V} \subset U_i \end{array} \right\}.$$

If $x \in K$, then $x \in U_i$ for some index $i \in \{1, \dots, n\}$ and, by Lemma A.3, there is an open set $V \subset X$ such that $\bar{V}$ is compact and $x \in V \subset \bar{V} \subset U_i$. Thus $\mathcal{V}$ is an open cover of K. Since K is compact there exist finitely many open sets $V_1, \dots, V_\ell \in \mathcal{V}$ such that $K \subset V_1 \cup \dots \cup V_\ell$. For $i = 1, \dots, n$ define

$$K_i := \bigcup_{1 \le j \le \ell,\, \bar{V}_j \subset U_i} \bar{V}_j.$$

Then $K \subset K_1 \cup \dots \cup K_n$ and K_i is a compact subset of U_i for each i. Hence it follows from Urysohn's lemma (Theorem A.1) that, for each i, there exists a compactly supported continuous function $g_i \colon X \to \mathbb{R}$ such that

$$0 \le g_i \le 1, \quad \operatorname{supp}(g_i) \subset U_i, \quad g_i|_{K_i} \equiv 1.$$

Define

$$\begin{aligned} f_1 &:= g_1, \\ f_2 &:= (1 - g_1) g_2, \\ f_3 &:= (1 - g_1)(1 - g_2) g_3, \\ &\vdots \\ f_n &:= (1 - g_1) \cdots (1 - g_{n-1}) g_n. \end{aligned}$$

Then $\operatorname{supp}(f_i) \subset \operatorname{supp}(g_i) \subset U_i$ for each i and

$$1 - \sum_{i=1}^n f_i = \prod_{i=1}^n (1 - g_i).$$

Since $K \subset K_1 \cup \dots \cup K_n$ and the factor $1 - g_i$ vanishes on K_i, this implies $\sum_{i=1}^n f_i(x) = 1$ for all $x \in K$. This proves Theorem A.4. □

Appendix B

The product topology

Let $(X, \mathcal{U}_X)$ and $(Y, \mathcal{U}_Y)$ be topological spaces, denote the product space by

$$X \times Y := \{(x, y) \mid x \in X,\ y \in Y\},$$

and let $\pi_X : X \times Y \to X$ and $\pi_Y : X \times Y \to Y$ be the projections onto the first and second factor. Consider the following universality property for a topology $\mathcal{U} \subset 2^{X \times Y}$ on the product space.

(F) *Let $(Z, \mathcal{U}_Z)$ be any topological space and let $h: Z \to X \times Y$ be any map. Then $h: (Z, \mathcal{U}_Z) \to (X \times Y, \mathcal{U})$ is continuous if and only if the maps*

$$f := \pi_X \circ h : (Z, \mathcal{U}_Z) \longrightarrow (X, \mathcal{U}_X), \tag{B.1a}$$

$$g := \pi_Y \circ h : (Z, \mathcal{U}_Z) \longrightarrow (Y, \mathcal{U}_Y) \tag{B.1b}$$

are continuous.

Theorem B.1. (i) There is a unique topology $\mathcal{U}$ on $X \times Y$ that satisfies (P).

(ii) Let $\mathcal{U} \subset 2^{X \times Y}$ be as in (i). Then $W \in \mathcal{U}$ if and only if there are open sets $U_i \in \mathcal{U}_X$ and $V_i \in \mathcal{U}_Y$, indexed by any set I, such that $W = \bigcup_{i \in I}(U_i \times V_i)$.

(iii) Let $\mathcal{U} \subset 2^{X \times Y}$ be as in (i). Then $\mathcal{U}$ is the smallest topology on $X \times Y$ with respect to which the projections π_X and π_Y are continuous.

(iv) Let $\mathcal{U} \subset 2^{X \times Y}$ be as in (i). Then the inclusion

$$\iota_x : (Y, \mathcal{U}_Y) \longrightarrow (X \times Y, \mathcal{U}), \quad \iota_x(y) := (x, y) \quad \text{for } y \in Y,$$

is continuous for every $x \in X$ and the inclusion

$$\iota_y : (X, \mathcal{U}_X) \longrightarrow (X \times Y, \mathcal{U}), \quad \iota_y(x) := (x, y) \quad \text{for } x \in X,$$

is continuous for every $y \in Y$.

Definition B.2. The *product topology* on $X \times Y$ is defined as the unique topology that satisfies (P) or, equivalently, as the smallest topology on $X \times Y$ such that the projections π_X and π_Y are continuous. It is denoted by

$$\mathcal{U}_{X\times Y} \subset 2^{X\times Y}.$$

Proof of Theorem B.1. The proof has five steps.

Step 1. *If* $\mathcal{U} \subset 2^{X\times Y}$ *is a topology satisfying* (P), *then the projections* π_X *and* π_Y *are continuous.*

Take $h := \mathrm{id}\colon X \times Y \to X \times Y$ so that $f = \pi_X \circ h = \pi_X$ and $g = \pi_Y \circ h = \pi_Y$.

Step 2. *We prove uniqueness in* (i).

Let $\mathcal{U}, \mathcal{U}' \subset 2^{X\times Y}$ be two topologies satisfying (P) and consider the map

$$h := \mathrm{id}\colon (X \times Y, \mathcal{U}) \longrightarrow (X \times Y, \mathcal{U}').$$

Since $f = \pi_X\colon (X \times Y, \mathcal{U}) \to (X, \mathcal{U}_X)$ and $g = \pi_Y\colon (X \times Y, \mathcal{U}) \to (Y, \mathcal{U}_Y)$ are continuous by Step 1, and $\mathcal{U}'$ satisfies (P), it follows that h is continuous and hence $\mathcal{U}' \subset \mathcal{U}$. Interchange the roles of $\mathcal{U}$ and $\mathcal{U}'$ to obtain $\mathcal{U}' = \mathcal{U}$.

Step 3. *We prove* (ii) *and existence in* (i).

Define $\mathcal{U} \subset 2^{X\times Y}$ as the collection of all sets of the form $W = \bigcup_{i\in I}(U_i \times V_i)$, where I is any index set and $U_i \in \mathcal{U}_X$, $V_i \in \mathcal{U}_Y$ for $i \in I$. Then $\mathcal{U}$ is a topology and the projections $\pi_X\colon (X \times Y, \mathcal{U}) \to (X, \mathcal{U}_X)$ and $\pi_Y\colon (X \times Y, \mathcal{U}) \to (Y, \mathcal{U}_Y)$ are continuous. We prove that $\mathcal{U}$ satisfies (P). To see this, let $(Z, \mathcal{U}_Z)$ be any topological space, let $h\colon Z \to X \times Y$ be any map, and define $f := \pi_X \circ h$ and $g := \pi_Y \circ h$ as in (B.1). If h is continuous, then so are f and g. Conversely, if f and g are continuous, then $h^{-1}(U \times V) = f^{-1}(U) \cap g^{-1}(V)$ is open in Z for all $U \in \mathcal{U}_X$ and all $V \in \mathcal{U}_Y$, and hence it follows from the definition of $\mathcal{U}$ that $h^{-1}(W)$ is open for all $W \in \mathcal{U}$. Thus h is continuous.

Step 4. *We prove* (iii).

Let $\mathcal{U}$ be the topology in (i) and let $\mathcal{U}'$ be any topology on $X \times Y$ with respect to which π_X and π_Y are continuous. If $U \in \mathcal{U}_X$ and $V \in \mathcal{U}_Y$, then $U \times V = \pi_X^{-1}(U) \cap \pi_Y^{-1}(V) \in \mathcal{U}'$. Hence $\mathcal{U} \subset \mathcal{U}'$ by (ii). Since π_X and π_Y are continuous with respect to $\mathcal{U}$, it follows that $\mathcal{U}$ is the smallest topology on $X \times Y$ with respect to which π_X and π_Y are continuous.

Step 5. *We prove* (iv).

Fix an element $x \in X$ and consider the map $h := \iota_x : Y \to X \times Y$. Then the map $f := \pi_X \circ h : Y \to X$ is constant and $g := \pi_Y \circ h : Y \to Y$ is the identity. Hence f and g are continuous and so is h by condition (P). An analogous argument shows that ι_y is continuous for all $y \in Y$. $\square$

Appendix C

The inverse function theorem

This appendix contains a proof of the inverse function theorem. The result is formulated in the setting of continuously differentiable maps between open sets in a Banach space. Readers who are unfamiliar with bounded linear operators on Banach spaces may simply think of continuously differentiable maps between open sets in finite-dimensional normed vector spaces. The inverse function theorem is used on page 81 in the proof of Lemma 2.19, which is a key step in the proof of the transformation formula for the Lebesgue measure (Theorem 2.17). Assume throughout that $(X, \|\cdot\|)$ is a Banach space. When $\Phi\colon X \to X$ is a bounded linear operator denote its operator norm by

$$\|\Phi\| := \|\Phi\|_{\mathcal{L}(X)} := \sup_{x \in X \setminus \{0\}} \frac{\|\Phi x\|}{\|x\|}.$$

For $x \in X$ and $r > 0$ denote by $B_r(x) := \{y \in X \mid \|x - y\| < r\}$ the open ball of radius r about x. For $x = 0$ abbreviate $B_r := B_r(0)$.

Theorem C.1 (inverse function theorem). *Fix an element $x_0 \in X$ and two real numbers $r > 0$ and $0 < \alpha < 1$. Let $\psi\colon B_r(x_0) \to X$ be a continuously differentiable map such that*

$$\|d\psi(x) - \mathbb{1}\|_{\mathcal{L}(X)} \le \alpha \quad \text{for all } x \in B_r(x_0). \tag{C.1}$$

Then

$$B_{(1-\alpha)r}(\psi(x_0)) \subset \psi(B_r(x_0)) \subset B_{(1+\alpha)r}(\psi(x_0)). \tag{C.2}$$

Moreover, the map ψ is injective, its image is open, the map ψ^{-1} is continuously differentiable, and

$$d\psi^{-1}(y) = d\psi(\psi^{-1}(y))^{-1} \quad \text{for all } y \in \psi(B_r(x_0)).$$

Proof. Assume without loss of generality that $x_0 = \psi(x_0) = 0$.

Step 1. *ψ is a homeomorphism onto its image and $\psi(B_r) \subset B_{(1+\alpha)r}$.*

Define
$$\phi := \mathrm{id} - \psi \colon B_r \longrightarrow X.$$
Then $\|d\phi(x)\| \le \alpha$ for all $x \in B_r$. Hence
$$\|\phi(x) - \phi(y)\| \le \alpha\|x - y\|. \tag{C.3}$$
for all $x, y \in B_r$, and so
$$(1-\alpha)\|x - y\| \le \|\psi(x) - \psi(y)\| \le (1+\alpha)\|x - y\|. \tag{C.4}$$
The second inequality in (C.4) shows that $\psi(B_r) \subset B_{(1+\alpha)r}$ and the first inequality in (C.4) shows that ψ is injective and ψ^{-1} is Lipschitz continuous.

Step 2. $B_{(1-\alpha)r} \subset \psi(B_r)$.

Let $\xi \in B_{(1-\alpha)r}$ and define $\varepsilon > 0$ by $\|\xi\| =: (1-\alpha)(r-\varepsilon)$. Then, by (C.3) with $y = 0$, we have $\|\phi(x)\| \le \alpha\|x\|$ for all $x \in B_r$. If $\|x\| \le r - \varepsilon$ this implies $\|\phi(x) + \xi\| \le r - \varepsilon$. Thus the map $x \mapsto \phi(x) + \xi$ is a contraction of the closed ball $\bar{B}_{r-\varepsilon}$. By the contraction mapping principle, it has a unique fixed point x and the fixed point satisfies $\psi(x) = x - \phi(x) = \xi$. Hence $\xi \in \psi(B_r)$.

Step 3. *$\psi(B_r)$ is open.*

Let $x \in B_r$ and define $y := \psi(x)$. Choose $\varepsilon > 0$ such that $B_\varepsilon(x) \subset B_r$. Then, by Step 2, $B_{(1-\alpha)\varepsilon}(\psi(x)) \subset \psi(B_\varepsilon(x)) \subset \psi(B_r)$.

Step 4. *ψ^{-1} is continuously differentiable.*

Let $x_0 \in B_r$ and define $y_0 := \psi(x_0)$ and $\Psi := d\psi(x_0)$. Then $\|\mathbb{1} - \Psi\| \le \alpha$, so Ψ is invertible, $\Psi^{-1} = \sum_{k=0}^{\infty}(\mathbb{1} - \Psi)^k$, and $\|\Psi^{-1}\| \le (1-\alpha)^{-1}$. We prove that ψ^{-1} is differentiable at y_0 and $d\psi^{-1}(y_0) = \Psi^{-1}$. Let $\varepsilon > 0$. Since ψ is differentiable at x_0 and $d\psi(x_0) = \Psi$, there is a constant $\delta > 0$ such that, for all $x \in B_r$ with $\|x - x_0\| < \delta(1-\alpha)^{-1}$, we have
$$\|\psi(x) - \psi(x_0) - \Psi(x - x_0)\| \le \varepsilon(1-\alpha)^2\|x - x_0\|.$$
Shrinking δ, if necessary, we may assume, by Step 3, that $B_\delta(y_0) \subset \psi(B_r)$. Now suppose $\|y - y_0\| < \delta$ and denote $x := \psi^{-1}(y) \in B_r$. Then, by (C.4),
$$\|x - x_0\| \le (1-\alpha)^{-1}\|y - y_0\| < \delta(1-\alpha)^{-1},$$

and hence

$$\begin{aligned}\|\psi^{-1}(y) - \psi^{-1}(y_0) - \Psi^{-1}(y - y_0)\| &= \|\Psi^{-1}(y - y_0 - \Psi(x - x_0))\| \\ &\le \frac{1}{1-\alpha} \|\psi(x) - \psi(x_0) - \Psi(x - x_0)\| \\ &\le \varepsilon(1-\alpha) \|x - x_0\| \\ &\le \varepsilon \|y - y_0\|.\end{aligned}$$

Hence ψ^{-1} is differentiable at y_0 and $d\psi^{-1}(y_0) = \Psi^{-1} = d\psi(\psi^{-1}(y_0))^{-1}$. Thus $d\psi^{-1}$ is continuous by Step 1. This proves Theorem C.1. □

Bibliography

[1] R. B. Ash, *Basic probability theory.* John Wiley & Sons, New York, 1970. Reprint, Dover Publications, Mineola, N,Y., 2008. John Wiley & Sons 1970, Dover Publications, 2008.
MR 0268923 Zbl 1234.60001 Zbl 0206.18001 (reprint)
http://www.math.uiuc.edu/~r-ash/BPT/BPT.pdf

[2] V. I. Bogachev, *Measure Theory.* Volumes I and II. Springer, Berlin etc., 2006.
MR 2267655 Zbl 1120.28001

[3] W. F. Eberlein, Notes on integration I: The underlying convergence theorem. *Comm. Pure Appl. Math.* **10** (1957), 357–360.
MR 0088537 Zbl 0077.31302

[4] D. H. Fremlin, *Measure theory.* Volumes 1–5. Torres Fremlin, Colchester, 2000–2008.
MR 2462519 (I) MR 2462280 (II) MR 2459668 (III) MR 2462372 (IV)
Zbl 1162.28001 (I) Zbl 1165.28001 (II) Zbl 1165.28002 (III)
Zbl 1166.28001 (IV) Zbl 1166.28002 (V)
https://www.essex.ac.uk/maths/people/fremlin/mt.htm

[5] D. Gilbarg and N. S. Trudinger, *Elliptic partial differential equations of second order.* Second edition. Springer, Berlin etc., 1983.
MR 0737190 Zbl 0562.35001

[6] P. R. Halmos, *Measure theory.* University Series in Higher Mathematics. D. Van Nostrand, New York, and Macmillan & Co., London, 1950. Reprint, Graduate Texts in Mathematics, 18. Springer, Berlin etc., 1974.
MR 0033869 Zbl 0040.16802 Zbl 0283.28001 (reprint)

[7] C. E. Heil, *Lecture notes on real analysis.* Georgia Institute of Technology, Atlanta, Fall 2007.
http://people.math.gatech.edu/~heil/6337/fall07/section3.4.pdf
http://people.math.gatech.edu/~heil/6337/fall07/section3.5b.pdf

[8] H. König, New versions of the Radon–Nikodým theorem. *Arch. Math. (Basel)* **86** (2006), no. 3, 251–260.
MR 2215314 Zbl 1091.28002
http://www.math.uni-sb.de/service/preprints/preprint132.pdf

[9] T. W. Körner, *A companion to analysis.* A second first and first second course in analysis. Graduate Studies in Mathematics, 62. American Mathematical Society, Providence, R.I., 2004.
MR 2015825 Zbl 1061.26003
https://www.dpmms.cam.ac.uk/~twk/

[10] U. Lang, *Mass & Integral*, Vorlesungsmanuskript. ETH, Zürich, September 2006.
http://www.math.ethz.ch/~lang/mi.pdf

[11] U. Lang, Existence and uniqueness of Haar integrals. Preprint. ETH Zürich, January 2015.
http://www.math.ethz.ch/~lang/HaarIntegral.pdf

[12] D. Ma, The lexicographic order and the double arrow space. Online survey, 2009.
http://dantopology.wordpress.com/2009/10/07/

[13] P. Malliavin, *Integration and probability.* With the collaboration of H. Airault, L. Kay, and G. Letac. Edited and translated from the French by L. Kay. With a foreword by M. Pinsky. Graduate Texts in Mathematics, 157. Springer, New York, 1995.
MR 1335234 Zbl 0874.28001

[14] J.R. Munkres, *Topology.* Second edition. Prentice Hall, Upper Saddle River, N.J., 2000.
Zbl 0951.54001

[15] G. Nagy, *Real Alalysis.* Version 2.0. Section IV.6, Duals of L^p spaces. Lecture notes. Kansas State University, Manhattan, KS.
http://www.math.ksu.edu/~nagy/real-an/4-06-dual-lp.pdf

[16] G. K. Pedersen, The existence and uniqueness of the Haar integral on a locally compact topological group. Preprint, University of Copenhagen, Copenhagen, 2000.
http://www.math.ku.dk/kurser/2004-2/mat3re/haarintegral.pdf

[17] W. Rudin, *Real and complex analysis.* Third edition. McGraw-Hill, New York, 1987.
MR 0924157 Zbl 0925.00005

[18] D. A. Salamon, Das Riemannsche Integral. Lecture Notes, ETH, Zürich, 2014.
http://www.math.ethz.ch/~salamon/PREPRINTS/ana1-int.pdf

[19] D. A. Salamon, Mehrfache Integrale. Lecture notes, ETH, Zürich, 2014.
http://www.math.ethz.ch/~salamon/PREPRINTS/ana2-int.pdf

[20] D. A. Salamon, *Funktionentheorie.* Grundstudium Mathematik. Birkhäuser, Basel, 2012.
Zbl 1236.30001
http://www.math.ethz.ch/~salamon/PREPRINTS/cxana.pdf

[21] W. F. Trench, *Introduction to real analysis.* Prentice Hall/Pearson Education, Upper Saddle River, N.J., 2003.
Zbl 1204.00023
http://ramanujan.math.trinity.edu/wtrench/misc/index.shtml

Index

– M –